Differential Equations With Applications

Class Notes With Detailed Examples
Version 2

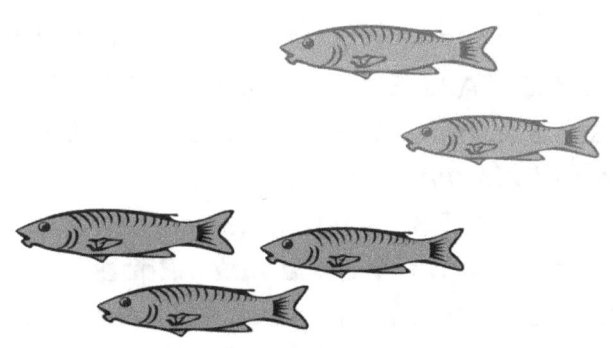

Copyright © 2019 by
Jigarkumar Patel and Kathryn Paulk

Please email any comments to Dr. Patel at
jigarkumar.patel@utdallas.edu

TABLE OF CONTENTS

PREFACE 1

INTRODUCTION 2

FIRST-ORDER DE 4
    FIRST-ORDER LINEAR DE 6
    SEPARABLE DE 13
    SUBSTITUTION METHODS 16
        Bernoulli's DE 17
        Homogeneous DE 22
        Special Substitution DE 25
    EXACT EQUATIONS 35
        Integrating Factor 40
    STRATEGIES FOR FIRST-ORDER DE AND OTHER STRATEGIES 56
    EXISTENCE AND UNIQUENESS THEOREM 58

SECOND-ORDER DE 59
    SECOND-ORDER LINEAR DE WITH CONSTANT COEFFICIENTS 60
        Characteristic Equation With Real and Distinct Roots 61
        Existence and Uniqueness Theorem 64
        Wronskian Determinant 66
        Characteristic Equation With Complex Roots 73
        Characteristic Equation With Repeated Roots 76
    HIGHER-ORDER HOMOGENEOUS DE WITH CONST. COEFF. 78
    NON-HOMOGENEOUS DE WITH CONST. COEFF. 87
        Method of Undetermined Coefficients 88
        Method of Variation of Parameters 100
    EULER'S DE 110
    REDUCTION OF ORDER 121
    MECHANICAL AND ELECTRICAL VIBRATIONS 133
        Spring Mass System 134
        Electrical Vibrations 145

SERIES SOLUTION OF DE 149
POWER SERIES (REVIEW) 150
POWER SERIES NEAR AN ORDINARY POINT 162
TYPES OF SINGULAR POINTS 177
SERIES SOL'N NEAR A REGULAR SINGULAR POINT 186
BESSEL'S DE 211

LAPLACE TRANSFORM 218
IMPROPER INTEGRATION (REVIEW) 219
DEFINITION OF LAPLACE TRANSFORM 225
GAMMA FUNCTION 237
SOLVE IVP USING LAPLACE TRANSFORM (GENERAL IDEA) 246
STEP FUNCTION 255
IMPULSE FUNCTION 287
CONVOLUTION INTEGRAL 300

SYSTEMS OF FIRST-ORDER DE 310
SYSTEMS OF FIRST-ORDER LINEAR DEs 311
EIGENVALUES AND EIGENVECTORS 316
HOMOGENEOUS SYSTEM WITH CONST. COEFF. 328
NON-HOMOGENEOUS SYSTEM WITH CONST. COEFF. 344
Fundamental Matrix 344
How to Solve Non-Homogeneous Systems (General Idea) 352
THE PHASE PLANE 368
LOCALLY LINEAR SYSTEMS (GENERAL IDEA) 402
COMPETING SPECIES 415
PREDATOR-PREY SYSTEMS 429

A. INTEGRATION BY PARTS (TABULAR METHOD) 1
B. COMPLEX NUMBERS 1
C. LINEAR INDEPENDENCE 1
D. MATRICES REVIEW 1

PREFACE

This book is a collection of a student's class notes from the Math 2420 class (Differential Equations With Applications) at the University of Texas at Dallas during the spring of 2017. The notes have been reviewed and updated by the professor, Jigarkumar Patel. Additional notes and examples were added to make this collection of notes more complete.

The content of this book parallels Dr. Patel's lectures. Many detailed examples, organized from simple to complex are included. Theory, remarks, and guidelines are included as needed. This book should be considered a reference for students taking a course in Differential Equations. It has been formatted to be readable on electronic, hand-held devices. Prerequisites for this class include: Differential Calculus, Integral Calculus, and Linear Algebra.

Jigarkumar Patel received a Ph.D. in Applied Mathematics and an M.S. in Applied Mathematics from the University of Texas at Dallas (UTDallas) in Richardson TX. Before attending UTDallas, he earned a B.S. and an M.S. degree in Pure Mathematics and a B.Ed. degree from Gujarat University in Ahmedabad, Gujarat INDIA. Dr. Patel is currently an Associate Professor of Instruction at UTDallas.

Kathryn Paulk is the student who audited Dr. Patel's class and took the notes. She received an M.S. in Industrial Engineering and a B.S. in Mechanical Engineering from the University of Pittsburgh, in Pittsburgh PA.

INTRODUCTION

The purpose of this book is to enhance the reader's understanding of differential equations, using step-by-step procedures and detailed examples. This book should be considered a supplement to course lectures and textbook(s). It is assumed that the reader has taken previous courses in Differential and Integral Calculus. Familiarity with Linear Algebra is also recommended. Statements of major theorem and results are included in this book. However, the proofs are not included as the focus of this book is primarily on examples and applications.

The scope of this book is based on the Differential Equations With Applications course at The University of Texas at Dallas. The course begins with a general definition of the Differential Equation (DE). DEs may be challenging or impossible to solve due to the complex relationship between variables and derivatives of various orders. Therefore, we start with first-order DEs which are less challenging.

First-order DEs may include a complex relationship between variables, making them tricky to solve. Special types of first-order DEs, where the solution techniques are known, are discussed first. Substitution and integrating-factor methods, to convert special types of DEs into known DEs, are also discussed. Remarks about graphical and numerical solutions are included.

Second-order DEs are more difficult to solve because of the complex relationship between variables (dependent and independent) and derivatives (first and second-order). Therefore, the focus is on the simple case, the second-order linear DE. The following cases of second-order DEs, in increasing complexity, are discussed:
- Second-order linear DEs with constant coefficients.
- Second-order non-linear DEs with variable coefficients.

Many real-world applications of DEs involve a discontinuous force or impulse force. These problems are difficult to solve with techniques used previously. An effective transform method (Laplace Transform) is used to solve these types of problems.

Higher-order DEs with constant coefficients are briefly discussed. An alternate method to convert them to a system of first-order linear DEs is included. The last, and perhaps the most interesting part of this book includes techniques to find equilibrium solutions for non-homogeneous linear systems of DEs with constant coefficients. In this section, the phase portrait, type, and stability of equilibrium solutions are discussed in detail. These techniques can also be used to solve certain nonlinear systems. The course concludes with some interesting applications (competing species and predator-prey systems) which are modeled as non-linear systems.

FIRST-ORDER DE

A simple equation is the relationship between a variable and numbers (e.g. $2x = 10$). In this case, the solution is simply a number ($x = 5$). A first-order differential equation (DE) is a relationship between a function and its first derivative (e.g. $y = y'$). The solution is a function ($y = e^x$). In fact, there may be many solutions ($y = e^x + C$).

This chapter starts with the definition of a DE of order n. The solution may be difficult or impossible to find because of the complex relationship between derivatives and variables. Hence, we focus on a less challenging case, the first-order DE form ($y' = f(x, y)$). We realize that even these DEs may be very challenging to solve if the function $(y = f(x, y))$ is complex. Therefore, we look into special subclasses, of a first-order DE, and introduce methods to solve them.

General methods, to solve first-order linear, exact, Bernoulli, and homogeneous DEs are discussed in detail. To convert a complex first-order DE into a known form, special techniques, such the substitution method and the integrating factor method, are discussed.

This chapter concludes with a few brief remarks on numerical solutions, graphical solutions, and the Existence and Uniqueness Theorem for DEs.

Definition:

A <u>differential equation</u> (DE) is a relationship between the derivative and function of several variables.

Notation: $\quad \frac{d^n y}{dx^n} = f\left(x, y, \frac{dy}{dx}, \frac{d^2 x}{dy^2}, \ldots \frac{d^{n-1} x}{dy^{n-1}}\right) \quad$ ➔ Order = n

Examples

- $\frac{d^2 y}{dx^2} = x^2 + y^3 + 2y^2 \left(\frac{dy}{dx}\right)^2 \quad$ ➔ Order = 2
- $\frac{d^3 y}{dx^3} = x^3 y^2 + \left(\frac{dy}{dx}\right)^7 \left(\frac{d^2 y}{dx^2}\right) + 9 \quad$ ➔ Order = 3

Remarks:
- Highest number of derivative in the DE is called the order of the DE.
- It could be very challenging to solve the DE due to the complexity of the right-hand side.
- The solution of the DE (if exists) is called the general solution (G.S.)
- Since n^{th} order DEs are challenging to solve, we focus on the simplest version.

First Order DE is the DE of the form: $\quad \frac{dy}{dx} = f(x, y)$

First-Order Linear DE

First-Order Differential Equation (DE) – Some Examples:
- $\frac{dy}{dx} = x^2 + y^3 + 7xy + \sin(x^3 + y^4)$
- $y' = 2xy^8 + 7y^5 + 9x^7$
- $\frac{dy}{dx} = y^3 + 2y + 7$

Initial Value Problem (IVP) → Initial data is provided.
A DE with initial conditions is called an Initial Value Problem (IVP).

- IVP Notation: $\frac{dy}{dx} = f(x, y), \quad y(x_0) = y_0$
  Where x_0 and y_0 are numbers.

- The solution of an IVP (if exists) is called the Particular Solution (P.S.)

> Previously
> $y = f(x)$
> Now $y = y(x)$

It is clear that computation of an analytic solution (as a function or curve) of a DE or IVP could be very challenging. So, we concentrate on special cases, such as the First-Order <u>LINEAR</u> DE.

First-Order Linear DE – Some Examples:
- A DE of the form: $\frac{dy}{dx} + p(x) \cdot y = g(x)$
  is called a "First-Order Linear DE."

- IVP form: $\frac{dy}{dx} + p(x) \cdot y = g(x), \quad y(x_0) = y_0$

General Idea of how to solve an Initial Value Problem (IVP)

Step	General Idea of how to solve an IVP $\frac{dy}{dx} + p(x) \cdot y = g(x)$, $y(x_0) = y_0$
1.	Write the problem in standard form. The coefficient of $\frac{dy}{dx} = 1$ The first term is $\frac{dy}{dx}$. The second term must be in the form: $p(x) \cdot y$
2.	Compare with standard form to find: $p(x)$, $g(x)$, x_0, y_0
3.	Construct the Integrating Factor, $\mu(x)$ $I(x) = \mu(x) = e^{\int p(x)dx}$
4.	Multiply both sides of the DE by the integrating factor. Notice the left side is product rule. Express it as: $\mu(x)\frac{dy}{dx} + \mu(x) \cdot p(x) \cdot y = \mu(x) \cdot g(x)$ Integrate both sides wrt "x". $\mu(x) \cdot y = \int g(x) \cdot \mu(x) + C$ $y = \frac{\int g(x) \cdot \mu(x) + C}{\mu(x)}$ G.S. (Could just do this!)
5.	Use x_0, y_0 to get C Then write P.S.

Example: First-Order Linear DE

Step	Example: Solve the IVP. $\frac{dy}{dx} + 6x^2 y = 24x^2$, $y(0) = 3$
1.	Write the problem in standard form. $$\frac{dy}{dx} + 6x^2 y = 24x^2$$
2.	Find: $p(x), g(x), x_0, y_0$ $p(x) = 6x^2$, $g(x) = 24x^2$, $x_0 = 0$, $y_0 = 3$
3.	Construct the Integrating Factor: $\mu(x) = e^{\int p(x)dx}$ $\mu(x) = e^{\int 6x^2 dx} = e^{2x^3}$
4.	Just use the equation: $y = \frac{\int g(x) \cdot \mu(x) + C}{\mu(x)}$ G.S. $y = \frac{\int 24x^2 \cdot e^{2x^3} dx + C}{e^{2x^3}}$ U-SUB: $u = 2x^3$ $du = 6x^2 dx$ $y = \frac{4 \int e^u du + C}{e^{2x^3}} = \frac{4e^u + C}{e^{2x^3}} = \frac{4e^{2x^3} + C}{e^{2x^3}}$ G.S.
5.	Use x_0, y_0 to get C. Then write the P.S. $3 = \frac{4e^0 + C}{e^0} = 4 + C$ → $C = -1$ $y = \frac{4e^{2x^3} - 1}{e^{2x^3}}$ P.S.

Step	Example: Solve the IVP. $t\dfrac{dy}{dt} + 2y = 16te^{2t}$, $y(1) = 3$
1.	Standard form: $\dfrac{dy}{dt} + \left(\dfrac{2}{t}\right)y = 16e^{2t}$
2.	$p(t) = \dfrac{2}{t}$, $\quad g(t) = 16e^{2t}$, $\quad t_0 = 1$, $\quad y_0 = 3$
3.	Construct the Integrating Factor: $\quad \mu(x) = e^{\int p(x)dx}$ $\mu(t) = e^{\int p(t)dt} = e^{\int \frac{2}{t} dt} = e^{2\ln t} = t^2$
4.	$y = \dfrac{\int g(x) \cdot \mu(x) + C}{\mu(x)} \quad$ G.S. $y = \dfrac{\int t^2 \, 16 e^{2t} dt + C}{t^2}$ $16 \int t^2 e^{2t} dt$ Use Integration by Parts 	Du	$\int dv$	 |---|---| 	$t^2 \quad +$	e^{2t}	 	$2t \quad -$	$\left(\dfrac{1}{2}\right)e^{2t}$	 	$2 \quad +$	$\left(\dfrac{1}{4}\right)e^{2t}$	 	$0 \quad -$	$\left(\dfrac{1}{8}\right)e^{2t}$	 $16 \int t^2 e^{2t} dt =$ $= 16\left[t^2\left(\dfrac{1}{2}\right)e^{2t} - 2t\left(\dfrac{1}{4}\right)e^{2t} + 2\left(\dfrac{1}{8}\right)e^{2t}\right]$ $= 8t^2 e^{2t} - 8t e^{2t} + 4e^{2t}$ $y = \dfrac{8t^2 e^{2t} - 8t e^{2t} + 4e^{2t} + C}{t^2} \quad$ G.S.
5.	$3 = \dfrac{8e^2 - 8e^2 + 4e^2 + C}{1} = 4e^2 + C$ $C = 3 - 4e^2$ $y = \dfrac{8t^2 e^{2t} - 8t e^{2t} + 4e^{2t} + 3 - 4e^2}{t^2} \quad$ P.S.

Example: Find the G.S. of $\frac{dy}{dx} + \cot x \, y = \csc x$

Step	Example: First-Order Liner DE Find the G.S. of: $\frac{dy}{dx} + \cot x \, y = \csc x$
1.	Write the problem in standard form. $\frac{dy}{dx} + \cot x \, y = \csc x$
2.	Find: $p(x), g(x), x_0, y_0$ $p(x) = \cot x \,, g(x) = \csc x$
3.	Construct the Integrating Factor: $\mu(x) = e^{\int p(x)dx}$ $\mu(x) = e^{\int \cot x \, dx} = e^{\ln|\sin x|} = \sin x$
4.	Just use the equation: $y = \frac{\int g(x) \cdot \mu(x) + C}{\mu(x)}$ G.S. $y = \frac{\int \sin x \cdot \csc x \, dx + C}{\sin x} = \frac{\int 1 \, dx + C}{\sin x} = \frac{x + C}{\sin x}$ G.S.
5.	If initial conditions are given, compute the particular solution (P.S.) using x_0, y_0 to get C. In this DE, initial conditions are not given, therefore, only the general solution (G.S.) can be computed.

Step	Example: First-Order Liner DE Find the General Solution (G.S.) of: $ty' + y = 9t \sin 3t$
1.	Write the problem in standard form. $\dfrac{dy}{dt} + \left(\dfrac{1}{t}\right) y = 9 \sin 3t$
2.	Find: $p(x), g(x), x_0, y_0$ $p(t) = \dfrac{1}{t}, \quad g(t) = 9 \sin 3t$
3.	Construct the Integrating Factor: $\mu(x) = e^{\int p(x)dx}$ $\mu(t) = e^{\int p(t)dt} = e^{\int \frac{1}{t} dt} = e^{\ln t} = t$
+	Just use the equation: $y = \dfrac{\int g(x) \cdot \mu(x) + C}{\mu(x)}$ G.S. $y = \dfrac{\int t \, 9 \sin 3t \, dt + C}{t}$ Find: $9 \int t \cdot \sin 3t \, dt$

	Du		$\int dv$
	t	+	$\sin 3t$
	1	−	$-\left(\dfrac{1}{3}\right) \cos 3t$
	0	+	$-\left(\dfrac{1}{9}\right) \sin 3t$

$y = \dfrac{9\left[-\dfrac{t}{3} \cos 3t + \dfrac{1}{9} \sin 3t\right] + C}{t} = \dfrac{-3t \cos 3t + \sin 3t + C}{t}$ G.S. |
| 4. | If initial conditions are given, compute the particular solution (P.S.) using x_0, y_0 to get C.

In this DE, initial conditions are not given, therefore, only the general solution (G.S.) can be computed. |

Extra Practice:

Find the General Solution of the following DEs

1. $\dfrac{dy}{dx} + 3y = 10e^{7x}$

2. $\dfrac{dy}{dx} + 3y = 9x$

3. $\dfrac{dy}{dx} + \dfrac{4}{x}y = \sin x, \quad x > 0$

4. $\dfrac{dy}{dx} - \dfrac{5}{x}y = 10x^6, \quad x > 0$

5. $\dfrac{dy}{dx} - \cot x\, y = \csc x \quad , \quad 0 \leq x \leq \pi$ \qquad Hint: $\mu(x) = \csc x$

6. $x\dfrac{dy}{dx} + 3y = 81\, xe^{3x}$

7. $x\dfrac{dy}{dx} + y = \dfrac{2}{x-2}$

8. $\dfrac{dy}{dx} = 4y + 16\, e^{2x}$

9. $\dfrac{dy}{dx} - \dfrac{3}{x}y = \ln x$

10. $\dfrac{dy}{dx} + \dfrac{y}{x} = \dfrac{1}{\sqrt{x-2}}$

Separable DE

Now, we will discuss another special type of First-Order DE whose analytic solution can be computed easily.

DE form: $g(x) + h(y)\frac{dy}{dx} = 0$ is called a "Separable DE."

IVP Form: $g(x) + h(y)\frac{dy}{dx} = 0$, $y(x_0) = y_0$

General Idea of how to solve a Separable DE

$g(x) + h(y)\frac{dy}{dx} = 0$

$h(y)\frac{dy}{dx} = -g(x)$

$h(y)\, dy = -g(x)\, dx$ (Variables can be separated.)

$\int h(y)\, dy = -\int g(x)\, dx$

$H(y) = G(x) + C$

Here, $H(y)$ is an anti-derivative of $h(y)$
and $G(x)$ is an anti-derivative of $-g(x)$.

Example #1:   Find the Particular Solution (P.S.) of the IVP

$dy + (3x^2 + 4x + 5)dx = 0$, $y(1) = 7$

Solution: Note: $h(y)$ is missing. → special case

$dy = -(3x^2 + 4x + 5)dx$ Now, variables are separated.

$\int dy = -\int (3x^2 + 4x + 5)\, dx$

$y = -x^3 - 2x^2 - 5x + C$ G.S.

$y(1) = 7$ → $7 = -1 - 7 + C$ → $C = 15$

$y = -x^3 - 2x^2 - 5x + 15$ P.S.

Example #2: Find the G.S. of $(x^2 + x)\, dy = -y\, dx$

Solution:

$\dfrac{1}{y} dy = \dfrac{-1}{(x^2+x)} dx$

$\int \dfrac{1}{y} dy = \int \dfrac{-1}{(x^2+x)} dx$

$\ln y = \int \dfrac{-1}{(x^2+x)} dx = \int \dfrac{-1}{x(x+1)} dx$

$\dfrac{-1}{x(x+1)} = \dfrac{A}{x} + \dfrac{B}{x+1}$

$-1 = A(x+1) + B(x)$

$x = 0 \rightarrow -1 = A$

$x = -1 \rightarrow -1 = B(-1) \rightarrow B = 1$

$\dfrac{-1}{x(x+1)} = \dfrac{-1}{x} + \dfrac{1}{x+1}$

$\ln y = \int \dfrac{-1}{x(x+1)} dx$

$\ln y = \int \dfrac{-1}{x} + \dfrac{1}{(x+1)} dx = -\ln|x| + \ln|x+1| + C$

$\ln y = -\ln|x| + \ln|x+1| + \ln C$

$\ln y = \ln\left(\dfrac{C(x+1)}{x}\right)$ Note: "$\ln$" is a one-one function.

$y = \dfrac{C(x+1)}{x}$ General Solution (G.S.)

Note: Any of the last four equations may be considered the G.S.

Example: Find the G.S. of $e^{1-x^2} y' = \dfrac{2x}{y \ln y}$

Note: If not First-Order DE, try separable.

Solution:

$$e^{1-x^2} \dfrac{dy}{dx} = \dfrac{2x}{y \ln y}$$

$$y \ln y \, dy = \dfrac{2x}{e^{1-x^2}} dx$$

$$\int y \ln y \, dy = \int \dfrac{2x}{e^{1-x^2}} dx$$

$$\int y \ln y \, dy = \int 2x \cdot e^{-(1-x^2)} dx$$

$$\int y \ln y \, dy = \int 2x \cdot e^{x^2-1} dx$$

Left Side: Integrate by Parts:

Du	$\int dv$
$\ln y$ +	y
$\dfrac{1}{y}$ −	$\left(\dfrac{1}{2}\right) y^2$

$I = \ln y \left(\dfrac{1}{2}\right) y^2 - \int \left(\dfrac{1}{y}\right)\left(\dfrac{1}{2}\right) y^2 \, dy$

$I = \ln y \left(\dfrac{1}{2}\right) y^2 - \dfrac{1}{2} \int y \, dy$

$I = \ln y \left(\dfrac{1}{2}\right) y^2 - \dfrac{1}{2}\left(\dfrac{1}{2}\right) y^2$

$I = \dfrac{1}{2} y^2 \ln y - \dfrac{1}{4} y^2$

Right Side: U-sub.

$u = x^2 - 1$

$du = 2x \, dx$

$\int e^u \, du = e^u$

$\qquad = e^{x^2-1}$

$$\dfrac{1}{2} y^2 \ln y - \dfrac{1}{4} y^2 = e^{x^2-1} + C$$

Substitution Methods

Question: What if the DE is not First-Order Linear or Separable?

Answer: Try one of the following substitution methods.

SUBSTITUTION METHODS TO SOLVE DE

Substitution methods allow us to convert certain types of DEs into First-Order Linear or Separable form. Once DEs are converted to a known form, we can compute the General Solution (G.S.) or the Particular Solution (P.S.).

BERNOULLI DE:

A DE of the form: $\dfrac{dy}{dx} + P(x) \cdot y = g(x) \cdot y^n$

is called a Bernoulli DE.

Remarks:

- Can be converted to a First-Order DE by using the substitution:

$$v = y^{1-n}$$

- If $n = 0$, then it's just a First-Order linear DE.
- If $n = 1$, then it's just a First-Order linear and separable DE.

Bernoulli's DE

General Idea: To convert a Bernoulli DE to a First-Order linear DE.

Given a Bernoulli DE: $\quad \dfrac{dy}{dx} + p(x) \cdot y = g(x) \cdot y^n$

First, divide both sides of the equation by y^n.

$$\dfrac{1}{y^n}\dfrac{dy}{dx} + p(x)\dfrac{1}{y^{n-1}} = g(x)$$

$$\dfrac{1}{y^n}\dfrac{dy}{dx} + p(x)\boxed{\dfrac{1}{y^{n-1}}} = g(x)$$

Substitute: $v = \dfrac{1}{y^{n-1}} = y^{1-n}$

$$\dfrac{dv}{dx} = (1-n)y^{1-n-1}\dfrac{dy}{dx}$$

$$\dfrac{dv}{dx} = \dfrac{1-n}{y^n}\dfrac{dy}{dx}$$

So...

$$\dfrac{dy}{dx} = \dfrac{y^n}{1-n} \cdot \dfrac{dv}{dx}$$

$$\dfrac{1}{y^n}\left(\dfrac{y^n}{1-n} \cdot \dfrac{dv}{dx}\right) + p(x)(v) = g(x)$$

$$\left(\dfrac{1}{1-n}\right)\dfrac{dv}{dx} + p(x)v = g(x)$$

First-Order Linear DE ☺

Example: Find the G.S. of: $\quad y' - \dfrac{2}{x}y = -5x^2 y^2$

Not First-Order Linear ☹   Not Separable ☹

Solution:

Is it Bernoulli? YES! → Convert it to a First-Order Linear DE

→ Divide both sides by y^2

$$\left(\dfrac{1}{y^2}\right)\dfrac{dy}{dx} - \left(\dfrac{1}{y}\right)\dfrac{2}{x} = -5x^2$$

$$\left(\dfrac{1}{y^2}\right)\dfrac{dy}{dx} - \left(\dfrac{1}{y}\right)\dfrac{2}{x} = -5x^2 \quad \to \text{Substitute: } v = \dfrac{1}{y}$$

$$\boxed{\begin{array}{l} v = y^{-1} \\[4pt] \dfrac{dv}{dx} = -y^{-2}\dfrac{dy}{dx} \\[4pt] \dfrac{dy}{dx} = -y^2 \dfrac{dv}{dx} \end{array}}$$

$$\left(\dfrac{1}{y^2}\right)\left(-y^2 \dfrac{dv}{dx}\right) - (v)\dfrac{2}{x} = -5x^2$$

$$-\dfrac{dv}{dx} - \left(\dfrac{2}{x}\right)v = -5x^2$$

$$\dfrac{dv}{dx} + \left(\dfrac{2}{x}\right)v = 5x^2 \quad \to \text{First-Order Linear DE}$$

$\mu(x) =$ Integrating Factor

$$\mu(x) = e^{\int p(x)dx} = e^{\int \frac{2}{x}dx} = e^{2\ln|x|} = x^2$$

$$v = \dfrac{\int x^2(5x^2)\,dx + C}{x^2} = \dfrac{5\int x^4\,dx + C}{x^2} = \dfrac{x^5 + C}{x^2}$$

$$\dfrac{1}{y} = \dfrac{x^5 + C}{x^2}$$

$$y = \dfrac{x^2}{x^5 + C} \qquad \text{G.S.}$$

Example: Find the G.S. of $\dfrac{dy}{dx} - 6y = -36xy^2$

Solution:

$\dfrac{1}{y^2}\dfrac{dy}{dx} - \dfrac{6}{y} = -36x$

$\dfrac{1}{y^2}\dfrac{dy}{dx} - \left(\dfrac{1}{y}\right)6 = -36x$

$\dfrac{1}{y^2}\left(-y^2\dfrac{dv}{dx}\right) - (v)6 = -36x$

$-\dfrac{dv}{dx} - 6v = -36x$

$\dfrac{dv}{dx} + 6v = 36x$ → First-Order Linear DE

$\mu(x) = e^{\int p(x)dx} = e^{\int 6\,dx} = e^{6x}$

$v = \dfrac{\int \mu(x)\cdot g(x)dx + C}{\mu(x)} = \dfrac{\int e^{6x}(36x)\,dx + C}{e^{6x}}$

$v = \dfrac{36\int e^{6x}(x)\,dx + C}{e^{6x}}$

$\boxed{v = \dfrac{1}{y} = y^{-1}}$

$\dfrac{dv}{dx} = (-1)(y)^{-2}\dfrac{dy}{dx}$

$\dfrac{dy}{dx} = -y^2\dfrac{dv}{dx}$

Du		$\int dv$
x	+	e^{6x}
1	−	$\left(\dfrac{1}{6}\right)e^{6x}$
0	+	$\left(\dfrac{1}{36}\right)e^{6x}$

$I = x\left(\dfrac{1}{6}\right)e^{6x} - \left(\dfrac{1}{36}\right)e^{6x}$

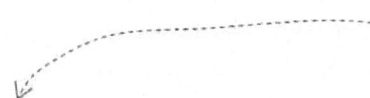

$v = \dfrac{36\left[x\left(\dfrac{1}{6}\right)e^{6x} - \left(\dfrac{1}{36}\right)e^{6x}\right] + C}{e^{6x}} = \dfrac{6xe^{6x} - e^{6x} + C}{e^{6x}}$

$y = \dfrac{1}{v} = \dfrac{e^{6x}}{6xe^{6x} - e^{6x} + C}$

Example: Find the G.S. of: $\quad \dfrac{dy}{dx} - 6y = 8e^{10x} y^{-2}$

Solution: → Bernoulli

$$y^2 \dfrac{dy}{dx} - 6y^3 = 8e^{10x}$$

$$\boxed{\begin{array}{l} v = y^3 \\[4pt] \dfrac{dv}{dx} = 3y^2 \dfrac{dy}{dx} \\[4pt] \dfrac{1}{3y^2} \dfrac{dv}{dx} = \dfrac{dy}{dx} \end{array}}$$

$$y^2 \left(\dfrac{1}{3y^2} \dfrac{dv}{dx} \right) - 6(v) = 8e^{10x}$$

$$\dfrac{1}{3} \dfrac{dv}{dx} - 6v = 8e^{10x}$$

$$\dfrac{dv}{dx} - 18v = 24e^{10x}$$

$$\boxed{\begin{array}{l} \text{Integrating Factor:} \\[4pt] \mu(x) = e^{\int p(x)dx} = e^{\int -18\, dx} \\[4pt] \mu(x) = e^{-18x} \end{array}}$$

$$v = \dfrac{\int \mu(x) \cdot g(x) dx + C}{\mu(x)}$$

$$v = \dfrac{\int e^{-18x} (24e^{10x}) dx + C}{e^{-18x}}$$

$$v = \dfrac{24 \int e^{-8x} dx + C}{e^{-18x}}$$

$$v = \dfrac{24 \left(\dfrac{-1}{8} \right) e^{-8x} + C}{e^{-18x}}$$

$$v = \dfrac{-3 e^{-8x} + C}{e^{-18x}}$$

$$y = \sqrt[3]{v}$$

$$y = \sqrt[3]{\dfrac{-3 e^{-8x} + C}{e^{-18x}}} \qquad \text{G.S.}$$

	So far, the following strategies have been used to solve some First-Order DEs
Step 1	Check if the DE can be expressed as a First-Order Linear DE in this form: $\frac{dy}{dx} + p(x)y = g(x)$ If yes: • Compute: $\mu(x) = e^{\int p(x)dx}$ • General Solution is: $y = \dfrac{\int \mu(x) \cdot g(x)dx + C}{\mu(x)}$
Step 2	Check if given DE is separable or not. If it is separable, then: • $\frac{dy}{dx} = g(x) \cdot h(y)$ • $\int \frac{1}{h(y)} dy = \int g(x)\, dx$
Step 3	Check for Substitution. Consider: • Bernoulli DE form: $\frac{dy}{dx} + p(x)y = g(x)\, y^n$, $y \neq 0, 1$

Homogeneous DE

Question:

What if the DE is not First-Order Linear, Separable, or Bernoulli?

Answer: Check to see if it is a First-Order Homogeneous DE.

Note: Homogeneous → $f(tx, ty) = t^k f(x, y)$

Homogeneous → The joint degree of each term is the same.

Step	General Idea: How to solve a DE that is **NOT** First-Order, Separable, or Bernoulli.
1.	Check if the degree of each term is the same.
2.	Divide all terms by the highest power of x.
3.	Use the substitution: $v = \frac{y}{x}$ Converts homogeneous DE to a First-Order Separable DE.
4.	Perform algebra and separate variables.
5.	Compute the G.S. (and the P.S. if IVP).

Simple Example: Check if $\frac{dy}{dx} = \frac{x^2 + xy + y^2}{2x^2 - 3y^2}$ is homogeneous.

Solution:

$f(x, y) = \frac{x^2 + xy + y^2}{2x^2 - 3y^2}$

$f(xt, ty) = \frac{(tx)^2 + (tx)(ty) + (ty)^2}{2(tx)^2 - 3(ty)^2} = \left(\frac{t^2}{t^2}\right) \frac{x^2 + xy + y^2}{2x^2 - 3y^2} = \frac{x^2 + xy + y^2}{2x^2 - 3y^2}$

→ Homogeneous!

Example: Find the G.S. of: $\quad x^2 \dfrac{dy}{dx} = x^2 + xy + y^2$

Solution:

$$\dfrac{dy}{dx} = \dfrac{x^2 + xy + y^2}{x^2}$$

→ Homogeneous because degree of all terms is 2

$$\dfrac{dy}{dx} = 1 + \dfrac{y}{x} + \left(\dfrac{y}{x}\right)^2$$

→ Substitute $v = \left(\dfrac{y}{x}\right)$ to get a separable DE.

$$\boxed{\begin{aligned} v &= \dfrac{y}{x} \\ y &= xv \\ \dfrac{dy}{dx} &= x\dfrac{dv}{dx} + v \quad \text{(Product rule)} \end{aligned}}$$

$$v\dfrac{dv}{dx} + v = 1 + v + (v)^2$$

$$x\dfrac{dv}{dx} = 1 + v^2 \qquad\qquad \text{Separable DE.}$$

$$\dfrac{1}{1+v^2} dv = \dfrac{1}{x} dx$$

$$\int \dfrac{1}{1+v^2} dv = \int \dfrac{1}{x} dx$$

$$\tan^{-1} v = \ln|x| + C$$

$$\tan^{-1}\left(\dfrac{y}{x}\right) = \ln|x| + C \qquad\qquad \text{G.S.}$$

Example: Find the G.S. of: $\dfrac{dy}{dx} = \dfrac{x^2 - y^2}{xy}$

Solution:

$$\dfrac{dy}{dx} = \dfrac{1 - \left(\dfrac{y}{x}\right)^2}{\left(\dfrac{y}{x}\right)}$$

→ Let $v = \dfrac{y}{x}$

$$\boxed{\begin{array}{l} v = \dfrac{y}{x} \qquad \text{We need } \dfrac{dy}{dx} \\[4pt] y = vx \\[4pt] \dfrac{dy}{dx} = \dfrac{dv}{dx}(x) + v(1) \\[4pt] \dfrac{dy}{dx} = x\dfrac{dv}{dx} + v \end{array}}$$

$$x\dfrac{dv}{dx} + v = \dfrac{1 - v^2}{v}$$

$$x\dfrac{dv}{dx} = \dfrac{1 - 2v^2}{v}$$

$$\dfrac{v}{1 - 2v^2}\, dv = \dfrac{1}{x}\, dx$$

$$\int \dfrac{v}{1 - 2v^2}\, dv = \int \dfrac{1}{x}\, dx$$

$$\boxed{\begin{array}{l} u = 1 - 2v^2 \\ du = -4v\, dv \end{array}}$$

$$\left(-\dfrac{1}{4}\right) \int \dfrac{1}{u}\, du = \int \dfrac{1}{x}\, dx$$

$$\left(-\dfrac{1}{4}\right) \ln|u| = \ln|x|$$

$$\left(-\dfrac{1}{4}\right) \ln|1 - 2v^2| = \ln|x|$$

$$\left(-\dfrac{1}{4}\right) \ln\left|1 - 2\left(\dfrac{y}{x}\right)^2\right| = \ln|x| + C \qquad \text{G.S.}$$

Extra Example: Try this: $\dfrac{dy}{dx} = \dfrac{x^2 + y^2}{xy}$

Special Substitution DE

Question: What if the DE is NOT First-Order Linear, Separable, Bernoulli, or a Homogeneous DE?

Answer: Try Special Substitution for: $\frac{dy}{dx} = f(x, y)$

Step	General Idea How to solve: $\frac{dy}{dx} = f(x,y)$ With Special Substitution
1.	Find a special combination of variables (x and y) from $f(x,y)$ and call it $v = g(x,y)$. (see example next page) The ideal choice for $g(x,y)$ may be a complicated expression.
2.	Compute: $\frac{dv}{dx} = g_x(x,y) + g_y(x,y) \cdot \frac{dy}{dx}$ Then solve for: $\frac{dy}{dx}$
3.	Replace $g(x,y)$ by: "v" Replace $\frac{dy}{dx}$ by: $\frac{1}{g_y(x,y)} \cdot \frac{dv}{dx} - \frac{g_x(x,y)}{g_y(x,y)}$
4.	Use algebra to convert the DE into a known form. (e.g. First-Order Linear, Separable, Bernoulli, or Homogeneous DE)
5.	Use a known method to compute the G.S. Then, use back substitution to get the G.S. of the original problem. If initial conditions are give, compute the P.S.

Special Substitution Example: Find the G.S. of $\dfrac{dy}{dx} = \sin(x+y)$

Solution:

Let $v = x + y$

$\dfrac{dv}{dx} - 1 = \sin v$

$\dfrac{dv}{dx} = 1 + \sin v$ → Separable DE

> $v = x + y$
> We want $\dfrac{dy}{dx}$
> $\dfrac{dv}{dx} = 1 + \dfrac{dy}{dx}$
> $\dfrac{dy}{dx} = \dfrac{dv}{dx} - 1$

$\dfrac{1}{1+\sin v}\, dv = 1\, dx$

$\displaystyle\int \dfrac{1}{1+\sin v}\, dv = \int 1\, dx$

$\displaystyle\int \dfrac{1}{1+\sin v}\left(\dfrac{1-\sin v}{1-\sin v}\right) dv = x$

$\displaystyle\int \dfrac{1-\sin v}{1-\sin^2 v}\, dv = x$

$\displaystyle\int \dfrac{1-\sin v}{\cos^2 v}\, dv = x$

$\displaystyle\int \dfrac{1}{\cos^2 v} - \dfrac{\sin v}{\cos v}\cdot\dfrac{1}{\cos v}\, dv = x$

$\displaystyle\int \tan^2 v - \tan v \sec v\, dv = x$

$\tan v - \sec v = x + C$

$\tan(x+y) - \sec(x+y) = x + C$ G.S.

Special Substitution Example:

Find the G.S. of $\quad yy' + x = \sqrt{x^2 + y^2}$

Solution:

$\quad$ Let $\quad v = \sqrt{x^2 + y^2}$

$$\boxed{\begin{array}{l} v^2 = x^2 + y^2 \\[4pt] \text{We want } \dfrac{dy}{dx} \text{ or } y\dfrac{dy}{dx} \\[4pt] 2v\dfrac{dv}{dx} = 2x + 2y\dfrac{dy}{dx} \\[4pt] v\dfrac{dv}{dx} = x + y\dfrac{dy}{dx} \\[4pt] y\dfrac{dy}{dx} = v\dfrac{dv}{dx} - x \end{array}}$$

$\left(v\dfrac{dv}{dx} - x\right) + x = \sqrt{x^2 + y^2}$

$v\dfrac{dv}{dx} = \sqrt{v^2}$

$v\dfrac{dv}{dx} = v$

$\dfrac{dv}{dx} = 1$

$dv = dx$

$\int dv = \int dx$

$v = x + C$

$\sqrt{x^2 + y^2} = x + C \qquad\qquad\qquad \text{G.S.}$

Special Substitution Example:

Find the G.S. of $\quad \dfrac{dy}{dx} = \dfrac{y+2}{x+1} + \sec\left(\dfrac{y-2x}{x+1}\right)$

Solution:

$$\boxed{\begin{array}{l} \text{Let } v = \dfrac{y-2x}{x+1} \\[6pt] (x+1)\dfrac{dv}{dx} + v = \dfrac{dy}{dx} - 2 \\[6pt] \dfrac{dy}{dx} = (x+1)\dfrac{dv}{dx} + v + 2 \end{array}}$$

Now:

$$(x+1)\dfrac{dv}{dx} + v + 2 = \dfrac{y+2}{x+1} + \sec\left(\dfrac{y-2x}{x+1}\right)$$

We need a DE in "v" and "x"

$$\boxed{\begin{array}{l} v = \dfrac{y-2x}{x+1} \\[6pt] v(x+1) = y - 2x \\[6pt] y = v(x+1) + 2x \end{array}}$$

$$(x+1)\dfrac{dv}{dx} + v + 2 = \dfrac{[v(x+1) + 2x] + 2}{x+1} + \sec\left(\dfrac{[v(x+1) + 2x] - 2x}{x+1}\right)$$

$$(x+1)\dfrac{dv}{dx} + v + 2 = \dfrac{v(x+1) + 2(x+1)}{x+1} + \sec\left(\dfrac{v(x+1)}{x+1}\right)$$

$$(x+1)\dfrac{dv}{dx} + v + 2 = v + 2 + \sec(v)$$

$$(x+1)\dfrac{dv}{dx} = \sec(v)$$

$$\int (x+1)\, dv = \int \sec(v)\, dx \qquad\qquad \text{(Continued...)}$$

$$\int \frac{1}{\sec(v)} \, dv = \int \frac{1}{(x+1)} \, dx$$

$$\int \cos v \, dv = \int \frac{1}{x+1} \, dx$$

$$\sin v = \ln|x+1| + C \qquad\qquad \text{Recall: } v = \frac{y-2x}{x+1}$$

$$\sin\left(\frac{y-2x}{x+1}\right) = \ln|x+1| + C \qquad\qquad \text{G.S.}$$

Another solution:

$$\frac{dv}{dx}(x+1) + v + 2 = \frac{(2x+v(x+1)) + 2}{x+1} + \sec v$$

$$\frac{dv}{dx}(x+1) + v + 2 = \frac{2x + v(x+1) + 2}{x+1} + \sec v$$

$$\frac{dv}{dx}(x+1) + 2 = \frac{2x+2}{x+1} + \sec v$$

$$\frac{dv}{dx}(x+1) = \frac{2x+2 - 2(x+1)}{x+1} + \sec v$$

$$\frac{dv}{dx}(x+1) = \frac{0}{x+1} + \sec v$$

$$\frac{dv}{dx}(x+1) = \sec v$$

$$\cos v \, dv = \frac{1}{(x+1)} \, dx$$

$$\int \cos v \, dv = \int \frac{1}{(x+1)} \, dx$$

$$\sin v = \ln|x+1| + C \qquad\qquad \text{Recall: } v = \frac{y-2x}{x+1}$$

$$\sin\left(\frac{y-2x}{x+1}\right) = \ln|x+1| + C \qquad\qquad \text{G.S.}$$

Example: Find the G.S. of $(y + \sqrt{xy})\, dx = x\, dy$

Solution:

Let $v = \sqrt{xy}$

$y + \sqrt{xy} = x\dfrac{dy}{dx}$

$\dfrac{v^2}{x} + v = x\left(\dfrac{2v\frac{dv}{dx}x - v^2}{x^2}\right)$

$\dfrac{v^2}{x} + v = \dfrac{2v\frac{dv}{dx}x - v^2}{x}$

$v^2 + vx = 2v\dfrac{dv}{dx}x - v^2$

$v + x = 2\dfrac{dv}{dx}x - v$

$2v + x = 2\dfrac{dv}{dx}x$

$\dfrac{v}{x} + \dfrac{1}{2} = \dfrac{dv}{dx}$

$\dfrac{dv}{dx} - \left(\dfrac{1}{x}\right)v = \dfrac{1}{2}$ $\quad\Rightarrow\quad p(x) = \dfrac{1}{x}$

> $v^2 = xy$
> We want $\dfrac{dy}{dx}$
> $y = \dfrac{v^2}{x}$
> $\dfrac{dy}{dx} = \dfrac{2v\frac{dv}{dx}x - v^2(1)}{x^2}$

(Integrating Factor)

$\mu(x) = e^{\int p(x)\, dx} = e^{-\int\left(\frac{1}{x}\right) dx} = e^{-\ln x} = \dfrac{1}{x}$

$v = \dfrac{\int\left(\frac{1}{x}\right)\left(\frac{1}{2}\right) dx + C}{\left(\frac{1}{x}\right)} = \dfrac{\left(\frac{1}{2}\right)\ln|x| + C}{\left(\frac{1}{x}\right)}$

$v = \sqrt{xy} = \left(\dfrac{1}{2}\right)x\ln|x| + xC$ \qquad G.S.

Example:

Find the G.S. of $xy\,dy = (y^2 + x)dx$

Solution:

Let $v = (y^2 + x)$

$\dfrac{dy}{dx} = \dfrac{(y^2 + x)}{xy}$

$2y\dfrac{dy}{dx} = 2\dfrac{(y^2 + x)}{x}$

$\dfrac{dv}{dx} - 1 = \dfrac{2v}{x}$

$\dfrac{dv}{dx} - \left(\dfrac{2}{x}\right)v = 1$

$v = y^2 + x$ We want $\dfrac{dy}{dx}$

$\dfrac{dv}{dx} = 2y\dfrac{dy}{dx} + 1$

$2y\dfrac{dy}{dx} = \dfrac{dv}{dx} - 1$

Integrating Factor

$\mu(x) = e^{\int p(x)dx} = e^{\int \left(-\frac{2}{x}\right)dx} = e^{-2\ln x} = \dfrac{1}{x^2}$

$v = \dfrac{\int \left(\frac{1}{x^2}\right)(1)dx + C}{\left(\frac{1}{x^2}\right)} = \dfrac{\left(-\frac{1}{x}\right) + C}{\left(\frac{1}{x^2}\right)} = -x + Cx^2$

$v = -x + Cx^2$

$y^2 + x = -x + Cx^2$

$y^2 = -2x + Cx^2$ G.S.

Strategies to compute the G.S. (or P.S.) of a First-Order DE
Check for First-Order Linear DE. If YES, then: • $\dfrac{dy}{dx} + p(x)y = g(x)$ • $\mu(x) = e^{\int p(x)dx}$ Integrating Factor • $y = \dfrac{\int \mu(x) \cdot g(x) dx + C}{\mu(x)}$
Check for Separable DE. If YES, then: • $\dfrac{dy}{dx} = g(x) \cdot h(y)$ • $\int \dfrac{1}{h(y)} dy = \int g(x)\, dx$
Check for EXACT DE. (Next Topic)
Consider Substitution: • Bernoulli DE　　　　　　　　　→ Use: $v = y^{1-n}$ • First Order Homogeneous DE → Use: $v = \dfrac{y}{x}$ • Special Substitution → Convert to First-Order Linear 　or Separable DE　　　　　　　→ Use: $v = g(x,y)$

Before introducing a new technique, consider the following example: Find the G.S. of $(2x + y^2) + 2xy\frac{dy}{dx} = 0$

Solution:

Note that this DE is not First-Order Linear, Separable, Bernoulli, or Homogeneous.

Consider:

$\psi(x,y) = x^2 + xy^2$ (Computing $\psi(x,y)$ will be discussed later.)

So:

$\psi_x(x,y) = 2x + y^2$ (This is part of the given DE.)

$\psi_y(x,y) = 2xy$ (This is also part of the given DE.)

The DE can be rewritten as:

$\psi_x + \psi_y \frac{dy}{dx} = 0$ Product Rule

$\frac{d}{dx}(\psi(x,y)) = 0$ Integrate wrt "x"

$\psi(x,y) + C = 0$ ➔ $x^2 + xy^2 + C = 0$ G.S.

Result: The Second Order, mixed partial derivatives of $\psi(x,y)$ are the same. In other words:

$$\psi_{xy}(x,y) = \psi_{yx}(x,y)$$

or $$\frac{\partial^2 \psi}{\partial x\, \partial y} = \frac{\partial^2 \psi}{\partial y\, \partial x}$$

Remark: A First-Order DE can be written in one of the following two forms. Both forms are equivalent and they represent the same DE.

Forms of a First-Order DE	
Form (1)	$\frac{dy}{dx} = f(x,y)$
Form (2)	$M(x,y) + N(x,y)\frac{dy}{dx} = 0$ $M(x,y)dx + N(x,y)dy = 0$

Examples of a First-Order DE in both forms.	
$\frac{dy}{dx} = \frac{2x^2 + 3y}{4x + 5y}$	Form (1)
$\underbrace{(2x^2 + 3y)}_{M} - \underbrace{(4x + 5y)}_{N}\frac{dy}{dx} = 0$	Form (2)
$\frac{dy}{dx} = 2x^2 + y^3 + 9$	Form (1)
$\underbrace{(2x^2 + y^3 + 9)}_{M} - \underbrace{(1)}_{N}\frac{dy}{dx} = 0$	Form (2)

Exact Equations

> **EXACT DE** If M, N, M_y and N_x are continuous functions on the rectangular region R, where $R: = \{(x,y) | \alpha < x < \beta,\ \mu < y < \gamma\}$
> Then, $\quad M(x,y) + N(x,y)\frac{dy}{dx} = 0 \quad$ is EXACT
> $\quad\quad$ iff $\quad M_y(x,y) = N_x(x,y)$

Consider: $\quad M(x,y) + N(x,y)\frac{dy}{dx} = 0$

IF EXACT → Then, there exists a solution: $\psi(x,y)$

$\quad\quad$ s.t. $\quad M(x,y) = \psi_x(x,y) \quad$ AND $\quad N(x,y) = \psi_y(x,y)$

Step	General Idea how to solve an EXACT DE
1.	Write DE in std. form: $M(x,y) + N(x,y)\frac{dy}{dx} = 0$
2.	Compute M_y and N_x $\quad$ Check: $M_y = N_x$ If yes → EXACT → Solution exists.
3.	Label: $M(x,y) = \psi_x(x,y)$ and $N(x,y) = \psi_y(x,y)$
4.	Integrate ψ_x wrt x $\quad$ (or Integrate ψ_y wrt y) $\int \psi_x(x,y)dx = \psi(x,y) + f(y)$
5.	Differentiate wrt y partially and equate with $\psi_y(x,y)$ → $f'(y) =$ expression in "y" Integrate wrt "y" and get $f(y)$
6.	Compute G.S.

* wrt = "with respect to"

Step	EXACT DE Example: Find the G.S. of $(2x + y^2) + 2xy \frac{dy}{dx} = 0$
1.	$M(x, y) = 2x + y^2 \qquad N(x, y) = 2xy$
2.	$M_y = 2y$ and $N_x = 2y$ $M_y = N_x$ → EXACT → There is a solution $\psi(x, y)$ s.t. $\psi_x = M(x, y)$ and $\psi_y = N(x, y)$
3.	$\psi_x = M(x, y) = 2x + y^2$ $\psi_y = N(x, y) = 2xy$
4.	$\psi(x, y) = \int \psi_x(x, y)\, dx = \int 2x + y^2\, dx$ $\psi(x, y) = x^2 + xy^2 + f(y)$ Note: Constant of integration is a function of "y"
5.	Now find $f(y)$ The constant of integration. $\psi_y(x, y) = 0 + 2xy + f'(y)$ $2xy = 2xy + f'(y)$ $0 = f'(y)$ Here, constant of integration = 0 Integrate wrt y → $f(y) = C$
6.	General Solution: $\psi(x, y) = x^2 + xy^2 + C$ Note: $\psi(x, y)$ is an arbitrary function. For simplicity, set $\psi(x, y) = 0$ Therefore: $x^2 + xy^2 + C = 0 \qquad$ G.S.

Step	EXACT DE Example: Find the G.S. of: $(y \cos x + 2xe^y)dx + (\sin x + x^2 e^y + 3y^2 - 5)\, dy = 0$
1.	$M = (y \cos x + 2x\, e^y) \quad\quad\Rightarrow\quad M_y = \cos x + 2x\, e^y$ $N = \sin x + x^2 e^y + 3y^2 - 5 \quad\Rightarrow\quad N_x = \cos x + 2x\, e^y$ $M_y = N_x \;\Rightarrow\;$ EXACT $\;\Rightarrow\;$ A solution exists. $\psi(x,y)$
2.	$\psi_x = M = y \cos x + 2xe^y$ $\psi_y = N = \sin x + x^2 e^y + 3y^2 - 5$ Just pick one to work with.
3.	Let's pick ψ_x to find: $\psi = \int \psi_x\, dx$ $\psi_x = y \cos x + 2xe^y$ $\int \psi_x\, dx = \int y \cos x + 2xe^y\, dx$ $\psi = y \sin x + x^2 e^y + f(y)$ $f(y) =$ Integration Constant
4.	$\psi_y = \sin x + x^2 e^y + f'(y)$ Differentiate both sides wrt y $\sin x + x^2 e^y + 3y^2 - 5 = \sin x + x^2 e^y + f'(y)$ $3y^2 - 5 = f'(y)$
5.	$\int 3y^2 - 5\, dy = \int f'(y)\, dy$ Integrate both sides wrt y $3xy^2 - 5y + C = f(y)$
6.	$y = \psi = y \sin x + x^2 e^y + f(y)$ $y = y \sin x + x^2 e^y + 3xy^2 - 5y + C$ G.S.

Step	EXACT DE Example: Find the G.S. of: $\left(\frac{y}{x} + 6x\right)dx + (\ln x - 3y^2 + 2y + 4)dy = 0$
1.	Check to see if the DE is exact. $M = \frac{y}{x} + 6x$ → $M_y = \frac{1}{x}$ $N = \ln x - 3y^2 + 2y + 4$ → $N_x = \frac{1}{x}$ → EXACT ☺ EXACT → ∃ a solution $\psi(x, y)$
2.	Find ψ_x and ψ_y $M(x, y) = \psi_x(x, y) = \frac{y}{x} + 6x$ $N(x, y) = \psi_y(x, y) = \ln x - 3y^2 + 2y + 4$
3.	Find: $\psi(x, y)$ $\psi(x, y) = \int \psi_x(x, y)\, dx = \int \frac{y}{x} + 6x\, dx$ $\psi(x, y) = y \ln x + 3x^2 + f(y)$; $f(y) =$ Const. of Integration
4.	Step 4: Differentiate both sides. Must find $f(y)$. → Differentiate wrt y $\psi_y(x, y) = \ln x + f'(y)$ Recall: $N(x, y) = \psi_y(x, y)$ $\ln x - 3y^2 + 2y + 4 = \ln x + f'(y)$ $-3y^2 + 2y + 4 = f'(y)$

Continued …

Continued ...

Step	EXACT DE Example: Find the G.S. of: $\left(\frac{y}{x} + 6x\right)dx + (\ln x - 3y^2 + 2y + 4)dy = 0$
5.	Integrate both sides to find $f(y)$. $\int -3y^2 + 2y + 4 \ dy = \int f'(y) \ dy$ $-y^3 + y^2 + 4y + C = f(y)$
6.	Write the General Solution (G.S.) $\psi(x,y) = y \ln x + 3x^2 + f(y)$ $\psi(x,y) = y \ln x + 3x^2 + y^3 + y^2 + 4y + C$ Use: $\psi(x,y) = 0$ So... $y \ln x + 3x^2 + y^3 + y^2 + 4y + C = 0$ G.S.

Integrating Factor

INTEGRATING FACTOR:

Question: What if the DE is not EXACT? → $M_y \neq N_x$

Answer: In some cases, it is possible to convert the DE into exact form by multiplying it by an integrating factor (μ). In general, $\mu = \mu(x, y)$ is a function of variables x and y. In some special cases, μ is a function of a single variable.

Notes:

- Computing the integrating factor is as difficult as solving the DE most of the time.

- It is possible to get more than one integrating factor for a first-order DE because there are two different ways to compute the integrating factor, $\mu(x, y)$.

- One way to compute the integrating factor is to solve:

$$\frac{d\mu}{dx} = \frac{M_y - N_x}{N} \mu$$

Where $\frac{M_y - N_x}{N} = f(x)$ only → Separable DE in μ and x

- Another way to compute the integrating factor is to solve:

$$\frac{d\mu}{dy} = \frac{N_x - M_y}{M} \mu$$

Where $\frac{N_x - M_y}{M} = f(y)$ only → Separable DE in μ and y

Case	INTEGRATING FACTOR SUMMARY of all possible cases
	For a DE in the form: $M(x,y) + N(x,y)\frac{dy}{dx} = 0$
i.	Compute: $\frac{M_y - N_x}{N}$
	Check if it is a factor of ONE variable. Then: $p(x) = \frac{M_y - N_x}{N}$
	$\frac{d\mu}{dx} = p(x) \cdot \mu$ $\quad\rightarrow\quad$ $\frac{1}{\mu} d\mu = p(x)dx$
	$\int \frac{1}{\mu} d\mu = \int p(x)\, dx$ $\quad\rightarrow\quad$ $\ln \mu = \int p(x)\, dx$
	$\mu = e^{\ln\mu} = e^{\int p(x)\, dx}$ $\quad$ Integrating Factor.
ii.	Compute: $\frac{N_x - M_y}{M}$
	Check if it is a factor of ONE variable. Then: $p(y) = \frac{N_x - M_y}{M}$
	$\frac{d\mu}{dy} = p(y) \cdot \mu$ $\quad\rightarrow\quad$ $\frac{1}{\mu} d\mu = p(y)dy$
	$\int \frac{1}{\mu} d\mu = \int p(y)\, dy$ $\quad\rightarrow\quad$ $\ln \mu = \int p(y)\, dy$
	$\mu = e^{\ln\mu} = e^{\int p(y)\, dy}$ $\quad$ Integrating Factor
iii.	If both: $\frac{M_y - N_x}{N} = p(x,y)$ $\quad$ AND $\quad$ $\frac{N_x - M_y}{M} = p(x,y)$
	Then the integrating factor is a function of two variables (if it exists). This is a more difficult situation! ☹ Some very special cases of this form will be discussed later.

Example: Find the G.S. of: $(3xy + y^2) + (x^2 + xy) y' = 0$
$\phantom{\text{Example: Find the G.S. of:}\quad}$ M N

Solution:

$M = 3xy + y^2 \;\Rightarrow\; M_y = 3x + 2y$

$N = x^2 + xy \;\Rightarrow\; N_x = 2x + y$ $\qquad$ Note: $M_y \neq N_x$

$\Rightarrow$ Not Exact ☹

Compute the integrating factor. (We may have to check both.)

$\dfrac{d\mu}{dx} = \dfrac{M_y - N_x}{N} \mu$

$\dfrac{d\mu}{dx} = \dfrac{3x + 2y - 2x - y}{x^2 + xy} \mu = \dfrac{x + y}{x(x+y)} \mu = \dfrac{1}{x} \mu$

$\dfrac{d\mu}{dx} = \dfrac{1}{x} \mu$ $\qquad$ It is a Separable DE

$\dfrac{1}{\mu} d\mu = \dfrac{1}{x} dx$

$\int \dfrac{1}{\mu} d\mu = \int \dfrac{1}{x} dx$

$\ln |\mu| = \ln |x|$ $\qquad \Rightarrow \;\mu = x \qquad$ Integrating factor!!!

Multiply both sides of the original D.E. by integrating factor.

$(3xy + y^2) + (x^2 + xy) y' = 0$

$x(3xy + y^2) + x(x^2 + xy) y' = x \cdot 0$

$(3x^2 y + xy^2) + (x^3 + x^2 y) y' = 0$
 M N

This will be exact. ☺ $\quad$ (Continued…)

$M_y = 3x^2 + 2xy$

$N_x = 3x^2 + 2xy$ \hspace{2em} Now, $M_y = N_x$ → Exact ☺

$\psi_x = M = 3x^2y + xy^2$

$\psi_y = N = x^3 + x^2y$ \hspace{2em} Let's use ψ_y to find ψ

$\psi_y = x^3 + x^2y$

$\int \psi_y \, dy = \int x^3 + x^2y \, dy$

$\psi = x^3y + \left(\frac{1}{2}\right)x^2y^2 + f(x)$ \hspace{2em} $f(x) =$ Constant of Integration

$\psi_x = 3x^2y + \left(\frac{1}{2}\right)2xy^2 + f'(x)$

$\psi_x = 3x^2y + xy^2 + f'(x)$

$3x^2y + xy^2 = 3x^2y + xy^2 + f'(x)$

$0 = f'(x)$

$f'(x) = 0$ → $f(x) = C$

$\psi = x^3y + \left(\frac{1}{2}\right)x^2y^2 + f(x)$

$\psi = x^3y + \left(\frac{1}{2}\right)x^2y^2 + C$ \hspace{2em} Set $\psi = 0$

$x^3y + \left(\frac{1}{2}\right)x^2y^2 + C = 0$ \hspace{2em} G.S.

Example:

Find the G.S. of: $(x+y)\sin y + (x\sin y + \cos y)y' = 0$

Solution:

$M = (x+y)\sin y$ → $M_y = (1)\sin y + (x+y)\cos y$

$N = x\sin y + \cos y$ → $N_x = \sin y$

$M_y \neq N_x$ → Not Exact ☹

Compute: $\dfrac{M_y - N_x}{N} = \dfrac{\sin y + (x+y)\cos y - \sin y}{x\sin y + \cos y}$

$= \dfrac{(x+y)\cos y}{x\sin y + \cos y} \neq p(x)$

Compute: $\dfrac{N_x - M_y}{M} = \dfrac{\sin y - \sin y - (x+y)\cos y}{(x+y)\sin y}$

$= \dfrac{-(x+y)\cos y}{(x+y)\sin y} = \dfrac{-\cos y}{\sin y}$

$= -\cot y = p(y)$

$\mu = e^{\int p(y)dy} = e^{\int -\cot y\, dy} = e^{-\ln|\sin y|} = \dfrac{1}{|\sin y|} = \dfrac{1}{\sin y}$

Multiply the DE by the integrating factor to make it exact.

$\left(\dfrac{1}{\sin y}\right)(x+y)\sin y + \left(\dfrac{1}{\sin y}\right)(x\sin y + \cos y)y' = \left(\dfrac{1}{\sin y}\right)0$

$(x+y) + \left(\dfrac{x\sin y + \cos y}{\sin y}\right)y' = 0$

$(x+y) + (x + \cot y)y' = 0$ (Continued...)

$(x + y) + (x + \cot y)y' = 0$

$M = x + y$ ➔ $M_y = 1$

$N = x + \cot y$ ➔ $N_y = 1$

$M_y = N_x$ ➔ EXACT ➔ There is a solution: $\psi(x, y)$

$M = \psi_x(x, y) = x + y$

$N = \psi_y(x, y) = x + \cot y$ Work with either ψ_x or ψ_y

$\psi(x, y) = \int \psi_x \, dx$

$\psi(x, y) = \int x + y \, dx = \dfrac{x^2}{2} + xy + f(y)$

$\psi_y = 0 + x + f'(y)$

Differentiate both sides of $\psi(x, y)$ wrt y

$\quad x + \cot y = x + f'(y)$

$\quad f'(y) = \cot y$

Integrate $f'(y)$ to find $f(y)$.

$\quad \int f'(y) \, dy = \int \cot y \, dy$

$\quad f(y) = \ln|\sin y| + C$

$\psi(x, y) = \dfrac{x^2}{2} + xy + \ln|\sin y| + C = 0$ G.S.

Example:

Find the G.S. of: $(x+y)\sin y + (x\sin y + \cos y)y' = 0$

Using $\mu = \mu(y)$

Solution:

Here, we are told the Integrating Factor (I.F.) is a function of y so there is NO need to verify this! Very useful hint!!! ☺

We know that if we multiply both sides of the equation by μ then it will be EXACT. We can use this fact to help solve for μ

Recall: $\frac{d}{dx}(u,v,w) = uvw' + uwv' + vwu'$

Product Rule for 3 factors.

$M = (x+y)\sin y \cdot \mu(y)$

$M_y = (x+y)\sin y \, u' + (x+y)\cos y \, \mu + \sin y \, \mu$

$N = (x\sin y + \cos y) \cdot \mu(y)$

$N_x = \sin y \cdot \mu$

(Continued...)

$M_y = N_x$ \qquad\qquad Because the DE is EXACT

$(x+y)\sin y \; u' + (x+y)\cos y \; \mu + \sin y \; \mu = \sin y \cdot \mu$

$(x+y) u' + \dfrac{(x+y)\cos y}{\sin y} \mu + \mu = \mu$

$(x+y) u' + (x+y) \cot y \; \mu = 0$

$u' + \cot y \; \mu = 0$

$u' = -\cot y \; \mu$

$\dfrac{1}{\mu} \cdot \dfrac{d\mu}{dy} = -\cot y$

$\dfrac{1}{\mu} d\mu = -\cot y \; dy$

$\int \dfrac{1}{\mu} d\mu = -\int \cot y \; dy$

$\ln|\mu| = -\ln|\sin y|$

$\mu = e^{-\ln|\sin y|} = \dfrac{1}{|\sin y|} = \dfrac{1}{\sin y}$ \qquad Assume it's positive.

Now we know the integrating factor.

We have already solved this problem. The second half of this problem is the same as the second half of the previous problem.

Example: (Revisit a problem done previously.)

Find the G.S. of: $(3xy + y^2) + (x^2 + xy)y' = 0$ Using $\mu = \mu(x)$

Solution:

Find μ then multiply both sides of the DE to get an exact DE

$M = (3xy + y^2)\mu \;\Rightarrow\; M_y = (3x + 2y)\mu$ Note: $\mu = \mu(x)$

$N = \mu(x^2 + xy) \;\Rightarrow\; N_x = \mu'(x^2 + xy) + \mu(2x + y)$

$M_y = N_x$ Because it's EXACT

$(3x + 2y)\mu = \mu'(x^2 + xy) + \mu(2x + y)$

$\mu'(x^2 + xy) = (3x + 2y - 2x - y)\mu$

$\mu' x(x + y) = (x + y)\mu$

$\mu' x = \mu$

$\dfrac{d\mu}{dx} x = \mu$ Note: $\mu = \mu(x)$

$\dfrac{1}{\mu} d\mu = \dfrac{1}{x} dx$

$\int \dfrac{1}{\mu} d\mu = \int \dfrac{1}{x} dx$

$\ln|\mu| = \ln|x|$ $\Rightarrow \mu = x$ Integrating factor.

$x(3xy + y^2) + x(x^2 + xy)y' = x \cdot 0$ Now, the DE is exact

$(3x^2 y + xy^2) + (x^3 + x^2 y)y' = 0$

(Continued...)

$M = 3x^2y + xy^2 \rightarrow M_y = 3x^2 + 2xy$

$N = x^3 + x^2y \rightarrow N_x = 3x^2 + 2xy$ Note: $M_y = N_x$

→ EXACT (as expected)

Since it is exact, then:

$M = \psi_x = 3x^2y + xy^2$

$N = \psi_y = x^3 + x^2y$

$\psi = \int \psi_y \, dy = \int x^3 + x^2y \, dy$

$\psi = x^3y + \left(\frac{1}{2}\right)x^2y^2 + f(x)$

$\psi_x = 3x^2y^2 + xy^2 + f'(x)$

$3x^2y^2 + xy^2 + f'(x) = 3x^2y + xy^2$

$f'(x) = 0 \rightarrow f(x) = C$

$\psi = x^3y + \left(\frac{1}{2}\right)x^2y^2 + C = 0$ G.S.

Example: Find the G.S. of: $\left(3x + \dfrac{6}{y}\right)dx + \left(\dfrac{x^2}{y} + \dfrac{3y}{x}\right)dy = 0$

With: $\mu(x,y) = \mu(xy)$

Solution:

Multiply both sides by μ to make it EXACT.

$\mu\left(3x + \dfrac{6}{y}\right)dx + \mu\left(\dfrac{x^2}{y} + \dfrac{3y}{x}\right)dy = \mu \cdot 0$

$\mu(xy) \cdot F(x,y)\, dx + \mu(xy) \cdot G(x,y)\, dy = 0$

$M = \mu(xy) \cdot F(x,y) \;\;\Rightarrow\;\; M_y = \mu(xy)\dfrac{\partial F}{\partial y} + F \cdot \mu'(xy) \cdot x$

$N = \mu(xy) \cdot G(x,y) \;\;\Rightarrow\;\; N_x = \mu(xy)\dfrac{\partial G}{\partial x} + G \cdot \mu'(xy) \cdot y$

$M_y = N_x$ Because the DE is EXACT

$\mu(xy)\dfrac{\partial F}{\partial y} + F \cdot \mu'(xy) \cdot x = \mu(xy)\dfrac{\partial G}{\partial x} + G \cdot \mu'(xy) \cdot y$

$\mu\dfrac{\partial F}{\partial y} + F\mu' x = \mu\dfrac{\partial G}{\partial x} + G\mu' y$ Same eqn.

(Continued...)

Solve for μ'

$(\ )\mu' = (\ )\mu$

$(xF - yG)\mu' = \left(\dfrac{\partial G}{\partial x} - \dfrac{\partial F}{\partial y}\right)\mu$

$\dfrac{\mu}{\mu'} = \dfrac{\left(\dfrac{\partial G}{\partial x} - \dfrac{\partial F}{\partial y}\right)}{(xF - yG)}$

Compute Denominator: $(xF - yG)$

$F = 3x + \dfrac{6}{y}$ and $G = \dfrac{x^2}{y} + \dfrac{3y}{x}$

$(xF - yG) = 3x^2 + \dfrac{6x}{y} - x^2 - \dfrac{3y^2}{x}$

$\qquad\qquad = 2x^2 + \dfrac{6x}{y} - \dfrac{3y^2}{x}$

$\qquad\qquad = \dfrac{2x^3y + 6x^2 - 3y^3}{xy}$

Compute the Numerator: $\left(\dfrac{\partial G}{\partial x} - \dfrac{\partial F}{\partial y}\right)$

$\dfrac{\partial F}{\partial y} = -\dfrac{6}{y^2}$ and $\dfrac{\partial G}{\partial x} = \dfrac{2x}{y} - \dfrac{3y}{x^2}$

$\left(\dfrac{\partial G}{\partial x} - \dfrac{\partial F}{\partial y}\right) = \dfrac{2x}{y} - \dfrac{3y}{x^2} + \dfrac{6}{y^2}$

$\qquad\qquad\qquad = \dfrac{2x^3y - 3y^3 + 6x^2}{x^2y^2}$

$\dfrac{\mu}{\mu'} = \dfrac{\left(\dfrac{1}{x^2y^2}\right)}{\left(\dfrac{1}{xy}\right)} = \dfrac{xy}{x^2y^2} = \dfrac{1}{xy}$ \qquad (Continued...)

$$\frac{\mu}{\mu'} = \frac{1}{xy}$$

Let $z = xy$

$$\frac{\mu'(z)}{\mu(z)} = \frac{1}{z}$$

$$\frac{1}{\mu}\frac{d\mu}{dz} = \frac{1}{z} \qquad \text{Separable DE}$$

$$\int \frac{1}{\mu}\, d\mu = \int \frac{1}{z}\, dz$$

$$\ln|\mu| = \ln|z| \quad \rightarrow \quad \mu = z = xy$$

Original DE was: $\left(3x + \frac{6}{y}\right)dx + \left(\frac{x^2}{y} + \frac{3y}{x}\right)dy = 0$

If the original DE is multiplied by the integrating factor, it will be EXACT. So, multiply both sides of the DE by $\mu = xy$.

$$xy\left(3x + \frac{6}{y}\right)dx + xy\left(\frac{x^2}{y} + \frac{3y}{x}\right)dy = xy \cdot 0$$

$$\underbrace{(3x^2y + 6x)dx}_{M} + \underbrace{(x^3 + 3y^2)dy}_{N} = 0$$

$M = 3x^2y + 6x \quad \rightarrow \quad M_y = 3x^2$

$N = x^3 + 3y^2 \quad \rightarrow \quad N_x = 3x^2 \qquad$ Note: $M_y = N_x$

$\rightarrow$ EXACT as expected.

(Continued...)

$M = \psi_x = 3x^2y + 6x$

$N = \psi_y = x^3 + 3y^2$

$\psi = \int \psi_x \, dx = \int 3x^2y + 6x \, dx = x^3y + 3x^2 + f(y)$

$\psi_y = x^3 + f'(y)$

$\psi_y = x^3 + f'(y) = x^3 + 3y^2 \quad \rightarrow \quad f'(y) = 3y^2$

$f'(y) = 3y^2$

$\int f'(y) \, dy = \int 3y^2 \, dy$

$f(y) = y^3 + C$

$\psi = x^3y + 3x^2 + y^3 + C = 0$ \quad\quad G.S.

Example: Compute the Integrating factor of the form $\mu = \mu(x, y)$ to make the given DE Exact.

$$(7x^3 + 3x^2y + 4y) + (4x^3 + x + 5y)y' = 0$$
$$F G$$

Solution:

Since $\mu(x, y)$ is the integrating factor, then:

$$\mu F + \mu G y' = 0 \quad \text{is EXACT}$$

So:

$M = \mu F \;\; \rightarrow \;\; M_y = \mu F_y + F\mu'(x, y) \cdot (1) = \mu F_y + F\mu'$

$N = \mu G \;\; \rightarrow \;\; N_y = \mu G_x + G\mu'$

Since the DE is Exact:

$M_y = N_x$

$\mu F_y + F\mu' = \mu G_x + G\mu'$

$(F - G)\mu' = (G_x - F_y)\mu$

$$\frac{\mu'}{\mu} = \frac{G_x - F_y}{F - G} \text{Equation (1)}$$

$F = 7x^3 + 3x^2y + 4y \rightarrow F_y = 3x^2 + 4$

$G = 4x^3 + x + 5y \rightarrow G_x = 12x^2 + 1$

$G_x - F_y = 3(3x^2 - 1) $ Equation (2)

(Continued…)

$$F - G = (7x^3 + 3x^2y + 4y) - (4x^3 + x + 5y)$$
$$F - G = 7x^3 + 3x^2y + 4y - 4x^3 - x - 5y$$
$$F - G = 3x^3 + 3x^2y - x - y$$
$$F - G = 3x^2(x + y) - (x + y)$$
$$F - G = (3x^2 - 1) \cdot (x + y) \qquad \text{Equation (3)}$$

Substituting (2) and (3) into (1)

$$\frac{\mu'}{\mu} = \frac{G_x - F_y}{F - G}$$

$$\frac{\mu'}{\mu} = \frac{3(3x^2 - 1)}{(3x^2 - 1) \cdot (x + y)} = \frac{3}{x + y}$$

Let: $z = x + y$

$$\frac{\mu'(z)}{\mu(z)} = \frac{3}{z}$$

$$\frac{1}{\mu} \cdot \frac{d\mu}{dz} = \frac{3}{z}$$

$$\int \frac{1}{\mu} d\mu = \int \frac{3}{z} dz$$

$$\ln \mu = 3 \ln z \qquad \rightarrow \quad \mu = z^3 = (x + y)^3$$

Multiply original DE by the integrating factor (μ) to make it exact.

$$(7x^3 + 3x^2y + 4y) + (4x^3 + x + 5y)y' = 0$$
$$(x + y)^3(7x^3 + 3x^2y + 4y) + (x + y)^3(4x^3 + x + 5y)y' = 0$$

We know how to compute the G.S. in this case. ☺

Strategies for First-Order DE and Other Strategies

#	Strategies to compute the G.S. for a First-Order DE $\frac{dy}{dx} = f(x,y)$
1.	Check for First-Order Linear DE.
2.	Check for Separable DE.
3.	Check for Exact DE.
4.	Check for Bernoulli's DE.
5.	Check for Homogeneous DE.
6.	Check if $\frac{M_y - N_x}{N} = f(x)$ OR $\frac{N_x - M_y}{M} = f(y)$ If yes, compute the integrating factor to make the equation exact. Then compute the G.S.
7.	Check for special substitution. Find a combination of x and y from $f(x,y)$ and call it $z = g(x,y)$. Based on this substitution, write the DE in the form of $\frac{dz}{dx}$ and $h(x,z)$. Check if new DE is one of the above cases. If yes, then compute the G.S.
8.	Compute the integrating factor $\mu(x,y)$. If $\frac{\mu'}{\mu}$ can be expressed as a special combination of x and y (i.e. $x+y$, xy, $\frac{1}{xy}$, etc.) then compute the integrating factor to make the DE exact and compute the G.S.

If none of the previous strategies work, one of the additional strategies may be used.

	Some Additional Strategies to compute the G.S. for a First-Order DE $\frac{dy}{dx} = f(x,y)$
Graphical Method	<ul><li>This method was used in Integral Calculus.</li><li>The right side of the DE represents the slope of the solution of the DE.</li><li>One can sketch the Direction Field and, based on it, solution trajectories can be sketched.</li></ul>
Euler's Method	<ul><li>For Initial Value Problems (IVP).</li><li>This is a numerical method to estimate the solution of the IVP on the desired interval.</li><li>To solve: $\frac{dy}{dx} = f(x,y)$, $y(x_0) = y_0$ on $[x_0, x_n]$ Use the formula: $y_{i+1} = y_i + f(x_i, y_i) \cdot h$ Where: $h = \frac{x_n - x_0}{n}$ n = # of partition of interval $[x_0, x_n]$ $x_i = x_{i-1} + h$</li></ul>

Existence and Uniqueness Theorem

Question: How do we know if the DE has a solution?
Answer: Existence and Uniqueness Theorem.

\multicolumn{2}{c}{Existence and Uniqueness Theorem of a G.S. for a Linear DE $\frac{dy}{dx} = f(x,y)$}	
General Theorem	• Consider the IVP for: $\frac{dy}{dx} = f(x,y)$, $y(x_0) = y_0$ • If f and f' are continuous on a rectangular region $R = \{(x,y) \mid \alpha < x < \beta,\ \mu < y < \delta\}$ containing the point (x_0, y_0) • Then $\exists$ (there exists) a unique solution $y = \phi(x)$ of the given IVP. (See figure (1a))
For a First Order Linear DE	• Consider the Initial Value Problem (IVP) for: $\frac{dy}{dx} + p(x) \cdot y = g(x)$, $y(x_0) = y_0$ • If $p(x)$ and $g(x)$ are continuous on interval I containing the point (x_0, y_0) • Then $\exists$ (there exists) a unique solution $y = \phi(x)$ of the given IVP. (See figure (1b))

Figure (1a)

Figure (1b)

SECOND-ORDER DE

This chapter discusses special cases of the second-order linear DE and their solution techniques. We begin with the simplest case, the second-order linear homogeneous DE with constant coefficients (e.g. $ay'' + by' + cy = 0$). Notice that coefficients being constant and right side being zero are heavy restrictions on the DE.

For a second-order DE, it is necessary to compute two linearly independent solutions. The Wronkian determinant is introduced to check linear independence of the solutions. If one solution is known, a special method (reduction of order) may be used to find the other solution.

We make second-order DEs more interesting by releasing the restrictions, one-by-one. Methods to solve a second-order linear non-homogeneous DE are discussed with a series of examples. This idea can be extended naturally to solve higher-order DEs with constant coefficients which are discussed briefly. Euler's DE, a very special type of DE, can be converted to a second-order linear DE using proper substitution.

Second-order linear DEs have many interesting applications in the real world. They are vital to any serious investigation of the classical areas of mathematical physics. Oscillation of some basic mechanical and electrical systems are discussed at the end of this chapter.

SECOND-ORDER LINEAR DE WITH CONSTANT COEFFICIENTS

Second-Order DE ➔ Form:	$\frac{d^2y}{dx^2} = f\left(x, y, \frac{dy}{dx}\right)$

Need 2 initial conditions: $y(x_0) = y_0$ & $y'(x_0) = y_0'$

Examples:

- $\frac{d^2y}{dx^2} = x^2 + y^3 + \sin\left(x + \frac{dy}{dx}\right)$
- $\frac{d^2y}{dx^2} = \ln\left(x + y + \frac{dy}{dx}\right)$

It may be difficult to solve a general Second Order DE due to the complexity of the right side, so we will focus on special cases.

Differential Eqn.	Format
Second-Order DE	$\frac{d^2y}{dx^2} = f\left(x, y, \frac{dy}{dx}\right)$
Second-Order Linear Non-Homogeneous DE	$P(x)\frac{d^2y}{dx^2} + Q(x)\frac{dy}{dx} + R(x)y = G(x)$ OR $\frac{d^2y}{dx^2} + p(x)\frac{dy}{dx} + q(x)y = g(x)$ Where: $p(x) = \frac{Q(x)}{P(x)}, \quad q(x) = \frac{R(x)}{P(x)}, \quad g(x) = \frac{G(x)}{P(x)}$
2nd Order Linear Homogeneous DE	$\frac{d^2y}{dx^2} + p(x)\frac{dy}{dx} + q(x)y = 0$
2nd Order Linear Homogeneous DE with Constant Coefficients	$a\frac{d^2y}{dx^2} + b\frac{dy}{dx} + cy = 0$ Start with simplest case because solving 2nd Order Linear DE may be challenging.

Characteristic Equation With Real and Distinct Roots

Case (i)
Second-Order Linear Homogeneous DE with Constant Coefficients Where the Characteristic Equation has real and distinct roots.

- Form: $a\dfrac{d^2y}{dx^2} + b\dfrac{dy}{dx} + cy = 0$

- This type of DE can be solved! $ay'' + by' + cy = 0$

- Solution in the form: $y = e^{rx}$ ➔ $\dfrac{dy}{dx} = re^{rx}$ & $\dfrac{d^2y}{dx^2} = r^2 e^{rx}$

Substitute:

- $a(r^2 e^{rx}) + b(re^{rx}) + c(e^{rx}) = 0$
- $e^{rx}(ar^2 + br + c) = 0$
- $e^{rx} > 0$ ➔ $(ar^2 + br + c) = 0$
- Characteristic Equation:

  $ar^2 + br + c = 0$

This is similar to the solution for a First-Order Linear DE. Recall:

$\dfrac{dy}{dx} + ay = 0$

$\int \dfrac{1}{y}\, dy = \int -a\, dx$

$\ln y = -ax + b$

$y = e^{-ax} e^b$

$\quad e^b =$ a constant

$y = Ce^{-ax}$ $\qquad$ (*)

- The solution(s) are: $r_{1,2} = \dfrac{-b \pm \sqrt{b^2 - 4ac}}{2a}$

  Where r_1 and r_2 are real and distinct roots.

(*) It is reasonable to assume that a higher-order Linear DE, with constant coefficients, also has solutions in the format of $y = e^{rx}$.

Example: Solve the DE $\dfrac{d^2y}{dx^2} - \dfrac{dy}{dx} - 6y = 0$

Solution:

Let $\quad y = e^{rx} \quad \rightarrow \quad \dfrac{dy}{dx} = re^{rx} \quad$ and $\quad \dfrac{d^2y}{dx^2} = r^2 e^{rx}$

$(r^2 e^{rx}) - (re^{rx}) - 6(e^{rx}) = 0$

$e^{rx}(r^2 - r - 6) = 0$

$r^2 - r - 6 = 0 \qquad\qquad$ Characteristic Equation

$(r - 3)(r + 2) = 0 \quad \rightarrow \quad r_1 = 3 \quad$ and $\quad r_2 = -2$

Recall: $y = e^{rx} \quad \rightarrow \quad y_1 = e^{3x} \qquad y_2 = e^{-2x}$

$y = $ a linear combination of the two solutions.

$y = C_1 e^{3x} + C_2 e^{-2x} \qquad\qquad\qquad$ G.S.

Result:
- If y_1 and y_2 are solutions of the DE $\quad ay'' + by' + cy = 0$ then their linear combination $\quad y = C_1 y_1 + C_2 y_2 \quad$ is called the general solution (G.S.) of the DE.

Example: Find the P.S. for: $y'' + 9y' + 20y = 0$

Satisfying: $y(0) = 2$ and $y'(0) = -1$

Solution:

Let $y = e^{rx}$

Characteristic Equation is: $r^2 + 9r + 20 = 0$

$r^2 + 9r + 20 = 0$

$(r + 5)(r + 4) = 0$

$r = -4, -5$

$y = C_1 e^{-4x} + C_2 e^{-5x}$ G.S.

$y' = -4C_1 e^{-4x} - 5C_2 e^{-5x}$ (need this later)

$y(0) = 2$ → $2 = C_1 + C_2$ (1)

$y'(0) = -1$ → $-1 = -4C_1 - 5C_2$ (2)

$$\boxed{\begin{aligned} C_1 + C_2 &= 2 \\ -4C_1 - 5C_2 &= -1 \\ \\ 4C_1 + 4C_2 &= 8 \\ \underline{-4C_1 - 5C_2} &= \underline{-1} \\ -C_2 &= 7 \end{aligned}}$$

$C_2 = -7$

$C_1 = 9$

$y = 9e^{-4x} - 7e^{-5x}$ P.S.

Existence and Uniqueness Theorem

Question:
How do we know if a Second-Order Linear DE has a solution?
Answer: Existence and Uniqueness Theorem.

Existence and Uniqueness Theorem
For the Second-Order Linear DE

Consider the Initial Value Problem (IVP) for:

$$y'' + p(x) \cdot y' + q(x) = g(x)$$

With: $y(x_0) = y_0$, $y'(x_0) = y'_0$

If $p(x)$, $q(x)$ and $g(x)$ are continuous on the interval I containing the point (x_0, y_0) then $\exists$ (there exists) a unique solution
$$y = \phi(x) \quad \text{of the given IVP on I.}$$

<u>Example</u>: Find the largest interval on which the unique solution is guaranteed for the IVP: $\sin x \, y'' - xy' + 4y = 0$

With: $y\left(\frac{\pi}{2}\right) = 2$ and $y'\left(\frac{\pi}{2}\right) = 6$

<u>Solution</u>: First, re-format the given equation.

$$y'' - \frac{x}{\sin x} y' + \frac{4}{\sin x} y = 0 \quad \rightarrow \quad p(x) = -\frac{x}{\sin x} \text{ and } q(x) = \frac{4}{\sin x}$$

$p(x)$ & $q(x)$ are continuous everywhere except when $\sin x = 0$.

Note: $\sin x = 0$ when $x = n\pi$, where $n = integer$.

The initial point, $x_0 = \frac{\pi}{2} \in (0, \pi)$.

So, $(0, \pi)$ is the largest interval that contains the unique solution of the given IVP.

Example: Find the largest interval on which a unique solution is guaranteed for the IVP: $(x^2 - 5x)y'' - (x)y' + (x+7)y = 0$

With: $y(1) = 3$ and $y'(1) = 5$

Solution:

$$y'' - \left(\frac{x}{x^2-5x}\right)y' + \left(\frac{x+7}{x^2-5x}\right)y = 0$$

$$y'' - \frac{1}{(x-5)}y' + \frac{x+7}{x(x-5)}y = 0$$

$p(x)$ & $q(x)$ are continuous on: $(-\infty, 0)$, $(0, 5)$, $(5, \infty)$

But only $(0, 5)$ contains the initial value, $x_0 = 1$.

So, $(0, 5)$ is the largest interval containing the unique solution.

Example: Does this IVP have a solution?

$$y'' - \frac{2x}{x-3}y' + \frac{5x+4}{(x-2)(x+7)}y = 0$$

With: $y(2) = 5$ and $y'(2) = -7$

If yes, find the largest interval containing the solution.

Solution:

$$q(x) = \frac{5x+4}{(x-2)(x+7)}$$

$q(x)$ is discontinuous at the initial point, $x_0 = 2$.

So, there is NO solution.

Wronskian Determinant

Definition: Wronskian Determinant

If y_1 and y_2 are solutions of: $\quad y'' + p(x)y' + q(x)y = g(x)$

With: $\quad y(x_0) = y_0 \quad$ and $\quad y'(x_0) = y_0'$

Then the Wronskian Determinant is:

$$W(y_1, y_2) = \begin{vmatrix} y_1(x_0) & y_2(x_0) \\ y_1'(x_0) & y_2'(x_0) \end{vmatrix}$$

Example:

Compute the Wronskian for the pair of functions: e^{3x}, e^{-2x}

Solution:

$$W(y_1, y_2) = \begin{vmatrix} y_1 & y_2 \\ y_1' & y_2' \end{vmatrix}$$

$$W(e^{3x}, e^{-2x}) = \begin{vmatrix} e^{3x} & e^{-2x} \\ 3e^{3x} & -2e^{-2x} \end{vmatrix}$$

$$= (e^{3x})(-2e^{-2x}) - (e^{-2x})(3e^{3x})$$

$$= -2e^x - 3e^x$$

$$= -5e^x$$

Example: Compute the Wronskian Determinant.

$$W(\sin 3\theta, \cos 3\theta)$$

Solution:

$$W(\sin 3\theta, \cos 3\theta) = \begin{vmatrix} \sin 3\theta & \cos 3\theta \\ 3\cos 3\theta & -\sin 3\theta \end{vmatrix}$$

$$= -3\sin^2 3\theta - 3\cos^2 3\theta$$
$$= -3(\sin^2 3\theta + \cos^2 3\theta)$$
$$= -3$$

THEOREM

If y_1 and y_2 are two solutions of the DE
$$y'' + p(x)y' + q(x)y = 0$$
Then $y = C_1 y_1 + C_2 y_2$ is the G.S. of the DE

If and only if: The Wronskian Determinant, $W(y_1, y_2) \neq 0$.

Example: (a.) Compute the G.S. of: $y'' - 3y' - 10y = 0$

and (b.) Compute the Wronskian of the two solutions of the DE.

Solution:

(a.) Assume: $y = e^{rx}$ will be any solution.

Then: $y' = re^{rx}$ and $y'' = r^2 e^{rx}$

Substitute: $r^2 e^{rx} - 3re^{rx} - 10e^{rx} = 0$

$$e^{rx}(r^2 - 3r - 10) = 0$$

Note: $e^{rx} \neq 0$ In fact: $e^{rx} > 0$ for all x

Characteristic Equation: $r^2 - 3r - 10 = 0$

$(x + 2)(x - 5) = 0$

$x = -2, 5$ ➔ Two solutions: $y_1 = e^{5x}$ $y_2 = e^{-2x}$

G.S is a Linear Combination of the two solutions.

$$y = C_1 e^{5x} + C_2 e^{-2x} \qquad \text{G.S.}$$

(b.) Compute the Wronskian: Recall: $W(y_1, y_2)(x) = \begin{vmatrix} y_1 & y_2 \\ y_1' & y_2' \end{vmatrix}$

Here: $W = \begin{vmatrix} e^{5x} & e^{-2x} \\ 5e^{5x} & -2e^{-2x} \end{vmatrix}$

$W = -2e^{3x} - 5e^{3x} = -7e^{3x}$ Note: $W \neq 0$

Example: Compute the Wronskian of two functions:
$$f_1(x) = \sin^2 x \quad \text{and} \quad f_2(x) = 1 - \cos^2 x$$

Solution:

$f_1'(x) = 2 \sin x \cos x$

$f_2'(x) = -2 \cos x (-\sin x) = 2 \cos x \sin x$

$$W = \begin{vmatrix} y_1 & y_2 \\ y_1' & y_2' \end{vmatrix} = \begin{vmatrix} \sin^2 x & 1 - \cos^2 x \\ 2 \sin x \cos x & 2 \cos x \sin x \end{vmatrix}$$

$W = \sin^2 x \; 2 \cos x \sin x - 2 \sin x \cos x + \cos^2 x \; 2 \sin x \cos x$

$W = 2 \cos x \sin x \, (\sin^2 x - 1 + \cos^2 x)$

$W = 2 \cos x \sin x \, (1 - 1) = 0$

Note: $W = 0$ → The two functions are Linearly Dependent (L.D.) In other words, the functions are <u>NOT</u> Linearly Independent (L.I.)

<u>Linear Independence</u>: Let y_1 and y_2 be any two solutions of the DE in the format: $y'' + p(x)y' + q(x)y = 0$. Then:
- y_1 and y_2 are linearly independent if $W(y_1, y_2) \neq 0$
- y_1 and y_2 are linearly dependent if $W(y_1, y_2) = 0$

Recall:
- L.I. → Two different solutions.
- L.D. → One solution is a scalar multiple of the other.

<u>Recommendation:</u> Review Linear Independence (L.I.) and Linear Dependence (L.D.) in the Appendix.

THEOREM

Let y_1 and y_2 be the solutions of a Second-Order Linear DE with constant coefficients of the form:
$$ay'' + by' + cy = 0$$

Then;

The linear combination of the 2 solutions: $y = C_1 y_1 + C_2 y_2$ contains ALL Possible Solutions if and only if: $W \neq 0$

In other words...

If $W(y_1, y_2) \neq 0$ then: $y = C_1 y_1 + C_2 y_2$ is the G.S.

Abel's Theorem: An Alternative way to compute Wronskian

Let y_1 and y_2 be any two solutions of a DE
$$y'' + p(x)y' + q(x)y = 0$$
And $p(x)$ and $q(x)$ are continuous on some interval I.

Then; $\quad W(y_1, y_2)(x) = C e^{-\int p(x)\, dx}$

Where; $\quad$ C is a constant whose value depends on y_1 and y_2 but NOT on x

Example: For the DE: $2x^2y'' + 3xy' - y = 0$

(a.) Verify the solutions are: $y_1 = x^{\frac{1}{2}}$ and $y_2 = x^{-1}$

(b.) Compute the Wronskian using y_1 and y_2

(c.) Compute the Wronskian using Abel's Theorem.

Solution (a.)

Verify that y_1 and y_2 are solutions by substituting them into the original equation.

$$y_1 = x^{\frac{1}{2}} \rightarrow y_1' = \frac{1}{2}x^{-\frac{1}{2}} \quad \text{and} \quad y_1'' = -\frac{1}{4}x^{-\frac{3}{2}}$$

$$2x^2\left(-\frac{1}{4}x^{-\frac{3}{2}}\right) + 3x\left(\frac{1}{2}x^{-\frac{1}{2}}\right) - \left(x^{\frac{1}{2}}\right) = 0$$

$$\left(-\frac{1}{2}x^{\frac{1}{2}}\right) + \left(\frac{3}{2}x^{\frac{1}{2}}\right) - \left(x^{\frac{1}{2}}\right) = 0$$

$$x^{\frac{1}{2}}\left(-\frac{1}{2} + \frac{3}{2} - 1\right) = 0$$

TRUE $\rightarrow y_1$ is a solution.

$$y_2 = x^{-1} \rightarrow y_2' = -x^{-2} \quad \text{and} \quad y_2'' = 2x^{-3}$$

$$2x^2(2x^{-3}) + 3x(-x^{-2}) - (x^{-1}) = 0$$

$$4x^{-1} - 3x^{-1} - x^{-1} = 0$$

TRUE $\rightarrow y_2$ is a solution

Solution (b.)

Compute the Wronskian using y_1 and y_2

$$W = \begin{vmatrix} y_1 & y_2 \\ y_1' & y_2' \end{vmatrix}$$

$$W = \begin{vmatrix} x^{\frac{1}{2}} & x^{-1} \\ \frac{1}{2}x^{-\frac{1}{2}} & -x^{-2} \end{vmatrix} = -x^{-\frac{3}{2}} - \frac{1}{2}x^{-\frac{3}{2}}$$

$$W = -\frac{3}{2} x^{-\frac{3}{2}}$$

Solution (c.)

Compute the Wronskian using Abel's Theorem.

$$W = Ce^{-\int p(x)\,dx}$$

We need $p(x)$

$$2x^2 y'' + 3xy' - y = 0$$

$$y'' + \left(\frac{3}{2x}\right)y' - \left(\frac{1}{2x^2}\right)y = 0 \qquad \rightarrow p(x) = \frac{3}{2x}$$

$$W = Ce^{-\int \frac{3}{2x}\,dx} = Ce^{-\frac{3}{2}\ln x} = Ce^{\ln x^{-\frac{3}{2}}} = Cx^{-\frac{3}{2}}$$

$$W = Cx^{-\frac{3}{2}} \qquad \text{Same Answer!!!}$$

NOTE: $W \neq 0$ → Linear Combination of y_1 and y_2 = G.S.

Characteristic Equation With Complex Roots

Question:

What if the characteristic equation has complex roots?

Case (ii)
Second-Order Linear Homogeneous DE with Constant Coefficients Where the Characteristic Equation has Complex roots.

Characteristic Equation: $ar^2 + br + c = 0$

Roots: $\quad r_1 = \lambda + i\mu \quad$ and $\quad r_2 = \lambda - i\mu$

Linearly Independent (or fundamental) solutions:

$$y_1 = e^{\lambda x} \cdot \cos(\mu x) \quad \text{and} \quad y_2 = e^{\lambda x} \cdot \sin(\mu x)$$

General Solution (G.S.)

$$y = C_1 y_1 + C_2 y_2$$
$$y = C_1 e^{\lambda x} \cos(\mu x) + C_2 e^{\lambda x} \sin(\mu x)$$
$$y = e^{\lambda x}[C_1 \cos(\mu x) + C_2 \sin(\mu x)]$$
$$y = e^{\lambda x} \cos(\mu x - \delta) \qquad \text{Where: } \delta = \tan^{-1}\left(\frac{C_2}{C_1}\right)$$

Remarks:
- If $\lambda > 0$ Then the solution is growing oscillatory.
- If $\lambda < 0$ Then the solution is decaying oscillatory.
- If $\lambda = 0$ Then the solution is constant oscillatory.

Recommendation: Review "Complex Numbers" in the Appendix.

Example: Find the G.S. of DE $y'' + 2y' + 2y = 0$

Solution:

Let: $y = e^{rx}$ $y' = re^{rx}$ $y'' = r^2 e^{rx}$

Then:

$$r^2 e^{rx} + 2re^{rx} + 2e^{rx} = 0$$

$$e^{rx}(r^2 + 2r + 2) = 0$$

$$r^2 + 2r + 2 = 0$$

$$r_{1,2} = \frac{-b \pm \sqrt{b^2 - 4ac}}{2a} = \frac{-2 \pm \sqrt{4-8}}{2} = \frac{-2 \pm 2i}{2} = -1 \pm i$$

$y_1 = e^{-x} \cdot \cos(x)$ and $y_2 = e^{-x} \cdot \sin(x)$

Optional Check: $W \neq 0$ → Linearly Independent

$$W = \begin{vmatrix} (e^{-x} \cos x) & (e^{-x} \sin x) \\ (-e^{-x} \cos x - e^{-x} \sin x) & (-e^{-x} \sin x + e^{-x} \cos x) \end{vmatrix}$$

$$W = -e^{-2x} \sin x \cdot \cos x + e^{-2x} \cos^2 x$$
$$\quad\quad + e^{-2x} \sin x \cdot \cos x + e^{-2x} \sin^2 x$$

$$W = e^{-2x}(\sin^2 x + \cos^2 x) = e^{-2x} \neq 0$$

Therefore:

$$y = C_1 e^{-x} \cos x + C_2 e^{-x} \sin x$$

$$y = e^{-x}[C_1 \cos x + C_2 \sin x] \quad\quad \text{General Solution (G.S.)}$$

Example: Find the Particular Solution (P.S.) of $y'' + 9y = 0$

With: $y\left(\frac{\pi}{3}\right) = 2$ and $y'\left(\frac{\pi}{3}\right) = 7$

Solution:

Let: $y = e^{rx}$ $y' = re^{rx}$ $y'' = r^2 e^{rx}$

$$r^2 e^{rx} + 9 e^{rx} = 0$$

$$e^{rx}(r^2 + 9) = 0$$

$r = \pm 3i$ → $y_1 = \cos 3x$ $y_2 = \sin 3x$

Optional check: $W = \begin{vmatrix} \cos(3x) & \sin(3x) \\ -3\sin(3x) & 3\cos(3x) \end{vmatrix} = 3 \neq 0$ OK

$y = C_1 \cos 3x + C_2 \sin 3x$ G.S.

$y' = -3 C_1 \sin 3x + 3 C_2 \cos 3x$

$y\left(\frac{\pi}{3}\right) = 2$ → $2 = C_1 \cos \pi + C_2 \sin \pi$

$\phantom{y\left(\frac{\pi}{3}\right) = 2} \quad 2 = C_1(-1) + C_2(0)$ → $C_1 = -2$

$y'\left(\frac{\pi}{3}\right) = 7$ → $7 = -3(-2)\sin \pi + 3 C_2 \cos \pi$

$\phantom{y'\left(\frac{\pi}{3}\right) = 7} \quad 7 = 6(0) + 3 C_2 (-1)$ → $C_2 = -\frac{7}{3}$

$y = (-2) \cos 3x - \left(\frac{7}{3}\right) \sin 3x$ P.S.

Characteristic Equation With Repeated Roots

Question: What if the characteristic equation has only one root with multiplicity 2?

Case (iii)
Second-Order Linear Homogeneous DE with Constant Coefficients Where the Characteristic Equation has repeated roots.

Here: $r = r_1 = r_2$

$y_1 = e^{rx}$ and $y_2 = xe^{rx}$

$y = C_1 y_1 + C_2 y_2$ General Solution (G.S.)

Example: Find the G.S. of $y'' + 4y' + 4y = 0$

Solution:

Let: $y = e^{rx}$ → $y' = re^{rx}$ and $y'' = r^2 e^{rx}$

$r^2 e^{rx} + 4re^{rx} + 4e^{rx} = 0$

$e^{rx}(r^2 + 4r + 4) = 0$

$r^2 + 4r + 4 = 0$

$(r+2)(r+2) = 0$ → $r = -2$ (m2)

→ $y_1 = e^{-2x}$ and $y_2 = xe^{-2x}$

$y = C_1 e^{-2x} + C_2 xe^{-2x}$ General Solution (G.S.)

SUMMARY of all three cases for

Second-Order Linear Homogeneous DE with Constant Coefficients

With Characteristic Equation: $\quad ar^2 + br + c = 0$

With roots: $\quad r_{1,2} = \dfrac{-b \pm \sqrt{b^2 - 4ac}}{2a}$

Case	Roots: r_1 and r_2	General Solution (G.S.)
i.	Real and distinct	$y_1 = e^{r_1 x}$ $y_2 = e^{r_2 x}$ $y = C_1 e^{r_1 x} + C_2 e^{r_2 x}$ G.S.
ii.	Complex Roots $r_1 = \lambda + i\mu$ $r_2 = \lambda - i\mu$	$y_1 = e^{\lambda x} \cos \mu x$ $y_2 = e^{\lambda x} \sin \mu x$ $y = e^{\lambda x}(C_1 \cos \mu x + C_2 \sin \mu x)$ G.S.
iii.	Repeated Roots $r_1 = r_2 = r$	$y = e^{rx}$ $y_2 = xe^{rx}$ $y = C_1 e^{rx} + C_2 x e^{rx}$ $y = e^{rx}(C_1 + C_2 x)$ G.S.

NOTE: We can use the same idea to solve higher-order linear homogeneous DE with constant coefficients!!!

HIGHER-ORDER HOMOGENEOUS DE WITH CONST. COEFF.

Higher-Order Homogeneous DE with Constant Coefficients

Consider a nth Order Homogeneous DE with Constant Coefficients.

$$a_n \frac{d^n y}{dx^n} + a_{n-1} \frac{d^{n-1} y}{dx^{n-1}} + \ldots + a_2 \frac{d^2 y}{dx^2} + a_1 \frac{d^1 y}{dx^1} + a_0 y = 0$$

With n Initial Conditions: $\quad y(x_0) = y_0$

$$y'(x_0) = y'_0$$
$$y^{(2)}(x_0) = y_0^{(2)}$$
$$\ldots$$
$$y^{(n-1)}(x_0) = y_0^{(n-1)}$$

Let: $y = e^{rx}$ Be the Solution (as before)

Characteristic Equation: Has n roots.

So, we need to compute n Linearly Independent solutions.

$$a_n r^n + a_{n-1} r^{n-1} + \ldots + a_2 r^2 + a_1 r + a_0 = 0$$

Get the roots: $r_1, r_2, \ldots r_n$ This can be tricky!!!

See 4 CASES – next page…

4 CASES: $\quad a_n r^n + a_{n-1} r^{n-1} + \ldots + a_2 r^2 + a_1 r + a_0 = 0$

Roots: $r_1 \ldots r_n$	General Solution (G.S.)
CASE #1 All real and distinct.	$y_1 = e^{r_1 x} \quad \ldots \quad y_n = e^{r_n x}$ $y = C_1 y_1 + \ldots + C_n y_n \qquad$ G.S.
CASE #2 All real and some repeating. Let $r = r_1$ w. multiplicity k	$y_1 = e^{r_1 x}$ $y_2 = x e^{r_1 x}$ $y_3 = x^2 e^{r_1 x}$ $\qquad$ $\boxed{k \text{ Linearly Independent Solutions}}$ $\ldots$ $y_k = x^{k-1} e^{r_1 x}$
CASE #3 Complex roots NOT repeating. Suppose p pairs of complex roots. Here: n = 2p	$r_{11} = \lambda_1 + i\mu_1 \quad ; \quad r_{12} = \lambda_1 - i\mu_1$ $r_{21} = \lambda_2 + i\mu_2 \quad ; \quad r_{22} = \lambda_2 - i\mu_2$ $\ldots \qquad\qquad\qquad \ldots$ $r_{p1} = \lambda_p + i\mu_p \quad ; \quad r_{p2} = \lambda_p - i\mu_p$ And: $y_{11} = e^{\lambda_1 x} \cos(\mu_1 x) \; ; \; y_{12} = e^{\lambda_1 x} \sin(\mu_1 x)$ $\ldots \qquad\qquad\qquad \ldots$ $y_{p1} = e^{\lambda_p x} \cos(\mu_p x) \; ; \; y_{p2} = e^{\lambda_p x} \sin(\mu_p x)$
CASE #4 Complex roots repeating. Suppose one pair of complex roots $r = \lambda \pm i\mu$ w. multiplicity "s"	$y_{11} = e^{\lambda x} \cos \mu x \qquad ; \; y_{12} = e^{\lambda x} \sin \mu x$ $y_{21} = x e^{\lambda x} \cos \mu x \quad ; \; y_{22} = x e^{\lambda x} \sin \mu x$ $y_{31} = x^2 e^{\lambda x} \cos \mu x \; ; \; y_{32} = x^2 e^{\lambda x} \sin \mu x$ $\ldots \qquad\qquad\qquad \ldots$ $y_{s1} = x^{s-1} e^{\lambda x} \cos \mu x \; ; \; y_{s2} = x^{s-1} e^{\lambda x} \sin \mu x$ There are $2s$ Linearly Independent Soln's.

Example: Write the G.S. of the DE

whose characteristic equation is in factored form given by:

$r(r-2)^3(r-i)(r+i)(r+2+3i)^2(r+2-3i)^2(r-5) = 0$

Solution:

Note: Degree = 11 → 11 roots → Need to create 11 L.I. solutions

Set each factor to zero and solve for "r".

General Solution	From
$y = C_1 e^{0x} +$	$r = 0$
$C_2 e^{2x} + C_3 x e^{2x} + C_4 x^2 e^{2x} +$	$(r-2)^3 = 0$ $r = 2, 2, 2$
$C_5 \cos x + C_6 \sin x +$	$(r \pm i) = 0$ $r = \pm i$
$C_7 e^{-2x} \cos 3x + C_8 e^{-2x} \sin 3x +$	$(r+2+3i)^2 = 0$ $r = -2 \pm 3i$
$C_9 x e^{-2x} \cos 3x + C_{10} x e^{-2x} \sin 3x +$	$(r+2-3i)^2 = 0$ $r = -2 \pm 3i$
$C_{11} e^{5x}$	$(r-5) = 0$ $r = 5$

Example: Find the G.S. of $y^{(6)} - y''' = 0$

Solution:

Degree = 6

Let: $y = e^{rx}$ → $y''' = r^3 e^{rx}$ and $y^{(6)} = r^6 e^{rx}$

$r^6 e^{rx} - r^3 e^{rx} = 0$

$e^{rx}(r^6 - r^3) = 0$

$r^3(r^3 - 1) = 0$ Difference of two cubes.

$r^3(r - 1)(r^2 + r + 1) = 0$

$r = 0, 0, 0$ and $r = 1$ and

$r = \dfrac{-b \pm \sqrt{b^2 - 4ac}}{2a} = \dfrac{-1 \pm \sqrt{1-4}}{2} = -\dfrac{1}{2} \pm \left(\dfrac{\sqrt{3}}{2}\right) i$

$y = C_1 e^{0x} + C_2 x e^{0x} + C_3 x^2 e^{0x} + C_4 e^x +$

$\quad C_5 e^{-\frac{1}{2}x} \cos\left(\dfrac{x\sqrt{3}}{2}\right) + C_6 e^{-\frac{1}{2}x} \sin\left(\dfrac{x\sqrt{3}}{2}\right)$

$y = C_1 + C_2 x + C_3 x^2 + C_4 e^x +$

$\quad C_5 e^{-\frac{1}{2}x} \cos\left(\dfrac{x\sqrt{3}}{2}\right) + C_6 e^{-\frac{1}{2}x} \sin\left(\dfrac{x\sqrt{3}}{2}\right)$ G.S.

Example: Write the G.S. whose characteristic equation is given by:
$$r^3(r-1)(r+2)^2(r-i)(r+i)(r-2+5i)^2(r-2-5i)^2 = 0$$

Solution: Note: Degree = 12 → 12 roots

General Solution (G.S.)	Based on Roots
$y = C_1 + C_2 x + C_3 x^2 +$	$r^3 = 0$ $r_1 = r_2 = r_3 = 0$
$C_4 e^{(1)x} +$	$(r-1) = 0$ $r_4 = 1$
$C_5 e^{(-2)x} + C_6 x e^{(-2)x} +$	$(r+2)^2 = 0$ $r_5 = -2$ $r_6 = -2$
$C_7 \cos x + C_8 \sin x +$	$(r-i)(r+i) = 0$ $r_7 = i$ $r_8 = -i$
$C_9 e^{2x} \cos 5x + C_{10} e^{2x} \sin 5x +$	$(r-2+5i)^2 = 0$ $r_9 = 2+5i$ $r_{10} = 2-5i$
$C_{11} x e^{2x} \cos 5x + C_{12} x e^{2x} \sin 5x$	$(r-2-5i)^2 = 0$ $r_{11} = 2+5i$ $r_{12} = 2-5i$

Example: Find the P.S. of: $\quad y''' - y'' - 4y' + 4y = 0$

With: $\quad y(0) = 2, \quad y'(0) = 3, \quad y''(0) = 4$

Solution:

Let $\quad y = e^{rx} \quad$ be any solution.

Then: $\quad y' = re^{rx}$

$\quad y'' = r^2 e^{rx}$

$\quad y''' = r^3 e^{rx}$

$r^3 e^{rx} - r^2 e^{rx} - 4re^{rx} + 4e^{rx} = 0$

$re^{rx}(r^3 - r^2 - 4r + 4) = 0$

$r^3 - r^2 - 4r + 4 = 0$

$(r^3 - 4r) - (r^2 - 4) = 0$

$r(r^2 - 4) - (r^2 - 4) = 0$

$(r^2 - 4)(r - 1) = 0$

$(r + 2)(r - 2)(r - 1) = 0 \quad \rightarrow \quad r = 1, 2, -2$

$y = C_1 e^x + C_2 e^{2x} + C_3 e^{-2x} \quad$ G.S.

Use Initial Conditions (I.C.) to find $C_1, C_2, C_3 \quad$ (Continued...)

$$y = C_1 e^x + C_2 e^{2x} + C_3 e^{-2x} \qquad \text{G.S.}$$

To find the P.S. y, y', and y'' are needed.

$$\boxed{\begin{aligned} y &= C_1 e^x + C_2 e^{2x} + C_3 e^{-2x} \\ y' &= C_1 e^x + 2C_2 e^{2x} - 2C_3 e^{-2x} \\ y'' &= C_1 e^x + 4C_2 e^{2x} + 4C_3 e^{-2x} \end{aligned}}$$

Use the Initial Conditions to find the constants.

$y(0) = 2 \quad \rightarrow \quad 2 = C_1 + C_2 + C_3$

$y'(0) = 3 \quad \rightarrow \quad 3 = C_1 + 2C_2 - 2C_3$

$y''(0) = 4 \quad \rightarrow \quad 4 = C_1 + 4C_2 + 4C_3$

We have three equations and three unknowns.

Solve for the unknowns using algebra or row reduction.

(See next page to see row reduction computations.)

You will find: $\quad C_1 = \dfrac{7}{12} \qquad C_2 = \dfrac{3}{4} \qquad C_3 = -\dfrac{1}{12}$

$$y = \left(\dfrac{7}{12}\right) e^x + \left(\dfrac{3}{4}\right) e^{2x} - \left(\dfrac{1}{12}\right) e^{-2x} \qquad \text{P.S.}$$

Solving for the three constants (C_1, C_2, and C_3)
Using row reduction.

$$\begin{array}{ccc|c} 1 & 1 & 1 & 2 \\ 1 & 2 & -2 & 3 \\ 1 & 4 & 4 & 4 \end{array}$$

$$\begin{array}{ccc|c} 1 & 1 & 1 & 2 \\ 0 & 1 & -3 & 1 \\ 0 & 0 & 12 & -1 \end{array}$$

$$\begin{array}{ccc|c} 1 & 1 & 1 & 2 \\ 1 & 2 & -2 & 3 \\ 0 & 2 & 6 & 1 \end{array}$$

$$\begin{array}{ccc|c} 1 & 1 & 1 & 2 \\ 0 & 1 & 0 & 3/4 \\ 0 & 0 & 1 & -1/12 \end{array}$$

$$\begin{array}{ccc|c} 1 & 1 & 1 & 2 \\ 0 & 1 & -3 & 1 \\ 0 & 2 & 6 & 1 \end{array}$$

$$\begin{array}{ccc|c} 1 & 0 & 0 & 7/12 \\ 0 & 1 & 0 & 3/4 \\ 0 & 0 & 1 & -1/12 \end{array}$$

Example: Find the G.S. of: $y^{(6)} - y = 0$

Solution:

Let: $y = e^{rx} \rightarrow y^{(6)} = r^6 e^{rx}$ (Differentiate 6 times.)

$r^6 e^{rx} - e^{rx} = 0$

$e^{rx}(r^6 - 1) = 0$ Recall: $e^{rx} \neq 0$

$r^6 - 1 = 0$

$(r^3)^2 - 1 = 0$ Difference of two squares

$(r^3 - 1)(r^3 + 1) = 0$ Difference and sum of two cubes.

$(r - 1)(r^2 + r + 1) \cdot (r + 1)(r^2 - r + 1) = 0$

$$r = \frac{-1 \pm \sqrt{1-4}}{2} = -\frac{1}{2} \pm \frac{\sqrt{3}}{2} i$$

$$r = \frac{1 \pm \sqrt{1-4}}{2} = \frac{1}{2} \pm \frac{\sqrt{3}}{2} i$$

$r = 1, \; -1, \; -\frac{1}{2} \pm \frac{\sqrt{3}}{2} i, \; \frac{1}{2} \pm \frac{\sqrt{3}}{2} i$

$$y = C_1 e^x + C_2 e^{-x} + C_3 e^{-\frac{1}{2}x} \cos \frac{\sqrt{3}}{2} + C_4 e^{-\frac{1}{2}x} \sin \frac{\sqrt{3}}{2} +$$

$$+ C_5 e^{\frac{1}{2}x} \cos \frac{\sqrt{3}}{2} + C_6 e^{\frac{1}{2}x} \sin \frac{\sqrt{3}}{2}$$

NON-HOMOGENEOUS DE WITH CONST. COEFF.

NON-HOMOGENEOUS DE WITH CONSTANT COEFFICIENTS

<u>Question:</u> What if the right-hand-side (RHS) $\neq 0$?

Then, the DE is a Second-Order Linear <u>Non-Homogeneous</u> D.E. with constant coefficients, in the form:

$$a \frac{d^2y}{dx^2} + b \frac{dy}{dx} + cy = g(x)$$

Here, the General Solution is: $y = C_1 y_1 + C_2 y_2 + Y$

Remarks:

- y_1 and y_2 are the solutions (fundamental solutions) of the homogeneous version: $a \frac{d^2y}{dx^2} + b \frac{dy}{dx} + cy = 0$
- Y is called the "specific solution."
- Two Methods to compute the Specific Solution (Y)
    1. Method of Undetermined Coefficients:
       (For RHS = exponential, polynomials, sine, cosine, or combinations.)
    2. Method of Variation Parameter.

Method of Undetermined Coefficients

Method of Undetermined Coefficients to solve a Second-Order Linear Non-Homogeneous DE with Constant Coefficients $$a\frac{d^2y}{dx^2} + b\frac{dy}{dx} + cy = g(x)$$	
1.	Set the RHS = 0 Find: y_1 and y_2
2.	General solution: $y = C_1 y_1 + C_2 y_2 + Y$
3.	Based on: $g(x)$, $y_1(x)$, and $y_2(x)$ Assume Y is in a format, listed in the following table. Then, compute the unknown coefficients.

$g(x)$	Y
C	$x^s A$
e^{rx} or ae^{rx}	$x^s A e^{rx}$
$\cos rx$, $\sin rx$, $a \cos rx$, or combination.	$x^s [A \cos rx + B \sin rx]$
$ae^{rx} P_n(x)$	$x^s [e^{rx} A_n(x)]$
$ae^{rx} \cdot \begin{cases} \cos rx \\ \sin rx \end{cases}$	$x^s [A \cos rx + B \sin rx]$
$P_n(x) \cdot \begin{cases} \sin rx \\ \cos rx \end{cases}$	$x^s [A_n(x) \sin rx + B_n(x) \cos rx]$
$P_n(x) e^{rx} \cdot \begin{cases} \sin rx \\ \cos rx \end{cases}$	$x^s e^{rx} [A_n(x) \sin rx + B_n(x) \cos rx]$

Example: Find the G.S. of $y'' - 5y' + 6y = 3e^{4x}$

Solution – Step 1: Get Homogeneous Solution

$$r^2 - 5r + 6 = 0$$
$$(r - 2)(r - 3) = 0 \quad \rightarrow \quad r = 2, 3$$
$$y_1 = e^{2x} \quad y_2 = e^{3x}$$

Assume:
$y = e^{rx}$
$y' = re^{rx}$
$y'' = r^2 e^{rx}$

Solution -- Step 2: Write the General Solution.

$$y'' - 5y' + 6y = 3e^{4x}$$
$$y = C_1 e^{2x} + C_2 e^{3x} + Y \quad \text{G.S.}$$

Solution -- Step 3:

$$g(x) = 3e^{4x} \quad \rightarrow \quad \text{Use Table to find: } Y = x^s A e^{4x}$$

Note: $s = 0$ Because e^{4x} is linearly indep. from e^{2x} and e^{3x}

$Y = x^0 A e^{4x} = A e^{4x}$ Y is the "Specific Solution"

$Y' = 4A e^{4x}$

$Y'' = 16 A e^{4x}$ Must find "A"

$$y'' - 5y' + 6y = 3e^{4x}$$
$$16 A e^{4x} - 20 A e^{4x} + 6 A e^{4x} = 3e^{4x}$$
$$A e^{4x}(16 - 20 + 6) = 3e^{4x}$$
$$2A e^{4x} = 3e^{4x} \quad \rightarrow \quad A = \frac{3}{2} \quad \rightarrow \quad Y = \frac{3}{2} e^{4x}$$

$$y = C_1 e^{2x} + C_2 e^{3x} + \frac{3}{2} e^{4x} \quad \text{G.S.}$$

Example: Find the G.S. of: $y'' - 5y' + 6y = x^2 + 2x + 5$

Solution:

Step 1: Get the solution of the homogeneous part.

$r^2 - 5r + 6 = 0$

$(r-2)(r-3) = 0 \quad \rightarrow \quad r = 2, 3$

$\rightarrow y_1 = e^{2x}$ and $y_2 = e^{3x}$

Step 2: $y = C_1 e^{2x} + C_2 e^{3x} + Y$

Step 3: $g(x) = x^2 + 2x + 5 \quad \rightarrow \quad Y = x^s(Ax^2 + Bx + C)$

Pick: $s = 0$ Because the assumed Y is LI from both y_1 and y_2

$Y = Ax^2 + Bx + C$

$Y' = 2Ax + B$

$Y'' = 2A$

$y'' - 5y' + 6y = x^2 + 2x + 5$

$(2A) - 5(2Ax + B) + 6(Ax^2 + Bx + C) = x^2 + 2x + 5$

$()x^2 + ()x + () = x^2 + 2x + 5$

Match the coefficients

$(6A)x^2 + (-10A + 6B)x + (2A - 5B + 6C) = x^2 + 2x + 5$

(Continued ...)

Match the coefficients

$$(6A)x^2 + (-10A + 6B)x + (2A - 5B + 6C) = x^2 + 2x + 5$$

$$6A = 1 \quad \rightarrow \quad A = \frac{1}{6}$$

$$-\frac{10}{6} + 6B = 2 \quad \rightarrow \quad B = \frac{1}{6}\left(\frac{12}{6} + \frac{10}{6}\right) = \frac{22}{36} = \frac{11}{18}$$

$$\frac{2}{6} - \frac{55}{18} + 6C = 5 \quad \rightarrow \quad C = \frac{1}{6}\left(\frac{5*18}{18} - \frac{6}{18} + \frac{55}{18}\right)$$

$$C = \frac{1}{6}\left(\frac{139}{18}\right) = \frac{139}{108}$$

$$y = C_1 e^{2x} + C_2 e^{3x} + Y$$
$$y = C_1 e^{2x} + C_2 e^{3x} + Ax^2 + Bx + C$$

$$y = C_1 e^{2x} + C_2 e^{3x} + \frac{1}{6}x^2 + \frac{11}{18}x + \frac{139}{108} \qquad \text{G.S.}$$

Example: Find the G.S. of: $y'' - 5y' + 6y = \sin 3x$

Solution:

Notice the homogeneous part is the same as the last 2 examples.

Step 1: $r^2 - 5r + 6 = 0$

$(r-2)(r-3) = 0 \rightarrow r = 2, 3$

$\rightarrow y_1 = e^{2x}$ and $y_2 = e^{3x}$

Step 2: $y = C_1 e^{2x} + C_2 e^{3x} + Y$

Step 3: $g(x) = \sin 3x \rightarrow Y = x^s (A \cos 3x + B \sin 3x)$

Pick: $s = 0$ Because $\sin 3x$ is LI from both y_1 and y_2

$Y = A \cos 3x + B \sin 3x$

$Y' = -3A \sin 3x + 3B \cos 3x$

$Y'' = -9A \cos 3x - 9B \sin 3x$

$y'' - 5y' + 6y = \sin 3x$

$(-9A \cos 3x - 9B \sin 3x) - 5(-3A \sin 3x + 3B \cos 3x)$

$\quad \ldots + 6(A \cos 3x + B \sin 3x) = \sin 3x$

$(-9B + 15A + 6B) \sin 3x = \sin 3x$

$(-9A - 15B + 6A) \cos 3x = 0$

Note: There are no $\cos 3x$ terms on RHS   (Continued...)

$(-9B + 15A + 6B)\sin 3x = \sin 3x$

$(-9A - 15B + 6A)\cos 3x = 0$

Note: There are no $\cos 3x$ terms on RHS

$-3B + 15A = 1 \quad \rightarrow \quad A = \dfrac{1}{15}(1 + 3B)$

$-3A - 15B = 0$

$-\dfrac{3}{15}(1 + 3B) - 15B = 0$

$-\dfrac{1}{15}(1 + 3B) - 5B = 0$

$(1 + 3B) + 75B = 0 \quad \rightarrow \quad B = -\dfrac{1}{78}$

$\quad\quad\quad\quad\quad\quad\quad\quad\quad \rightarrow A = \dfrac{1}{15}\left(1 - \dfrac{3}{78}\right) = \dfrac{75}{15} = 5$

$y = C_1 e^{2x} + C_2 e^{3x} + Y$

$y = C_1 e^{2x} + C_2 e^{3x} + A\cos 3x + B\sin 3x$

$y = C_1 e^{2x} + C_2 e^{3x} + 5\cos 3x - \dfrac{1}{78}\sin 3x \quad\quad \text{G.S.}$

Example: Find the G.S. of: $y'' - 5y' + 6y = xe^{7x}$

Solution:

Step 1: $r^2 - 5r + 6 = 0$

$(r-2)(r-3) = 0$ ➔ $r = 2, 3$

➔ $y_1 = e^{2x}$ and $y_2 = e^{3x}$

Step 2: $y = C_1 e^{2x} + C_2 e^{3x} + Y$

Step 3: $g(x) = xe^{7x}$ ➔ $Y = x^s(Axe^{7x} + Be^{7x})$

Pick: $s = 0$

$Y = Axe^{7x} + Be^{7x}$

$Y' = 7Axe^{7x} + Ae^{7x} + 7Be^{7x}$

$Y'' = 49Axe^{7x} + 7Ae^{7x} + 49Be^{7x}$

$y'' - 5y' + 6y = xe^{7x}$

$(49Axe^{7x} + 7Ae^{7x} + 49Be^{7x}) - 5(7Axe^{7x} + Ae^{7x} + 7Be^{7x})$
$+ 6(Axe^{7x} + Be^{7x}) = xe^{7x}$

(Continued...)

$(49Ax + 7A + 49B) - 5(7Ax + A + 7B) + 6(Ax + B) = x$

$(49A - 35A + 6A)x + (7A + 49B - 5A + 35B + 6B) = x$

$(49A - 35A + 6A)x + (7A + 49B - 5A + 35B + 6B) = x$

$(20A)x + (2A + 90B) = x$

$20A = 1 \quad\quad\quad \rightarrow A = \frac{1}{20}$

$\frac{2}{20} + 90B = 0 \quad\quad \rightarrow B = \frac{1}{90}\left(-\frac{2}{90}\right) = -\frac{2}{8100} = -\frac{1}{4050}$

$y = C_1 e^{2x} + C_2 e^{3x} + Y$

$y = C_1 e^{2x} + C_2 e^{3x} + Axe^{7x} + Be^{7x}$

$y = C_1 e^{2x} + C_2 e^{3x} + \frac{1}{20}xe^{7x} - \frac{1}{4050}e^{7x}$ \quad\quad G.S.

Example: Find the G.S. of: $y'' + 4y = 6x \sin x$

Solution:

| Let: | $y = e^{rx}$ | $y' = re^{rx}$ | $y'' = r^2 e^{rx}$ |

Step 1: Consider the homogeneous solution.

$y'' + 4y = 0$

$r^2 e^{rx} + 4e^{rx} = 0$

$r^2 + 4 = 0 \;\Rightarrow\; r = \pm\sqrt{-4} = \pm 2i$ (Real part = 0)

$y_1 = e^{0x} \cos 2x$ and $y_2 = e^{0x} \sin 2x$

$y_1 = \cos 2x$ $\qquad\qquad\qquad y_2 = \sin 2x$

Step 2: Consider: $y'' + 4y = 6x \sin x$

$y = C_1 \cos 2x + C_2 \sin 2x + Y$

$g(x) = 6x \sin x$

Step 3: Y = Specific Solution

$Y = x^s [\,(Ax + B) \cos x + (Cx + D) \sin x\,]$

$Y = (Ax + B) \cos x + (Cx + D) \sin x$

> Note: $g(x) = 6x \sin x$
> Since $g(x)$ is L.I. to $\sin 2x$ & $\cos 2x$ $\Rightarrow s = 0$

(Continued…)

$$Y = (Ax + B)\cos x + (Cx + D)\sin x$$

$$Y = Ax\cos x + B\cos x + Cx\sin x + D\sin x$$

$$Y' = A\cos x - Ax\sin x - B\sin x +$$
$$+ C\sin x + Cx\cos x + D\cos x$$

$$Y' = (A + D)\cos x + (C - B)\sin x - Ax\sin x + Cx\cos x$$

$$Y'' = -(A + D)\sin x + (C - B)\cos x - A\sin x - Ax\cos x +$$
$$+ C\cos x - Cx\sin x$$

$$Y'' = (-A - D - A)\sin x + (C - B + C)\cos x$$
$$- Ax\cos x - Cx\sin x$$

$$Y'' + 4Y = 6x\sin x$$

$$(-2A - D)\sin x + (2C - B)\cos x - Ax\cos x - Cx\sin x +$$
$$+ 4Ax\cos x + 4B\cos x + 4Cx\sin x + 4D\sin x = 6x\sin x$$

$$(-2A + 3D)\sin x + (2C + 3B)\cos x + (3A)x\cos x +$$
$$+ (3C)x\sin x = 6x\sin x$$

$(3C) = 6 \rightarrow C = 2$	$(-2A + 3D) = 0 \rightarrow D = 0$
$(3A) = 0 \rightarrow A = 0$	$(2C + 3B) = 0 \rightarrow B = -\frac{4}{3}$

$$y = C_1\cos 2x + C_2\sin 2x + Y \quad \text{and} \quad Y = -\frac{3}{4}\cos x + 2x\sin x$$

$$y = C_1\cos 2x + C_2\sin 2x - \frac{3}{4}\cos x + 2x\sin x$$

Example: (Where $s \neq 0$) Find the G.S. of: $y'' - 5y' + 6y = 7e^{2x}$

Solution - Step 1: Consider: $y'' + 5y' + 6y = 0$

As usual: $y = e^{rx}, y' = re^{rx}, y'' = r^2 e^{rx}$

$r^2 - 5r + 6 = (r-3)(r-2) = 0$

$r = 2, 3$ → $y_1 = e^{2x}, y_2 = e^{3x}$

Solution - Step 2: Consider: $y'' - 5y' + 6y = 7e^{2x}$

$C_1 e^{2x} + C_2 e^{3x} + Y = 7e^{7x}$

$g(x) = 7e^{2x}$ → $Y = x^s(Ae^{2x}) = Axe^{2x}$

> Since e^{2x} is a solution of the Homogeneous part, select $s = 1$. So, then, it will be linearly independent.

$Y' = Ae^{2x} + 2Axe^{2x}$

$Y'' = 2Ae^{2x} + 2Ae^{2x} + 4Axe^{2x} = 4Ae^{2x} + 4Axe^{2x}$

$Y'' - 5Y' + 6Y = 7e^{2x}$

$4Ae^{2x} + 4Axe^{2x} - 5Ae^{2x} - 10Axe^{2x} + 6Axe^{2x} = 7e^{2x}$

$(4A - 5A)e^{2x} + (4A - 10A + 6A)xe^{2x} = 7e^{2x}$

$-(A)e^{2x} = 7e^{2x}$ → $A = -7$

Specific Solution: $Y = -7xe^{2x}$

G.S. $y = C_1 e^{2x} + C_2 e^{3x} - 7xe^{2x}$

Example: Find the G.S. of: $\quad y'' - 4y' + 4y = 8e^{2x}$

Solution -- Step 1: Consider: $\quad y'' - 4y' + 4y = 0$

$\quad r^2 - 4r + 4 = (r-2)(r-2) = 0 \quad \rightarrow \quad r = 2, 2$

$\quad y_1 = e^{2x} \quad\quad y_2 = xe^{2x}$

Step 2: G.S. $\quad y = C_1 e^{2x} + C_2 xe^{2x} + Y$

$\quad g(x) = 8e^{2x} \quad \rightarrow \quad Y = x^s A e^{2x} = x^2 A e^{2x}$

Pick $s = 2$ because $x^2 e^{2x}$ is L.I. from y_1 and y_2
Note:
- $s = 0$ gives a multiple of y_1
- $s = 1$ gives a multiple of y_2
- So, s must be 2.

$Y' = 2Axe^{2x} + 2Ax^2 e^{2x}$

$Y'' = 2Ae^{2x} + 4Axe^{2x} + 4Axe^{2x} + 4Ax^2 e^{2x}$

$Y'' - 4Y' + 4Y = 8e^{2x}$

$2Ae^{2x} + 8Axe^{2x} + 4Ax^2 e^{2x} - 8Axe^{2x} - 8Ax^2 e^{2x} +$
$\quad\quad\quad\quad\quad\quad\quad\quad\quad + 4x^2 Ae^{2x} = 8e^{2x}$

$(2A)e^{2x} + (8A - 8A)xe^{2x} + (4A - 8A + 4A)x^2 e^{2x} = 8e^{2x}$

$(2A)e^{2x} = 8e^{2x} \quad \rightarrow \quad 2A = 8 \quad \rightarrow \quad A = 4$

$y = C_1 e^{2x} + C_2 xe^{2x} + Y \;=\; C_1 e^{2x} + C_2 xe^{2x} + \mathbf{4x^2 e^{2x}}$

Method of Variation of Parameters

METHOD OF VARIATION OF PARAMETER (MVP)

$y'' + p(x)y' + q(x)y = g(x)$ If the RHS is this...	Then, use this method to find the Specific Solution **(Y)**
e^{ax}, $\sin x$, $\cos x$, $P_n(x)$	Method of Undetermined Coefficients
$\log x$, $\sec x$, or other complicated function	Method of Variation Parameter Works for all cases but is more complex.

Theorem: Method of Variation of Parameter (MVP)

Consider: $y'' + p(x)y' + q(x)y = g(x)$

   Where $p(x), q(x),$ and $g(x)$
   are continuous functions on some open interval (I).

Let y_1 and y_2 be the solution of the homogeneous part.
Then, the specific solution of the non-homogeneous DE is:

$$Y = -y_1(x) \int_{x_0}^{x} \frac{y_2(s) \cdot g(s)}{W(s)} \, ds + y_2(x) \int_{x_0}^{x} \frac{y_1(s) \cdot g(s)}{W(s)} \, ds$$

Where x_0 is any arbitrary point in the open interval I.

Remarks: For Simplicity...

- Let $x_0 = 0$

- Since $y_1(x), y_2(x), g(x),$ and $W(x)$ are ALL functions of x ...

$$Y = -y_1 \int \frac{y_2 \cdot g}{W} \, dx + y_2 \int \frac{y_1 \cdot g}{W} \, dx$$

	General Idea: Method of Variation Parameter (MVP) For solving: $y'' + p(x)y' + q(x)y = g(x)$
Step 1	Consider the Homogeneous Part and compute y_1 and y_2
Step 2	Consider the Non-Homogeneous D.E. $y = C_1 y_1 + C_2 y_2 + Y$
Step 3	Use: $Y = -y_1 \int \frac{y_2 \cdot g}{W} dx + y_2 \int \frac{y_1 \cdot g}{W} dx$

NOTE:

Use the Method of Undetermined Coefficients (MUC) if possible.

ADVANTAGES OF MVP:
- This method can be used for any Second-Order Linear DE with Non-Constant Coefficients.
- The MUC works with Second-Order Linear DE with Constant Coefficients and for DE equations that can be converted to that type of DE. (e.g. Substitution may be used to convert the DE.)

Example: Find the G.S. of $y'' - 5y' + 6y = 8e^{4x}$

Solve using Variation Parameter.

Solution – Step 1:

$$y'' - 5y' + 6y = 0$$
$$y = e^{rx} \quad y' = re^{rx} \quad y'' = r^2 e^{rx}$$
$$r^2 - 5r + 6 = (5-2)(r-3) = 0 \quad \rightarrow r = 2, 3$$
$$y_1 = e^{2x} \quad y_2 = e^{3x}$$

Solution -- Step 2:

$$y'' - 5y' + 6y = 8e^{4x}$$
$$y = C_1 e^{2x} + C_2 e^{3x} + Y$$

Solution -- Step 3:

$$Y = -y_1 \int \frac{y_2 \cdot g}{W} dx + y_2 \int \frac{y_1 \cdot g}{W} dx$$

$$y_1 = e^{2x} \quad y_2 = e^{3x} \quad g(x) = 8e^{4x}$$

(Continued...)

Recall: $y_1 = e^{2x}$ $y_2 = e^{3x}$ and $g(x) = 8e^{4x}$

$$W = \begin{vmatrix} y_1 & y_2 \\ y'_1 & y'_2 \end{vmatrix} = \begin{vmatrix} e^{2x} & e^{3x} \\ 2e^{2x} & 3e^{3x} \end{vmatrix}$$

$W = 3e^{3x} e^{2x} - 2e^{2x} e^{3x}$

$W = e^{5x} \neq 0$ ☺

$Y = -y_1 \int \frac{y_2 \cdot g}{W} dx + y_2 \int \frac{y_1 \cdot g}{W} dx$

> Integral boundaries are not needed as we are integrating functions of x wrt "x".

$Y = -e^{2x} \int \frac{e^{3x} \cdot 8e^{4x}}{e^{5x}} dx + e^{3x} \int \frac{e^{2x} \cdot 8e^{4x}}{e^{5x}} dx$

$Y = -8e^{2x} \int e^{2x} dx + 8e^{3x} \int e^x dx$

$Y = -8e^{2x} \left[\frac{e^{2x}}{2}\right] + 8e^{3x}[e^x]$

$Y = -4e^{4x} + 8e^{4x} = 4e^{4x}$

$y = C_1 e^{2x} + C_2 e^{3x} + 4e^{4x}$ G.S.

> Method of Undetermined Coefficients (MUC) would give the same result.

Example: Find the G.S. for $y'' + y = \csc x$

Solution – Step 1:

$$\boxed{y = e^{rx} \qquad y' = re^{rx} \qquad y'' = r^2 e^{rx}}$$

$y'' + y = 0$

$(r^2 + 1)e^{rx} = 0$

$r^2 + 1 = 0$

$r = \pm i$ $\rightarrow$ $y_1 = \cos x \qquad y_2 = \sin x$

Solution -- Step 2:

$y'' + y = \csc x$ \qquad From the problem statement.

$y = C_1 \cos x + C_2 \sin x + Y$ \qquad G.S.

Solution -- Step 3:

$$Y = -y_1 \int \frac{y_2 \cdot g}{W} \, dx + y_2 \int \frac{y_1 \cdot g}{W} \, dx$$

$$W = \begin{vmatrix} y_1 & y_2 \\ y'_1 & y'_2 \end{vmatrix} = \begin{vmatrix} \cos x & \sin x \\ -\sin x & \cos x \end{vmatrix}$$

$W = \cos^2 x + \sin^2 x = 1 \neq 0$ \qquad ☺ $\rightarrow$ L.I.

(Continued...)

$$Y = -y_1 \int \frac{y_2 \cdot g}{W} \, dx + y_2 \int \frac{y_1 \cdot g}{W} \, dx$$

$$Y = -\cos x \int \frac{\sin x \cdot \csc x}{1} \, dx + \sin x \int \frac{\cos x \cdot \csc x}{1} \, dx$$

$$Y = -\cos x \int 1 \, dx + \sin x \int \frac{\cos x}{\sin x} \, dx$$

$$Y = -x \cos x + \sin x \cdot \ln|\sin x|$$

General Solution:

$$y = C_1 \cos x + C_2 \sin x + Y$$

$$y = C_1 \cos x + C_2 \sin x - x \cos x + \sin x \cdot \ln|\sin x|$$

NOTE: For extra practice, try this: $y'' + y = \tan x$

EXTRA: More general form of Method of <u>Variation Parameter</u>

$$Y = \sum_{i=1}^{n} y_i(x) \int \frac{W_i(x) \cdot g(x)}{W(x)} \, dx$$

If the characteristic equation has order "3" then:

$$W(x) = \begin{vmatrix} y_1 & y_2 & y_3 \\ y'_1 & y'_2 & y'_3 \\ y''_1 & y''_2 & y''_3 \end{vmatrix} \quad \rightarrow \quad W_1 = \begin{vmatrix} 0 & y_2 & y_3 \\ 0 & y'_2 & y'_3 \\ 1 & y''_2 & y''_3 \end{vmatrix}$$

and $\quad W_2 = \begin{vmatrix} y_1 & 0 & y_3 \\ y'_1 & 0 & y'_3 \\ y''_1 & 1 & y''_3 \end{vmatrix} \quad W_3 = \begin{vmatrix} y_1 & y_2 & 0 \\ y'_1 & y'_2 & 0 \\ y''_1 & y''_2 & 1 \end{vmatrix}$

<u>Example</u>: Previous Solution: $y_1 = \cos x$, $y_2 = \sin x \quad g = \csc x$

$$W = \begin{vmatrix} y_1 & y_2 \\ y'_1 & y'_2 \end{vmatrix} = \begin{vmatrix} \cos x & \sin x \\ -\sin x & \cos x \end{vmatrix} = \cos^2 x + \sin^2 x = 1$$

$$W_1 = \begin{vmatrix} 0 & \sin x \\ 1 & \cos x \end{vmatrix} = -\sin x \quad \text{and} \quad W_2 = \begin{vmatrix} \cos x & 0 \\ -\sin x & 1 \end{vmatrix} = \cos x$$

$$Y = y_1 \int \frac{W_1 \cdot \csc x}{1} \, dx + y_2 \int \frac{W_2 \cdot \csc x}{1} \, dx$$

$$Y = \cos x \int \frac{(-\sin x) \cdot \csc x}{1} \, dx + \sin x \int \frac{(\cos x) \cdot \csc x}{1} \, dx$$

$$Y = \cos x \int (-1) \, dx + \sin x \int \frac{\cos x}{\sin x} \, dx$$

$$Y = -x \cos x + \sin x \ln|\sin x| \qquad \text{(Same Answer)}$$

Example: Find the G.S. of: $y''' - 2y'' = 90e^{3x}$

Solve using the more general form of Variation of Parameter Method

Solution: (Step 1)

$y''' - 2y'' = 0$

$y = e^{rx}$, $y' = re^{rx}$, $y'' = r^2 e^{rx}$, $y''' = r^3 e^{rx}$

Characteristic Equation:

$r^3 e^{rx} - 2r^2 e^{rx} = 0$

$r^2(r - 2)e^{rx} = 0$ ➔ $r = 0, 0, 2$

Solution: (Step 2)

Consider: $y''' - 2y'' = 90e^{3x}$

G.S. $y = C_1 e^{0x} + C_2 x e^{0x} + C_3 e^{2x} + Y$

$y = C_1 + C_2 x + C_3 e^{2x} + Y$

Solution: (Step 3)

$Y = y_1 \int \frac{W_1 \cdot g(x)}{W} dx + y_2 \int \frac{W_2 \cdot g(x)}{W} dx + y_3 \int \frac{W_3 \cdot g(x)}{W} dx$

$W = \begin{vmatrix} 1 & x & e^{2x} \\ 0 & 1 & 2e^{2x} \\ 0 & 0 & 4e^{2x} \end{vmatrix} = 4e^{2x}$ $W_1 = \begin{vmatrix} 0 & x & e^{2x} \\ 0 & 1 & 2e^{2x} \\ 1 & 0 & 4e^{2x} \end{vmatrix} = 2xe^{2x} - e^{2x}$

(Continued...)

Previously, we found: $y_1 = 1$, $y_2 = x$, $y_3 = e^{2x}$

Solution: (Step 3)

$$Y = y_1 \int \frac{W_1 \cdot g(x)}{W} dx + y_2 \int \frac{W_2 \cdot g(x)}{W} dx + y_3 \int \frac{W_3 \cdot g(x)}{W} dx$$

$$W = \begin{vmatrix} 1 & x & e^{2x} \\ 0 & 1 & 2e^{2x} \\ 0 & 0 & 4e^{2x} \end{vmatrix} = 4e^{2x} \qquad W_1 = \begin{vmatrix} 0 & x & e^{2x} \\ 0 & 1 & 2e^{2x} \\ 1 & 0 & 4e^{2x} \end{vmatrix} = 2xe^{2x} - e^{2x}$$

$$W_2 = \begin{vmatrix} 1 & 0 & e^{2x} \\ 0 & 0 & 2e^{2x} \\ 0 & 1 & 4e^{2x} \end{vmatrix} = -2e^{2x} \qquad W_3 = \begin{vmatrix} 1 & x & 0 \\ 0 & 1 & 0 \\ 0 & 0 & 1 \end{vmatrix} = 1$$

$$Y = 1 \int \frac{(2xe^{2x} - e^{2x}) \, 90e^{3x}}{4e^{2x}} dx - x \int \frac{(-2e^{2x}) \, 90e^{3x}}{4e^{2x}} dx + e^{2x} \int \frac{(1) \, 90e^{3x}}{4e^{2x}} dx$$

$$Y = \int \left(45xe^{3x} - \frac{45}{2}e^{3x}\right) dx - x \int 45e^{3x} dx + e^{2x} \int \frac{45}{2} e^{x} dx$$

Integrate 1st integral by parts.

	Du	$\int dv$
	x \quad +	$45\, e^{3x}$
	1 \quad −	$15\, e^{3x}$
	0 \quad +	$5\, e^{3x}$

$$Y = [15xe^{3x} - 5e^{3x}] - \left(\frac{15}{2}\right)e^{3x} - 15xe^{3x} + \frac{45}{2}e^{3x}$$

$$Y = \left[-\frac{10}{2} - \frac{15}{2} + \frac{45}{2}\right] e^{3x} = \left[\frac{20}{2}\right] e^{3x} = 10\, e^{3x}$$

G.S. $\quad y = C_1 + C_2 x + C_3 e^{2x} + 10\, e^{3x}$

Example: Find the G.S. of: $y''' - 2y'' = 90e^{3x}$

Previously solved using Method of Variation of Parameter (MVP)

Now, solve it using the Method of Undetermined Coefficients (MUC)

Solution: (Steps 1 & 2) Solve for homogeneous case as before.

G.S. $y = C_1 + C_2 x + C_3 e^{2x} + Y$

Solution: (Step 3) MUC has a different way to find "Y"

Assume: $Y = Ae^{3x}$ (L.I. from other terms in G.S.)

Then:
$$Y' = 3Ae^{3x}$$
$$Y'' = 9Ae^{3x}$$
$$Y''' = 27Ae^{3x}$$

Substitute:

$27Ae^{3x} - 18Ae^{3x} = 90e^{3x}$

$9Ae^{3x} = 90e^{3x}$

$9A = 90$

$A = 10$ → $Y = 10e^{3x}$

G.S. $y = C_1 + C_2 x + C_3 e^{2x} + 10e^{3x}$ (Same solution)

EULER'S DE

EULER's Differential Equation (DE)

Homogeneous EULER's DE is a DE of the form:
$$ax^2 y'' + bxy' + cy = 0$$

Remarks:
- The solutions of a Euler's DE are in the form: $y = x^r$
- A Euler's DE can be converted into a Second-Order Linear DE with constant coefficients, using the substitution: $z = \ln x$

The method to convert a Euler's DE into a Linear DE with constant coefficients is outlined on the following page.

To convert a Euler's DE in the form: $ax^2y'' + bxy' + cy = 0$
To a Second Order Linear DE with Constant Coefficients
The following **Substitutions** are needed:

Let	$z = \ln x$	$\dfrac{dz}{dx} = \dfrac{1}{x}$	$\dfrac{dx}{dz} = x$
$\dfrac{dy}{dx}$	$\dfrac{dy}{dz} = \dfrac{dy}{dx} \cdot \dfrac{dx}{dz} = \dfrac{dy}{dx} \cdot x$	$\Rightarrow$	$\dfrac{dy}{dx} = \dfrac{1}{x} \cdot \dfrac{dy}{dz}$
$\dfrac{d^2y}{dx^2}$	$\dfrac{d}{dx}\left[\dfrac{dy}{dx}\right] = \dfrac{d}{dx}\left[\dfrac{1}{x} \cdot \dfrac{dy}{dz}\right]$		
	$\dfrac{d^2y}{dx^2} = \dfrac{d}{dx}\left[\dfrac{1}{x} \cdot \dfrac{dy}{dz}\right] = -\dfrac{1}{x^2} \cdot \dfrac{dy}{dz} + \dfrac{1}{x} \cdot \dfrac{d}{dx}\left[\dfrac{dy}{dz}\right]$		
	$\dfrac{d^2y}{dx^2} = -\dfrac{1}{x^2} \cdot \dfrac{dy}{dz} + \dfrac{1}{x} \cdot \dfrac{d^2y}{dx \cdot dz}$		
	$\dfrac{d^2y}{dx^2} = -\dfrac{1}{x^2} \cdot \dfrac{dy}{dz} + \dfrac{1}{x} \cdot \left[\dfrac{dy}{dx} \cdot \dfrac{dy}{dz}\right]$		
	$\dfrac{d^2y}{dx^2} = -\dfrac{1}{x^2} \cdot \dfrac{dy}{dz} + \dfrac{1}{x} \cdot \left[\dfrac{dy}{dx}\right]\dfrac{dy}{dz}$		
	$\dfrac{d^2y}{dx^2} = -\dfrac{1}{x^2} \cdot \dfrac{dy}{dz} + \dfrac{1}{x} \cdot \left[\dfrac{1}{x} \cdot \dfrac{dy}{dz}\right]\dfrac{dy}{dz}$		
	$\dfrac{d^2y}{dx^2} = -\dfrac{1}{x^2} \cdot \dfrac{dy}{dz} + \dfrac{1}{x^2} \cdot \dfrac{d^2y}{dz^2}$		

$$ax^2[y''] + bx[y'] + c[y] = 0$$

$$ax^2\left[-\dfrac{1}{x^2} \cdot \dfrac{dy}{dz} + \dfrac{1}{x^2} \cdot \dfrac{d^2y}{dz^2}\right] + bx\left[\dfrac{1}{x} \cdot \dfrac{dy}{dz}\right] + cy = 0$$

$$-a\dfrac{dy}{dz} + a\dfrac{d^2y}{dz^2} + b\dfrac{dy}{dz} + cy = 0$$

$$\boxed{a\dfrac{d^2y}{dz^2} + (b-a)\dfrac{dy}{dz} + cy = 0}$$

General Idea – To solve a Euler's DE

$$ax^2 \cdot y'' + bx \cdot y' + c \cdot y = 0$$

Let: $y = x^r$ Be any solution.

Then: $y' = rx^{r-1}$ and $y'' = r(r-1)r^{r-2}$

$$ax^2 \cdot y'' + bx \cdot y' + c \cdot y = 0$$
$$ax^2[r(r-1)r^{r-2}] + bx[rx^{r-1}] + c[x^r] = 0$$
$$[\,ar(r-1) + br + c\,]x^r = 0$$
$$ar(r-1) + br + c = 0$$
$$ar^2 + (b-a)r + c = 0 \qquad \text{Characteristic Equation.}$$

CASE	Roots	General Solution (G.S.)
1	Roots are real and distinct. $r_1 \neq r_2$	$y = C_1 x^{r_1} + C_2 x^{r_2}$
2	Roots are repeated. $r_1 = r_2$	$y = C_1 x^{r_1} + C_2 x^{r_1} \ln x$ $y = x^{r_1}(C_1 + C_2 \ln x)$
3	Complex Roots. $r_{1,2} = \lambda \pm i\mu$	$y_1 = x^\lambda \cos(\mu \ln x)$ $y_2 = x^\lambda \sin(\mu \ln x)$ $y = x^\lambda [\,C_1 \cos(\mu \ln x) + C_2 \sin(\mu \ln x)\,]$

Three Euler DE examples, representing the three cases, follow.

Find the G.S. of the following DEs

1. $x^2 y'' + 6xy' + 6y = 0$ (Case 1)
2. $x^2 y'' + 7xy' + 9y = 0$ (Case 2)
3. $x^2 y'' + 3xy' + 2y = 0$ (Case 3)

Euler DE Example #1: $x^2 y'' + 6xy' + 6y = 0$ Find the G.S.

Solution:

Let: $y = x^r$ $y' = rx^{r-1}$ $y'' = r(r-1)x^{r-2}$

Then: $x^2[\, r(r-1)x^{r-2}\,] + 6x[\, rx^{r-1}\,] + 6[\, x^r\,] = 0$

$x^r[r(r-1) + 6r + 6] = 0$

$r(r-1) + 6r + 6 = 0$

$r^2 + 5r + 6 = 0$

$(r+3)(r+2) = 0$ → $r = -3, -2$

→ $y_1 = x^{-3}$ $y_2 = x^{-2}$

$y = C_1 x^{-3} + C_2 x^{-2}$ G.S.

Euler DE Example #2: $x^2 y'' + 7xy' + 9y = 0$ Find the G.S.

Solution:

Let: $y = x^r$ $y' = rx^{r-1}$ $y'' = r(r-1)x^{r-2}$

Then: $x^2[\,r(r-1)x^{r-2}\,] + 7x[\,rx^{r-1}] + 9[\,x^r\,] = 0$

$[\,r^2 - r + 7r + 9\,]x^r = 0$

$r^2 - r + 7r + 9 = 0$

$(r+3)^2 = 0$ → $r = -3, -3$

→ $y_1 = x^{-3}$ $y_2 = x^{-3} \ln|x|$

$y = C_1 x^{-3} + C_2 x^{-3} \ln|x|$ G.S.

Euler DE Example #3: $x^2 y'' + 3xy' + 2y = 0$ Find the G.S.

Solution:

Let: $y = x^r$ $y' = rx^{r-1}$ $y'' = r(r-1)x^{r-2}$

Then: $x^2 [\, r(r-1)x^{r-2}\,] + 3x [\, rx^{r-1}\,] + 2[\, x^r\,] = 0$

$r(r-1) + 3r + 2 = 0$

$r^2 + 2r + 2 = 0$

$r_{1,2} = \dfrac{-2 \pm \sqrt{4-8}}{2} = \dfrac{-2 \pm \sqrt{-4}}{2} = -1 \pm i$

$y_1 = x^{-1} \cos(\ln |x|)$

$y_2 = x^{-1} \sin(\ln |x|)$

$y = x^{-1} [\, C_1 \cos(\ln|x|) + C_2 \sin(\ln|x|)\,]$ G.S.

Example: Find the G.S. of $x^2 y'' - 2xy' + 2y = 20x^4 + 5$

This is a non-homogeneous Euler's DE

Recall: Method of Variation of Parameter does NOT require constant coefficients.

Solution

Step 1: (Compute the fundamental solutions.)

Consider: $x^2 y'' - 2xy\, y' + 2y = 0 \;\; \rightarrow$ Euler's

Let: $\quad y = x^r \qquad y' = r x^{r-1} \qquad y'' = r(r-1)x^{r-2}$

$x^2 [\, r(r-1)x^{r-2}\,] - 2x[\, rx^{r-1}\,] + 2[\, x^r\,] = 0$

$x^r [r^2 - r - 2r + 2] = 0$

$r^2 - 3r + 2 = 0$

$(r-1)(r-2) = 0 \;\; \rightarrow r = 1, 2$

$y_1 = x^1 \qquad y_2 = x^2$

Step 2: (Assume the form of a specific solution and its computation.)

Consider: $x^2 y'' - 2xy' + 2y = 20x^4 + 5$

$y = C_1 x + C_2 x^2 + Y \qquad\qquad$ G.S.

Where $Y = $ Specific Solution.

$Y = -y_1 \int \frac{y_2 \cdot g}{W} dx + y_2 \int \frac{y_1 \cdot g}{W} dx$

(Continued...)

$$y_1 = x \qquad y_2 = x^2 \qquad \text{and} \qquad g(x) = 20x^4 + 5$$

Check to make sure y_1 and y_2 are linearly independent.

$$W = \begin{vmatrix} x & x^2 \\ 1 & 2x \end{vmatrix} = 2x^2 - x^2 = x^2 \neq 0 \qquad ☺$$

Previously, we found:

$$y = C_1 x + C_2 x^2 + Y \qquad \text{G.S.}$$

Where Y = Specific Sol'n.

$$Y = -y_1 \int \frac{y_2 \cdot g}{W} \, dx + y_2 \int \frac{y_1 \cdot g}{W} \, dx$$

$$Y = -x \int \frac{x^2 \cdot (20x^4 + 5)}{x^2} \, dx + x^2 \int \frac{x \cdot (20x^4 + 5)}{x^2} \, dx$$

$$Y = -x \int (20x^4 + 5) \, dx + x^2 \int \left(20x^3 + \frac{5}{x}\right) dx$$

$$Y = -x [4x^3 + 5x] + x^2 [4x^4 + 5 \ln|x|]$$

$$Y = 4x^6 - 4x^4 - 5x^2 + 5x^2 \ln|x|$$

$$y = C_1 x + C_2 x^2 + 4x^6 - 4x^4 - 5x^2 + 5x^2 \ln|x| \qquad \text{G.S.}$$

Example: Find the general solution of the following DE using the Method of Undetermined Coefficients (MUC).
$$x^2 y'' + 8xy' + 10y = x^3$$

Note: The Method of Undetermined Coefficients (MUC) works for Second-Order Linear DE with Constant Coefficients. So, this DE must be converted to a Linear DE with Constant Coefficients using the substitution: $z = \ln x$.

Solution:
Let $z = \ln x$ $\rightarrow$ $x = e^z$, $x^3 = e^{3z}$

$$\frac{dz}{dx} = \frac{1}{x}$$

$$\frac{dy}{dz} = \frac{dy}{dx} \cdot \frac{dx}{dz} = \frac{dy}{dx} \cdot x \quad \rightarrow \quad \frac{dy}{dx} = \frac{1}{x}\frac{dy}{dz}$$

$$\frac{d^2 y}{dz^2} = -\frac{1}{x^2}\frac{dy}{dz} + \frac{1}{x^2}\frac{d^2 y}{dz^2} \quad \text{(Chain Rule)}$$

Substitute this into the DE

$$x^2 \left[-\frac{1}{x^2}\frac{dy}{dz} + \frac{1}{x^2}\frac{d^2 y}{dz^2} \right] + 8x \left[\frac{1}{x}\frac{dy}{dz} \right] + 10y = e^{3z}$$

$$\frac{dy}{dz} + \frac{d^2 y}{dz^2} + 8\frac{dy}{dz} + 10y = e^{3z} \quad \text{All "x" terms cancel.}$$

(Step 1) Set RHS = 0 to find the Characteristic Equation and Roots

$y'' + 7y' + 10y = 0$

$r^2 + 7r + 10 = 0$ $\rightarrow$ $r = -5, -2$

$\rightarrow$ $y_1 = e^{-5z}$, $y_2 = e^{-2z}$

(Step 1) Set RHS = 0 to find the Characteristic Equation and Roots

$y'' + 7y' + 10y = 0 \quad \rightarrow \quad r^2 + 7r + 10 = 0 \quad \rightarrow \quad r = -5, -2$

$$\rightarrow \quad y_1 = e^{-5z} \quad , \quad y_2 = e^{-2z}$$

(Step 2) Consider: $y'' + 7y' + 10y = e^{3z}$

G.S. $\quad y = C_1 e^{-5z} + C_2 e^{-2z} + Y$

(Step 3) Let: $Y = z^s A e^{3z}$

Note: $s = 0$ Because e^{3z} is L.I. to both e^{-5z} and e^{-2z}

$Y = A e^{3z} \qquad Y' = 3A e^{3z} \qquad Y'' = 9A e^{3z}$

Substitute Y, Y', and Y'' into the DE equation.

$[9Ae^{3z}] + 7[3Ae^{3z}] + 10[Ae^{3z}] = e^{3z}$

$9A + 21A + 10A = 1$

$40A = 1 \quad \rightarrow \quad A = \frac{1}{40} \quad \rightarrow \quad Y = \left(\frac{1}{40}\right) e^{3z}$

$y = C_1 e^{-5z} + C_2 e^{-2z} + Y \qquad\qquad$ G.S.

$y = C_1 e^{-5z} + C_2 e^{-2z} + \left(\frac{1}{40}\right) e^{3z} \qquad$ Recall: $z = \ln x$

$y = C_1 e^{-5 \ln x} + C_2 e^{-2 \ln x} + \left(\frac{1}{40}\right) e^{3 \ln x}$

$y = C_1 x^{-5} + C_2 x^{-2} + \left(\frac{1}{40}\right) x^3$

Extra Practice: Find G.S. of DE $x^2 y'' + 8xy' + 10y = x^3$
Hint: Compute two solutions of homogeneous part. Then use MVP
Should get the same answer.

REDUCTION OF ORDER

<u>REDUCTION OF ORDER</u> → NO restriction on coefficients!!!

If one solution of a Second Order Linear DE is given, then one can compute the second solution using one of the following two methods.

METHOD #1:   D'Alembert's Method.

   The two solutions of the DE are related.

   $$y_2 = y_1 \cdot V \quad ; \quad V = \text{Some unknown function}$$

In other words, the second solution can be obtained from the first solution by simply multiplying the first solution by some function of "x".

METHOD #2:   Set two definitions of Wronskian equal.

   (1.) Abel's Theorem: $W = C\, e^{-\int p(x)\, dx}$

   (2.) Definition: $W = \begin{vmatrix} y_1 & y_2 \\ y'_1 & y'_2 \end{vmatrix} = y_1 y'_2 - y_2 y'_1$

   $$C\, e^{-\int p(x)\, dx} = y_1 y'_2 - y_2 y'_1$$

Examples: Consider: $2x^2 y'' + 3xy' - y = 0$

1) Verify $y_1 = \frac{1}{x}$ is a solution.

2) Find second solution, using the above two REDUCTION OF ORDER methods.

Solution: Part (1) Verify $y_1 = \frac{1}{x}$ is a solution

$$y_1 = \frac{1}{x} \qquad y'_1 = -\frac{1}{x^2} \qquad y''_1 = \frac{2}{x^3}$$

$$2x^2 \left[\frac{2}{x^3}\right] + 3x\left[-\frac{1}{x^2}\right] - \left[\frac{1}{x}\right] = 0$$

$$\frac{4}{x} - \frac{3}{x} - \frac{1}{x} = 0 \quad \rightarrow \text{TRUE}$$

Solution: Part (2a)   Find 2nd solution using D'Alembert's Method

$$y_2 = y_1 V = \left(\frac{1}{x}\right) V \quad \rightarrow \quad y'_2 = \frac{1}{x} V' - \frac{1}{x^2} V$$

$$y''_2 = \frac{1}{x} V'' - \frac{1}{x^2} V' - \frac{1}{x^2} V' + \frac{2}{x^3} V$$

$$y''_2 = \frac{1}{x} V'' - \frac{2}{x^2} V' + \frac{2}{x^3} V$$

$$2x^2 y'' + 3xy' - y = 0$$

$$2x^2 \left[\frac{1}{x} V'' - \frac{2}{x^2} V' + \frac{2}{x^3} V\right] + 3x \left[\frac{1}{x} V' - \frac{1}{x^2} V\right] - \left[\left(\frac{1}{x}\right) V\right] = 0$$

$$2x V'' - 4V' + \frac{4}{x} V + 3V' - \frac{3}{x} V - \frac{1}{x} V = 0$$

$$2x V'' - V' = 0 \qquad \text{Note: All "V" terms cancel.}$$

Let: $u = V'$ and $u' = V''$

$$2x u' - u = 0 \quad \rightarrow \quad \text{1st Order Linear D.E.}$$

$$2x \frac{du}{dx} - u = 0$$

$$\int \frac{1}{u} du = \int \frac{1}{2x} dx$$

$$\ln|u| = \frac{1}{2} \ln|x| \quad \rightarrow \quad u = x^{\frac{1}{2}}$$

It's OK to ignore the constant of integration.

$$V' = \frac{dV}{dx} = x^{\frac{1}{2}} \quad \rightarrow \quad \int dV = \int x^{\frac{1}{2}} dx \quad \rightarrow \quad V = \frac{2}{3} x^{\frac{3}{2}}$$

Recall: $y_2 = y_1 V \quad \rightarrow \quad y_2 = \left(\frac{1}{x}\right) \left[\frac{2}{3} x^{\frac{3}{2}}\right] = \frac{2}{3} x^{\frac{1}{2}} = x^{\frac{1}{2}}$

$$y = C_1 \left(\frac{1}{x}\right) + C_2 x^{\frac{1}{2}} \qquad \text{General Solution (G.S.)}$$

Solution: Part (2b) Find 2nd solution using Compare W Method

$$2x^2 y'' + 3xy' - y = 0$$

$$2x^2 y'' + \frac{3x}{2x^2} y' - \frac{1}{2x^2} y = 0$$

$$y'' + \underbrace{\frac{3}{2x}}_{p(x)} y' - \underbrace{\frac{1}{2x^2}}_{q(x)} y = 0$$

$$W = C e^{-\int p(x)\, dx} = Ce^{-\int \frac{3}{2x}\, dx} = Ce^{-\frac{3}{2}\ln|x|} = Cx^{-\frac{3}{2}}$$

Also

$$W = y_1 y'_2 - y_2 y'_1 = \left(\frac{1}{x}\right) y'_2 + y_2 \left(\frac{1}{x^2}\right)$$

$$\boxed{y_1 = \frac{1}{x} \;\rightarrow\; y'_1 = -\frac{1}{x^2}}$$

$$W = W$$

$$\left(\frac{1}{x}\right) y'_2 + y_2 \left(\frac{1}{x^2}\right) = Cx^{-\frac{3}{2}}$$

$$y'_2 + y_2 \left(\frac{1}{x}\right) = Cx^{-\frac{1}{2}}$$

$$\boxed{\mu(x) = e^{\int p(x)\, dx} \\ \mu(x) = e^{\int \frac{1}{x}\, dx} = e^{\ln x} = x}$$

$$y_2 = \frac{\int x \cdot Cx^{-\frac{1}{2}}\, dx}{x} = \frac{C \int x^{\frac{1}{2}}\, dx}{x} = \frac{C\left[\frac{2}{3} x^{\frac{3}{2}}\right]}{x} = C\frac{2}{3} \cdot x^{\frac{1}{2}} = D x^{\frac{1}{2}}$$

$$y = C_1 \left(\frac{1}{x}\right) + C_2 x^{\frac{1}{2}} \qquad \text{G.S.} \quad \text{Same Answer!}$$

EXTRA PRACTICE: Try: $(x-1)y'' - xy' + y = 0$

(1) Verify $y_1 = e^x$ is a solution.

(2) Find second solution using Reduction of Order (Compare W)

Solution:

$y_1 = e^x \rightarrow y'_1 = e^x$ and $y''_1 = e^x$

$(x-1)y'' - xy' + y = 0$

$(x-1)e^x - xe^x + e^x = 0$

$(x-1) - x + 1 = 0$ $\rightarrow$ TRUE

$(x-1)y'' - xy' + y = 0$

$y'' - \frac{x}{x-1} y' + \frac{1}{x-1} y = 0$

$W = Ce^{\left(\int \frac{x}{x-1} dx\right)}$	$W = y_1 y'_2 - y_2 y'_1$
$W = Ce^{\left(\int 1 + \frac{1}{x-1} dx\right)}$	$W = e^x y'_2 - y_2 e^x$
$W = Ce^{[x + \ln|x-1|]}$	$W = e^x (y'_2 - y_2)$
$W = Ce^{x + \ln|x-1|}$	
$W = Ce^x e^{\ln|x-1|}$	
$W = Ce^x |x-1|$	

Continued...

Recall:

$\int \frac{x}{x-1} dx = \ln|x-1| + x$

$\frac{x}{x-1} = 1 + \frac{1}{x-1}$

$$\begin{array}{r} 1 \\ x-1 \overline{\smash{)} x} \\ -(x-1) \\ \hline 1 \end{array}$$

$W = W$

$Ce^x |x-1| = e^x(y'_2 - y_2)$

$C|x-1| = (y'_2 - y_2)$

$y'_2 - y_2 = C(x-1)$ \quad\quad $x > 1$

$y_2 = \frac{\int (e^{-x}) \cdot C(x-1) \, dx}{e^{-x}}$

$y_2 = \frac{C \int (xe^{-x} - e^{-x}) \, dx}{e^{-x}}$

$y_2 = \frac{C[-xe^{-x} - e^{-x} + e^{-x}]}{e^{-x}}$

$y_2 = C[-x] = (-C)x$

$y_2 = x$ \quad Ignore the constant.

$\mu(x) = e^{\int p(x) \, dx}$

$\mu(x) = e^{\int -1 \, dx} = e^{-x}$

$\int x e^{-x} \, dx$

Du	$\int dv$
x +	e^{-x}
1 −	$-e^{-x}$
0 +	e^{-x}

$= -xe^{-x} - e^{-x}$

EXTRA PRACTICE: Try: $(x-1)y'' - xy' + y = 0$

(1) Verify $y_1 = e^x$ is a solution.

(2) Find the second solution using Reduction of Order (D'Alembert's Method)   Hint: The answer should be: $y_2 = x$

Solution -- Part (1) Already done.

Solution – Part (2) Find 2nd sol'n using D'Alemberts Method

$y_1 = e^x$ → $y_2 = e^x V$

$y_2 = e^x V$

$y'_2 = e^x V + e^x V'$

$y''_2 = e^x V + e^x V' + e^x V'' + e^x V'$

$y''_2 = e^x V + 2e^x V' + e^x V''$

$(x-1)y'' - xy' + y = 0$

$(x-1)[e^x V + 2e^x V' + e^x V''] - x[e^x V + e^x V'] + [e^x V] = 0$

$(x-1)[V + 2V' + V''] - x[V + V'] + [V] = 0$

$V''[x-1] + V'[2x - 2 - x] + V[x - 1 - x + 1] = 0$

$V''[x-1] + V'[x-2] = 0$ All "V" terms cancel.

Let: $u = V'$ $u' = V''$

$u'[x-1] + u[x-2] = 0$ (Continued...)

$$u'[x-1] + u[x-2] = 0$$

$$\frac{du}{dx}[x-1] + u[x-2] = 0$$

$$\frac{du}{dx}[x-1] = -u[x-2]$$

$$-\frac{1}{u}du = \frac{x-2}{x-1}dx$$

$$-\int \frac{1}{u}du = \int \frac{x}{x-1}dx - \int \frac{2}{x-1}dx$$

$$-\ln|u| = \ln|x-1| + x - 2\ln|x-1|$$

$$-\ln|u| = x - \ln|x-1|$$

$$\ln|x-1| - \ln|u| = x$$

$$\ln\left|\frac{x-1}{u}\right| = x \qquad x > 1$$

$$\frac{x-1}{u} = e^x$$

$u = \frac{x-1}{e^x}$ and $u = V' \rightarrow V' = \frac{dv}{dx} = \frac{x-1}{e^x}$

$\int 1\, dv = \int \frac{x-1}{e^x} dx$

$V = \int \frac{x}{e^x} dx - \int e^{-x} dx = \int x e^{-x} dx + e^{-x}$

$V = -xe^{-x} - e^{-x} + e^{-x} = -xe^{-x}$

$y_2 = e^x V$

$y_2 = e^x[-xe^{-x}] = -x$

$y_2 = Cx$ Ignore the constant.

Example: Consider $x^2 y'' - x(x+2) y' + (x+2) y = 0$

This is a 2nd Order DE with non-constant coefficients. (Not Euler)

(a) Verify $y_1 = x$ is a solution of the given D.E.

(b) Use Reduction of Order to compute y_2

Solution – Part (a)

$y_1 = x \;\to\; y_1' = 1 \quad \text{and} \quad y_1'' = 0$

Substitute into D.E. equation:

$x^2(0) - x(x+2)(1) + (x+2)(x) = 0$

$0 \;-\; x^2 - 2x \;+\; x^2 + 2x = 0$

$0 = 0 \quad \text{TRUE} \;\to\; y_1 = x$ is a solution.

Solution – Part (b)

Can use either D'Alembert's or W-Comparison Method.

Using D'Alembert's method $\to\; y_2 = y_1 \cdot V = x \cdot V$

$y_2' = xV' + V$

$y_2'' = xV'' + V' + V' = xV'' + 2V'$

Substitute into D.E. equation.

$x^2 y'' \;-\; x(x+2) y' \;+\; (x+2) y = 0$

$x^2 (xV'' + 2V') - x(x+2)(xV' + V) + (x+2)(xV) = 0$

(Continued ...)

$$x^2(xV'' + 2V') - x(x+2)(xV' + V) + (x+2)(xV) = 0$$
$$x(xV'' + 2V') - (x+2)(xV' + V) + (x+2)(V) = 0$$
$$x^2 V'' + 2xV' - x^2 V' - xV - 2xV' - 2V + xV + 2V = 0$$
$$x^2 V'' - x^2 V' = 0 \quad \text{Note: V terms should all cancel out.}$$
$$V'' - V' = 0 \quad \text{OK to divide both sides by } x^2$$

$$\boxed{\text{Let:} \quad u = V', \quad u' = V''}$$

$$u' - u = 0$$
$$\frac{du}{dx} - u = 0 \quad \text{Separable.}$$
$$\int \frac{1}{u}\, du = \int 1\, dx$$
$$\ln|u| = x \quad \text{Assume: } u > 0$$
$$e^{\ln u} = e^x$$
$$u = e^x \quad \text{Recall: } u = V'$$
$$V' = e^x$$
$$\frac{dV}{dx} = e^x$$
$$\int 1\, dV = \int e^x\, dx$$
$$V = e^x$$

$$y_2 = y_1 \cdot V = x \cdot e^x$$

$$y = C_1 y_1 + C_2 y_2$$
$$y = C_1 x + C_2 x e^x \qquad \text{G.S.}$$

Same Example: Consider $x^2 y'' - x(x+2)y' + (x+2)y = 0$

This is a 2nd Order DE with non-constant coefficients. (Not Euler)

(a) Verify $y_1 = x$ is a solution of the given DE (Previously Done)

(b) Use Reduction of Order to compute y_2

Solution – Part (b)

Can use either D'Alembert's or W-Comparison Method.

Using W-Comparison method → $y_2 = y_1 \cdot V = x \cdot V$

$y'' - \frac{x(x+2)}{x^2} y' + \frac{(x+2)}{x^2} y = 0$ Rewrite DE equation

$y'' - \frac{(x+2)}{x} y' + \frac{(x+2)}{x^2} y = 0$

$y'' - \left(1 + \frac{2}{x}\right) y' + \frac{(x+2)}{x^2} y = 0$

Abel's Theorem	Definition of W
$W = C e^{-\int p(x) dx}$	$W = y_1 y_2' - y_2 y_1'$
$W = C e^{-\int -\left(1 + \frac{2}{x}\right) dx}$	$W = x y_2' - y_2$
$W = C e^{\int \left(1 + \frac{2}{x}\right) dx}$	
$W = C e^{x + 2 \ln|x|}$	
$W = C e^x e^{\ln x^2}$	
$W = C e^x x^2$	

Continued ...

$W = W$

$x y_2' - y_2 = C e^x x^2$

$y_2' - \left(\dfrac{1}{x}\right) y_2 = C x e^x$

> **Integrating Factor**
>
> $\mu(x) = e^{\int -\frac{1}{x} dx}$
>
> $\mu(x) = e^{-\int \frac{1}{x} dx}$
>
> $\mu(x) = e^{-\ln|x|} = e^{\ln\left|\frac{1}{x}\right|}$
>
> $\mu(x) = \dfrac{1}{x}$

$y_2 = \dfrac{\int \left(\frac{1}{x}\right) C x e^x \, dx}{\left(\frac{1}{x}\right)}$

$y_2 = Cx \int e^x \, dx = C x e^x$ \quad\quad **Same Answer!**

MECHANICAL AND ELECTRICAL VIBRATIONS

Mechanical and Electrical Vibrations (or Oscillations)

Many physical systems can be represented as second order linear D.E. with constant coefficients. Consider two types:
- Mechanical Vibrations
- Electrical Vibrations

Examples:
- Motion of a mass on a vibrating spring.
- Torsional oscillation of a shaft.
- Oscillation due to voltage drop in a simple circuit.

All of these systems can be described by the DE:

$ay'' + by' + cy = g(t)$, $y(0) = y_0$, $y'(0) = y'_0$

Where the independent variable "t" represents time.

Spring Mass System

SPRING-MASS SYSTEM:

- A mass weighs m lbs.
- Stretches a spring of original length l by L units.
- Additional stretch (beyond L) is u.
- Recall: $g = 9.8 \frac{m}{s^2} = 32 \frac{ft}{s^2}$
- At Equilibrium: $w = F_s \qquad mg = kL$

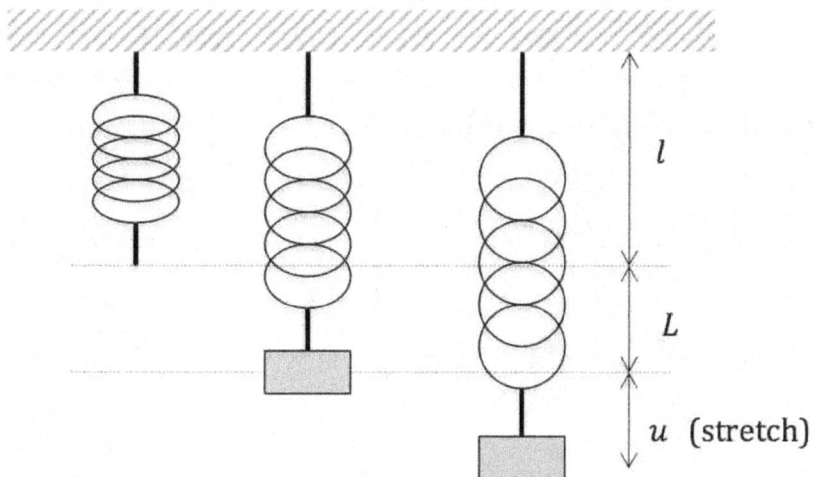

$F_s \propto L$ Force of Spring $= F_s = -kL$

$w = mg$ Gravitational Force (positive)

NEWTON'S LAW

Suppose the mass is displaced by "u" units (up or down). Consider Newton's Law.

> Newton's Law ➔
>
> $Force = ma$
>
> $f(t) = mu''$

#	Force acting on Block	Equations
1.	Gravity (positive)	$w = mg$
2.	Spring (negative)	$F_s \propto L$ $F_s \propto L + u$ $F_x = -k(L + u)$
3.	Damping (or Resistance) Always against the velocity direction.	$F_d \propto \dfrac{du}{dt}$ $F_d = -\mu \dfrac{du}{dt}$
4.	External	$f(t)$

Newton's Law: $Force = mass \cdot accelerataion$

$Force = m\,u'' = mass \cdot acceleration = \sum all\ forces$

$m\,u'' = w + F_s + F_d + f(t)$
$m\,u'' = w - k(L+u) - \mu\,u' + f(t)$
$m\,u'' = w - kL - ku - \mu\,u' + f(t)$

$$\boxed{m\,u'' + \mu\,u' + k\,u = f(t)}$$

$m > 0,\quad \mu > 0,\quad k > 0$

Initial Conditions: $u(0) = u_0,\quad u'(0) = v_0$

Where:
- $m =$ mass of block.
- $\mu =$ damping constant
- $k =$ spring constant.

If no external force, then:

$$\boxed{m\,u'' + \mu\,u' + k\,u = 0}$$

> **SPRING-MASS SYSTEM:** $mu'' + \mu u' + ku = f(t)$

$$m = \text{Mass of the block}$$
$$\mu = \text{Damping factor}$$
$$k = \text{Spring constant}$$
$$u(0) = u_0 = \text{Initial displacement}$$
$$u'(0) = u'_0 = v_0 = \text{Initial velocity}$$
$$f(t) = \text{External force (may be zero)}$$

If a spring mass system does not have damping, movement continues and does not slow down or stop. Such vibrations are often called "undamped" vibrations in a spring mass system. The governing DE in an undamped system has the following form:

$$mu'' + ku = 0 \;, \; u(0) = u_0, \; u'(0) = u'_0 = v_0$$

Typically, the movement in a spring-mass system has damping and the movement eventually stops. The amplitude of vibration decreases over time and the system rests at equilibrium. This system is called a "damped" system and the governing DE has the following form:

$$mu'' + \mu u' + ku = 0 \;, \; u(0) = u_0, \; u'(0) = u'_0 = v_0$$

UNDAMPED VIBRATIONS: ($\mu = 0$)

Equation: $m u'' + k u = 0$ With: $f(x) = 0$

Let: $u = e^{rt}$ be a solution.

Then: $u' = re^{rt}$

$u'' = r^2 e^{rt}$

Characteristic Equation: $mr^2 + k = 0$

Roots: $r^2 = -\frac{k}{m}$ → $r_{1,2} = \pm\sqrt{\frac{-k}{m}} = \pm\sqrt{\frac{k}{m}}\, i = \pm\omega_0 i$

General Solution: $u(t) = A\cos(\omega_0 t) + B\sin(\omega_0 t)$

It is possible to rewrite this General Solution (G.S.) in the following form, which is more convenient to use.

UNDAMPED VIBRATION:	$u(t) = R\cos(\omega_0 t + \delta)$	
ω_0	$\sqrt{\frac{k}{m}}$	Natural Frequency
R	$\sqrt{A^2 + B^2}$	Amplitude
δ	$\tan^{-1}\left(\frac{B}{A}\right)$	Phase
τ	$\frac{2\pi}{\omega_0} = 2\pi\sqrt{\frac{m}{k}}$	Period

DAMPED VIBRATIONS: ($\mu \neq 0$)

Equation: $mu'' + \mu u' + ku = 0$ With: $f(x) = 0$

$u(0) = u_0, \quad u'(0) = u'_0 = v_0$

Let: $u = e^{rt} \quad u' = re^{rt} \quad u'' = r^2 e^{rt}$

Char. Equation: $mr^2 + \mu r + k = 0 \rightarrow r_{1,2} = \dfrac{-\mu \pm \sqrt{\mu^2 - 4mk}}{2m}$

The three different cases, based on the roots (r_1 and r_2) are listed in the following table.

Root Description	General Solution (G.S.)	Type of Damping
$r_1 \neq r_2$ $r_1, r_2 \in \mathbb{R}$	$u(t) = A\,e^{r_1 t} + B\,e^{r_2 t}$ Slowly returns to equilibrium.	Over Damping
$r_1 = r_2 = \dfrac{-\mu}{2m}$ $r_1, r_2 \in \mathbb{R}$	If $\mu^2 - 4mk = 0$ $u(t) = (A + Bt)\,e^{\frac{-\mu t}{2m}}$ Returns to Equilibrium faster.	Critical Damping
$r_{1,2} = \lambda \pm i\mu$ $r = -\dfrac{\mu}{2m} \pm \dfrac{i\sqrt{4mk - \mu^2}}{2m}$ Quasi Frequency: $= \omega_d = \dfrac{\sqrt{4mk - \mu^2}}{2m}$ Quasi Period: $= \tau_d = 2\pi/\omega_d$	If $\mu^2 - 4mk < 0$ $u(t) = e^{\lambda t}(A\cos(\mu t) + B\sin(\mu t))$ $u(t) = e^{-\frac{\mu t}{2m}}(A\cos(\mu t) + B\sin(\mu t))$ $u(t) = R\,e^{-\frac{\mu t}{2m}}\cos(\mu t - \delta)$ Periodic Motion. (Oscillates) Most realistic case.	Under Damping

	Displacement: $mu'' + \mu u' + ku = 0$ $u(0) = u_0, \ u'(0) = u'_0 = v_0$
With No Damping ($\mu = 0$)	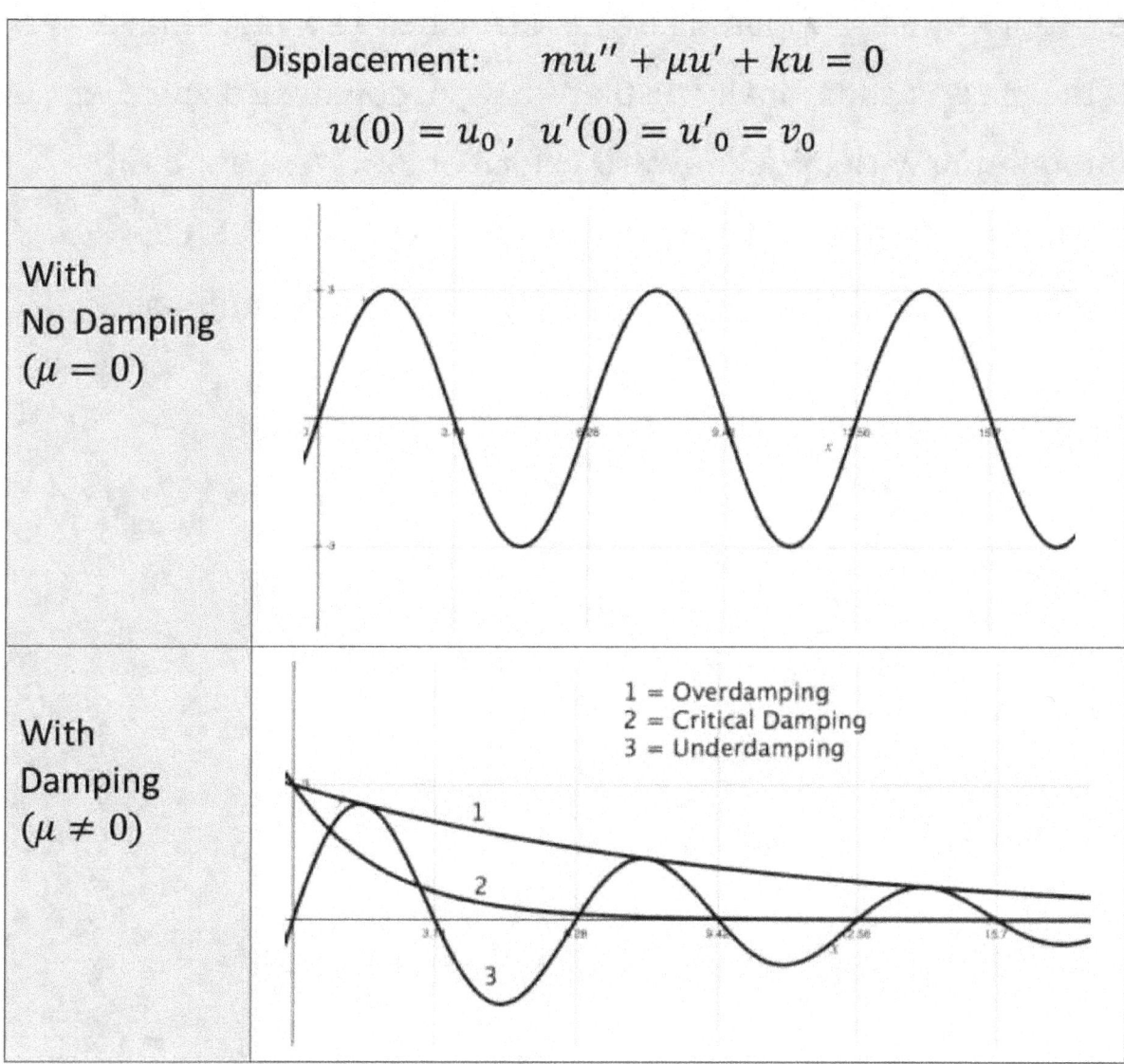
With Damping ($\mu \neq 0$)	1 = Overdamping 2 = Critical Damping 3 = Underdamping

Example: A mass weighing 64 lbs. stretches a spring 24 in. If the mass is additionally displaced 12 in. downward and then set into motion with initial upward velocity of 2 ft/s. Find the position of the block at any time t. (e.g. Get G.S. of D.E.) Then, find Period, Natural Frequency, Amplitude, and Phase.

Solution:

$$w = 64 \; lb.$$

$$m = \frac{w}{g} = \frac{64}{32} = 2$$

$$L = 24 \; in. = \frac{24}{12} = 2 \; ft.$$

$$u(0) = 12 \; in. = \frac{12}{12} = 1 \; ft.$$

$$u'(0) = -2 \quad \text{(upward)}$$

$$2u'' + 32 u = 0 \qquad \text{With: } u(0) = 1, \; u'(0) = -2$$

Characteristic Equation: $2r^2 + 32 = 0$

$$r^2 + 16 = 0$$

$$r = \pm\sqrt{-16}$$

$$r = \pm 4i$$

Continued…

Previously, found: $r = \pm 4i$

$u(t) = A\cos 4t + B\sin 4t$ G.S.

$t_0 = 0$ $u_0 = 1$ $u'(0) = -2$

$u(t) = A\cos(\omega_0 t) + B\sin(\omega_0 t)$
$1 = A\cos(0) + B\sin(0)$ ➔ $A = 1$

$u'(t) = -A\sin(\omega_0 t)\,\omega_0 + B\cos(\omega_0 t)\,\omega_0$
$u'(t) = -A4\sin(4t) + 4B\cos(4t)$

$-2 = -4\sin(0) + 4B\cos(0)$ ➔ $B = -\dfrac{2}{4} = -\dfrac{1}{2}$

$u(t) = \cos(4t) - \dfrac{1}{2}\sin(4t)$ P.S.

No Frequency	$\omega_0 = 4$
Period	$\tau = \dfrac{2\pi}{\omega} = \dfrac{2\pi}{4} = \dfrac{\pi}{2}$
No Damping	Will vibrate forever.

Example: A block weighing 8 lbs. stretches a spring by 4 in. Suppose the block is displaced an additional 6 inches downward and then released. The viscous damping is 4 lb. when velocity is 8 ft./s. Model the D.E. governing this motion.

Solution:

$$w = 8 \, lb. \; = mg \quad \Rightarrow \quad m = \frac{w}{g} = \frac{8}{32} = \frac{1}{4} \frac{lb.s^2}{ft.}$$

$$L = 4 \, in. = \frac{4}{12} = \frac{1}{3} \, ft.$$

$$u(0) = 6 \, in. = \frac{6}{12} = \frac{1}{2} \, ft.$$

$$u'(0) = 0 \qquad \text{No velocity when released} \;\Rightarrow\; v_0 = 0$$

$$w = F_s = kL \quad \Rightarrow \quad k = \frac{w}{L} = \frac{8}{\left(\frac{1}{3}\right)} = 24 \frac{lb}{ft.}$$

$$F_d = 4 \qquad \text{When } \frac{du}{dt} = 8 \frac{ft}{s}.$$

$$F_d = \mu \frac{du}{dt}$$

$$4 = \mu(8) \quad \Rightarrow \quad \mu = \frac{4}{8} = \frac{1}{2}$$

$$\boxed{m\,u'' + \mu\,u' + k\,u = 0}$$

$$\frac{1}{4} u'' + \frac{1}{2} u' + 24\,u = 0$$

DAMPING: The vibrations (or movement or displacement) of a system will eventually stop only if there is damping. The roots appear as an exponent in an exponential equation, which defines the movement. With a negative exponent, the movement (or displacement) approaches zero. Examples for the three types of damping are listed in the following table.

Over Damped $r_1 \neq r_2$ $r_1 < 0$ $r_2 < 0$	$u'' + 3u' + 2u = 0$ With: $u(0) = 2$ and $u'(0) = -1$ Characteristic Equation: $r^2 + 3u + 2 = 0$ $(u+1)(u+2) = 0 \quad \rightarrow r = -1, -2$ Solution: $u(t) = C_1 e^{-t} + C_2 e^{-2t}$
Critical Damping $r_1 = r_2 < 0$	$u'' + 4u' + 4u = 0$ With: $u(0) = 2$ and $u'(0) = 1$ Characteristic Equation: $r^2 + 4r + 4 = 0$ $(r+2)(r+2) = 0 \quad \rightarrow r_1 = r_2 = -2$ Solution: $u(t) = (C_1 + C_2 t)e^{-2t}$
Under Damping $r_{1,2} = p \pm qi$ $p < 0$	$u'' + 2u' + 2u = 0$ With: $u(0) = 1$ and $u'(0) = -1$ Characteristic Equation: $r^2 + 2r + 2 = 0$ $r_{1,2} = \dfrac{-2 \pm \sqrt{4-8}}{2} = -1 \pm i$ Solution: $u(t) = e^{-t}[C_1 \cos t + C_2 \sin t]$

NOTE: For each solution, C_1 and C_2 can be computed using the initial conditions.

Electrical Vibrations

ELECTRICAL VIBRATIONS (LCR Circuit)

<u>Electrical Circuit</u>: Flow of electric current in a simple circuit.

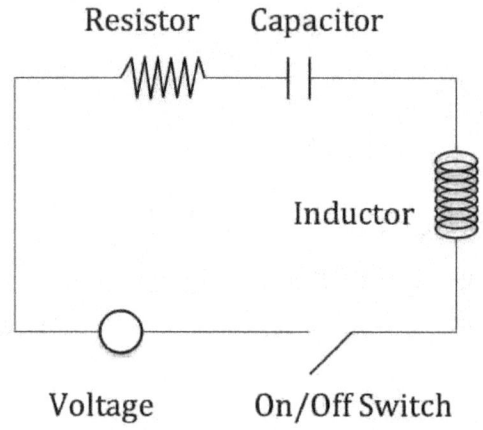

Item		Units
$I(t)$	Current	Amperes
R	Resistance	Ohms
C	Capacitance	Farads
L	Inductance	Henrys

Note: I = Current = $\frac{dQ}{dt}$ = Rate of change of the charge.

Three areas where voltage drop may occur:

1. Voltage drop across resistor = IR
2. Voltage drop across capacitor = $\frac{Q}{C}$
3. Voltage drop across inductor = $L\frac{dI}{dt}$

<u>Kirchhoff's Second Law</u>: In a closed, simple circuit

    Impressed Voltage = Sum of Voltage drops in circuit

$$L\frac{dI}{dt} + RI + \frac{1}{C}Q = E(t) \qquad Q(t_0) = Q_0$$

$$L\frac{d^2Q}{dt^2} + R\frac{dQ}{dt} + \frac{1}{C}Q = E(t) \qquad Q'(t_0) = I(t_0) = I_0$$

Example: Compute Current and Charge in a simple LCR Circuit.

Given: Resistance, $R = 30$ ohms

Capacitance, $C = 0.008$ farads

Inductance, $L = 1$ henry

Impressed Voltage, $E(t) = 10 \cos 2t$

With: $Q(0) = 0$

$Q'(0) = 0 = I(0)$

At $t = 0$, the charge and rate of change in the charge should be 0.

Solution:

$$L \frac{d^2Q}{dt^2} + R \frac{dQ}{dt} + \frac{1}{C} Q = E(t)$$

$$(1)Q'' + (30)Q' + (125)Q = 10 \cos 2t$$

(Step 1) Characteristic Equation: $\quad r^2 + 30r + 125 = 0$

$(r + 25)(r + 5) = 0 \qquad \rightarrow r = -25, -5$

(Step 2) Consider: $\quad Q'' + (30)Q' + (125)Q = 10 \cos 2t$

General Solution: $\quad Q(t) = C_1 e^{-5t} + C_2 e^{-25t} + \boldsymbol{Q_s}$

Where: Q_s = Specific Solution

(Step 3) Use Method of Undetermined Coefficients (MUC) to compute Q_s (Continued...)

(Step 3) Use MUC to compute the Specific Solution, Q_s

Let:
$$Q_s = A\cos 2t + B\sin 2t$$
$$Q'_s = -2A\sin 2t + 2B\cos 2t$$
$$Q''_s = -4A\cos 2t - 4B\sin 2t$$

$$Q'' + (30)Q' + (125)Q = 10\cos 2t$$

$$[-4A\cos 2t - 4B\sin 2t] + 30[-2A\sin 2t + 2B\cos 2t]$$
$$+ 125[A\cos 2t + B\sin 2t] = 10\cos 2t + 0\sin 2t$$

$$-4A\cos 2t - 4B\sin 2t - 60A\sin 2t + 60B\cos 2t$$
$$+ 125A\cos 2t + 125B\sin 2t = 10\cos 2t + 0\sin 2t$$

$$(121A + 60B)\cos 2t + (121B - 60A)\sin 2t = 10\cos 2t + 0\sin 2t$$

$$121A + 60B = 10$$
$$121B - 60A = 0 \quad\quad \rightarrow A = \frac{121}{60}B$$

$$121\left(\frac{121}{60}B\right) + 60B = 10 \quad \rightarrow B = 0.03289,\ A = 0.06633$$

$$Q_s = A\cos 2t + B\sin 2t$$
$$Q_s = (0.06633)\cos 2t + (0.03289)\sin 2t$$

Thus, the General Solution (G.S.) is:

$$Q(t) = C_1 e^{-5t} + C_2 e^{-25t} + \boldsymbol{Q_s}$$
$$Q(t) = C_1 e^{-5t} + C_2 e^{-25t} + (0.06633)\cos 2t + (0.03289)\sin 2t$$

To find the Particular Solution (P.S.) use the initial conditions.

Use:

$Q(t) = C_1 e^{-5t} + C_2 e^{-25t} + (0.0663)\cos 2t + (0.0329)\sin 2t$

$Q'(t) = -5C_1 e^{-5t} - 25C_2 e^{-25t} - (.1327)\sin 2t + (0.0658)\cos 2t$

$Q(0) = 0 \quad \rightarrow \quad C_1 + C_2 + 0.06633 = 0$

$Q'(0) = 0 \quad \rightarrow \quad -5C_1 - 25C_2 + 0.06578 = 0$

$C_1 = -0.0862 \quad$ and $\quad C_2 = 0.0199$

Particular Solution (P.S.)

$Q(t) =$ Charge at time "t"

$Q(t) = -0.0862\, e^{-5t} + .0199\, e^{-25t} + .0663\cos 2t + .0329\sin 2t$

Extra:

$I(t) =$ Current at time "t"

$I(t) = \frac{d}{dt}Q(t) = .431\, e^{-5t} - .497\, e^{-25t} - .133\sin 2t + .066\cos 2t$

SERIES SOLUTION of DE

In this chapter, we will learn how to solve a Second-Order Linear Homogeneous DE of the form: $P(x)y'' + Q(x)y' + R(x)y = 0$.

Previously, we have used methods to construct the linearly independent solutions of Homogeneous DEs with constant coefficients.

To deal with DEs with non-constant coefficients, it is necessary to extend our search for solutions beyond the familiar elementary functions of calculus. The main tool that we need, to represent such functions, is a power series. The basic idea is similar to method of undetermined coefficients. We assume the DE has a solution in form of a power series and, based on this assumption, we compute the coefficients of the power series.

POWER SERIES (REVIEW)

Power Series Review	Description
$\sum_{n=0}^{\infty} a_n (x - x_0)^n$ $= a_0 + a_1(x - x_0) + a_2 (x - x_0)^2 + \cdots$	Power Series centered at x_0
$\sum_{n=0}^{\infty} a_n x^n$ $= a_0 + a_1 x + a_2 x^2 + \cdots$	Power Series centered at 0. We usually assume this.
$\sum_{n=1}^{\infty} \frac{1}{n(n+1)} = \sum_{n=1}^{\infty} \left[\frac{A}{n} + \frac{B}{n+1} \right]$ $S_N = 1 + \frac{1}{2} - \frac{1}{N} - \frac{1}{N+1}$	Telescoping Series And Sum of first N terms.

POWER SERIES Review -- CONVERGENCE:

Every function can be expressed as a power series (using Taylor Series representation).

A Power Series is Convergent if:

$$\lim_{N \to \infty} \sum_{n=1}^{N} a_n (x - x_0)^n \quad \text{Exists as a real number.}$$

A Power Series is Absolutely Convergent if:

$$\lim_{N \to \infty} \sum_{n=1}^{N} |a_n| \cdot |x - x_0|^n \quad \text{Converges.}$$

CHECK for CONVERGENCE: Use the <u>Ratio</u> or <u>Root</u> Test.

For the Power Series: $\sum_{n=0}^{\infty} a_n (x - x_0)^n$

<u>Ratio Test</u> → Take limit of the ratio of 2 consecutive terms.

$$\lim_{n \to \infty} \left| \frac{a_{n+1} (x-x_0)^{n+1}}{a_n (x-x_0)^n} \right| = \lim_{n \to \infty} \left| \frac{a_{n+1}}{a_n} (x - x_0) \right|$$

$$= |x - x_0| \lim_{n \to \infty} \left| \frac{a_{n+1}}{a_n} \right| = |x - x_0| \cdot L$$

$|x - x_0| \cdot L < 1$ → CONVERGENT

$|x - x_0| \cdot L > 1$ → DIVERGENT

RADIUS of CONVERGENCE:

$$|x - x_0| < \frac{1}{L} = \delta = \text{Radius of Convergence}$$

INTERVAL of CONVERGE:

$$-\delta < x - x_0 < \delta$$
$$x_0 - \delta < x < x_0 + \delta$$

Interval of Convergence $= (x_0 - \delta, \; x_0 + \delta)$

- Each boundary must be checked separately. If the series is convergent at the boundary, then include it.
- The Ratio Test or the Root Test will not work for boundary point checks. Other testing techniques must be used.

POWER SERIES Review – EXAMPLES

Example (1)

Find Radius and Interval of Convergence for: $\sum_{n=0}^{\infty} x^n$

Solution:

Note: Centered at zero.

$$\lim_{n\to\infty} \left|\frac{x^{n+1}}{x^n}\right| = \lim_{n\to\infty} |x| = |x|$$

→ PS is Convergent when: $|x| < 1$

→ PS is Divergent when: $|x| > 1$

Interval of Convergence: $-1 < x < 1$

Must check each boundary separately.

Check: $x = 1$

$\sum_{n=0}^{\infty} (1)^n = 1 + 1 + 1 + \cdots$ → Divergent

Check: $x = -1$

$\sum_{n=0}^{\infty} (-1)^n = 1 - 1 + 1 \ldots$ → Limit DNE

Radius of Convergence $= 1$

Interval of Convergence: $(-1, 1)$ Boundaries not included.

POWER SERIES Review – EXAMPLES

Example (2)

Find Radius and Interval of Convergence for: $\sum_{n=0}^{\infty} n!\, x^n$

Solution:

$$\lim_{n \to \infty} \left| \frac{(n+1)!\, x^{n+1}}{n!\, x^n} \right| = \lim_{n \to \infty} \left| \frac{n!\, (n+1)\, x^n\, x}{n!\, x^n} \right|$$

$$= |x| \lim_{n \to \infty} (n+1) = \infty$$

This limit is < 1 ONLY when $x = 0$ ➔ ONLY one point.
Radius of Convergence $= 0$
Interval of Convergence: $\emptyset$ (empty set)

Example (3)

Find Radius and Interval of Convergence for: $\sum_{n=0}^{\infty} \frac{x^n}{n!}$

Solution:

$$\lim_{n \to \infty} \left| \frac{x^{n+1}}{(n+1)!} \cdot \frac{n!}{x^n} \right| = \lim_{n \to \infty} \left| \frac{x^n\, x}{(n+1)\, n!} \cdot \frac{n!}{x^n} \right|$$

$$= |x| \lim_{n \to \infty} \frac{1}{n+1} = 0$$

This limit is always < 1. Doesn't matter what x is.
This PS always converges.
Radius of Convergence $= \infty$
Interval of Convergence $= (-\infty, \infty)$

Note: These 3 Examples had series centered at 0 and were simple.

POWER SERIES Review – EXAMPLES

Example (4)

Find Radius and Interval of Convergence for: $\sum_{n=0}^{\infty} \frac{(x-3)^n}{2^n\, n}$

Solution:

$$\lim_{n\to\infty} \left| \frac{(x-3)^{n+1}}{2^{n+1}(n+1)} \cdot \frac{2^n \cdot n}{(x-3)^n} \right|$$

$$= \lim_{n\to\infty} \left| \frac{(x-3)^n (x-3)}{2^n \cdot 2 \cdot (n+1)} \cdot \frac{2^n \cdot n}{(x-3)^n} \right|$$

$$= \tfrac{1}{2} |x-3| \lim_{n\to\infty} \left| \frac{n}{n+1} \right|$$

$$= \tfrac{1}{2} |x-3| \lim_{n\to\infty} \frac{n}{n+1}$$

$$= \text{Indeterminate Form} \rightarrow \text{Use L'Hopital's Rule}$$

L.H. → Take the derivative of the numerator and denominator.

$$\text{L.H.} = \tfrac{1}{2} |x-3| \lim_{n\to\infty} \tfrac{1}{1} = \tfrac{1}{2} |x-3|$$

This PS is convergent when: $\tfrac{1}{2} |x-3| < 1$

$|x-3| < 2$ → Radius of Convergence = 2

Interval of Convergence:

$-2 < x-3 < 2$

$1 < x < 5$ Check the boundaries.

(Continued...)

Interval of Convergence:

$$-2 < x - 3 < 2$$
$$1 < x < 5 \qquad \text{Check the boundaries.}$$

Check: $x = 1$

$$\sum_{n=0}^{\infty} \frac{(x-3)^n}{2^n \, n} = \sum_{n=0}^{\infty} \frac{(1-3)^n}{2^n \, n} = \sum_{n=0}^{\infty} \frac{(-2)^n}{2^n \, n}$$

$$= \sum_{n=1}^{\infty} \frac{(-1)^n}{n}$$

= Alternating Harmonic Series

→ Convergent by Alternating Series Test

So, $x = 1$ is included.

Check: $x = 5$

$$\sum_{n=0}^{\infty} \frac{(x-3)^n}{2^n \, n} = \sum_{n=0}^{\infty} \frac{(5-3)^n}{2^n \, n} = \sum_{n=0}^{\infty} \frac{(2)^n}{2^n \, n}$$

$$= \sum_{n=1}^{\infty} \frac{1}{n}$$

P-Series Test → Divergent

So, $x = 5$ is NOT included.

Interval of Convergence: $[\,1, 5\,)$

POWER SERIES Review – EXAMPLES

Example:

Find Radius and Interval of Convergence for: $\sum_{n=1}^{\infty} \frac{3^n (2-x)^n}{n+1}$

Solution: Finish it yourself. One boundary point is included.

Remark: It is possible to add, subtract, multiply, and divide two Power Series ONLY if the generic term in both series contain the same power of x.

Example: Addition and Subtraction

$$\sum_{n=0}^{\infty} a_n(x-x_0)^n \pm \sum_{n=0}^{\infty} b_n(x-x_0)^n$$
$$= \sum_{n=0}^{\infty}(a_n + b_n)(x-x_0)^n$$

Example: Multiplication

$$\sum_{n=0}^{\infty} a_n(x-x_0)^n \cdot \sum_{n=0}^{\infty} b_n(x-x_0)^n$$
$$= \sum_{n=0}^{\infty} c_n(x-x_0)^n$$
$$= c_0 + c_1(x-x_0) + c_2(x-x_0)^2 + \cdots$$

Note:

$c_0 = a_0 b_0$, $c_1 = a_0 b_1 + a_1 b_0$, $c_2 = a_0 b_2 + a_1 b_1 + a_2 b_0$

$c_n =$ Cauchy Product

$c_n = a_n b_0 + a_{n-1} b_1 + \ldots + a_1 b_{n-1} + a_0 b_n$

For $n = 0, 1, 2, \ldots n$

Power Series Remarks:

If $f(x) = \sum_{n=0}^{\infty} a_n (x - x_0)^n$ is continuous and differentiable (infinitely many times)

- Then: $f', f'', f''', f^{(4)}, f^{(n)}$ can be calculated and they will all have the same radius of convergence, "ρ" (Rho).

If two power series have the same radius of convergence, then they can be added, subtracted, multiplied, and divided.

$$f(x) = \sum_{n=0}^{\infty} a_n (x - x_0)^n$$
$$f(x) = a_0 + a_1 (x - x_0) + a_2 (x - x_0)^2 + a_3 (x - x_0)^3 + \ldots$$

$$f'(x) = 0 + a_1 + 2a_2 (x - x_0) + 3a_3 (x - x_0)^2 + \ldots$$
$$f''(x) = \phantom{0 + a_1 + {}} 0 + 2a_2 + 6a_3 (x - x_0) + \ldots$$

$$f'(x) = \sum_{n=1}^{\infty} n \, a_n (x - x_0)^{n-1}$$
$$f''(x) = \sum_{n=2}^{\infty} n(n-1) \, a_n (x - x_0)^{n-2}$$
$$f'''(x) = \sum_{n=3}^{\infty} n(n-1)(n-2) \, a_n (x - x_0)^{n-3}$$

TAYLOR SERIES EXPANSION

Definition:

A function $f(x)$ that has a Taylor Series Expansion near $x = x_0$ (i.e. $f(x) = \sum_{n=0}^{\infty} a_n (x - x_0)^n$) is said to be <u>analytic</u> at $x = x_0$.

If a function has a Taylor Series Expansion

→ All coefficients (a_n) can be computed using,

$$a_n = \frac{f^{(n)}(x_0)}{n!}$$

Shift of Index:

$$\sum_{n=0}^{\infty} a_n (x - x_0)^n = \sum_{m=0}^{\infty} a_m (x - x_0)^m$$
$$= \sum_{i=0}^{\infty} a_i (x - x_0)^i$$

Hence: m, n, i are "dummy" indices.

Power Series Example:

Rewrite the following P.S. in terms of a P.S. starting with index "0."

$$\sum_{n=3}^{\infty} (n+2) a_n (x-x_0)^{n+5}$$

Solution:

Replace "n" by "$m+3$"

$$= \sum_{m+3=3}^{\infty} (m+3+2) a_{m+3} (x-x_0)^{m+3+5}$$
$$= \sum_{m=0}^{\infty} (m+5) a_{m+3} (x-x_0)^{m+8}$$
$$= \sum_{n=0}^{\infty} (n+5) a_{n+3} (x-x_0)^{n+8}$$

Power Series Example:

Rewrite the following P.S. in terms of a P.S. starting with index "0."

$$\sum_{n=-2}^{\infty} (n+3) a_n (x-x_0)^{n+3}$$

Solution:

Replace "n" by "$n-2$"

$$= \sum_{n=0}^{\infty} (n+1) a_{n-2} (x-x_0)^{n+1}$$

Power Series Example:

Rewrite the following P.S. with "n" power of generic term.

$$\sum_{n=2}^{\infty} n(n+5) a_{n+2} (x)^{n+3}$$

Solution:

Replace "n" by "$n-3$"

$$= \sum_{n=5}^{\infty} (n-3)(n+2) a_{n-1} (x)^n$$

Power Series Example:

Get "n" Power for: $\sum_{n=0}^{\infty} n(n-1)(n-5) a_{n+1} (x)^{n-2}$

Solution:

Replace "n" by "n+2"

$$= \sum_{n=-2}^{\infty} (n+2)(n+1)(n-3) a_{n+3} (x)^n$$

EQUATING POWER SERIES:

Two P.S. are equal if their coefficients coincide.

If $\quad \sum_{n=0}^{\infty} a_n (x - x_0)^n = \sum_{n=0}^{\infty} b_n (x - x_0)^n$

Then: $\quad a_n = b_n \quad$ for $n = 0, 1, 2, 3, \ldots$

EQUATING POWER SERIES (Special Case)

If $\quad \sum_{n=0}^{\infty} a_n (x - x_0)^n = 0$

Then:

$$\sum_{n=0}^{\infty} a_n (x - x_0)^n = \sum_{n=0}^{\infty} 0 \, (x - x_0)^n$$

Then: $\quad a_n = 0 \quad$ for $n = 0, 1, 2, 3, \ldots$

All coefficients are zero.

POWER SERIES NEAR AN ORDINARY POINT

Power Series Solution – Near an Ordinary Point
Recall this Example: $y' = y$ Find G.S.
Solution: Using Separable DE Method $\frac{dy}{dx} = y$ $\int \frac{1}{y} dy = \int 1\, dx$ $\ln y = x + C$ $y = e^{x+C} = e^x e^C = Ce^x$
Another Solution: Using First-Order Linear DE Method $y' - (1)y = 0$ $\mu(x) = e^{-\int 1\, dx} = e^{-x}$ (Integrating Factor) $y = \frac{\int 0 \cdot e^{-x}\, dx + C}{e^{-x}} = \frac{\int 0\, dx + C}{e^{-x}} = \frac{C}{e^{-x}} = Ce^x$
Another Solution: Using Characteristic Equation Method $y' - y = 0$ $r - 1 = 0$ → $r = 1$ → $y = Ce^{1x} = Ce^x$
Another Solution: Using Taylor Series (See ahead)

Example: Compute a_n that satisfy:

$\sum_{n=0}^{\infty} a_n (x)^n = \sum_{n=1}^{\infty} n\, a_n (x)^{n-1}$ Note: Similar to: $y = y'$

Solution:

Make "powers" the same. Change "n" by "n+1" on RHS.

$\sum_{n=0}^{\infty} a_n (x)^n = \sum_{n=0}^{\infty} (n+1)\, a_{n+1} (x)^n$

So: $a_n = (n+1)\, a_{n+1}$ → $a_{n+1} = \dfrac{a_n}{n+1}$ Recursive Eqn.

$n = 0$ → $a_1 = \dfrac{a_0}{0+1} = a_0$

$n = 1$ → $a_2 = \dfrac{a_1}{2} = \dfrac{a_0}{2 \cdot 1}$

$n = 2$ → $a_3 = \dfrac{a_2}{3} = \dfrac{a_1}{3 \cdot 2} = \dfrac{a_0}{3 \cdot 2 \cdot 1}$

$n = 3$ → $a_4 = \dfrac{a_3}{4} = \dfrac{a_2}{4 \cdot 3 \cdot} = \dfrac{a_1}{4 \cdot 3 \cdot 2} = \dfrac{a_0}{4 \cdot 3 \cdot 2 \cdot 1} = \dfrac{a_0}{4!}$

$n = 4$ → $a_5 = \dfrac{a_0}{5!}$

$f(x) = \sum_{n=0}^{\infty} a_n (x)^n = a_0 + a_1 x + a_2 x^2 + \ldots$

$f(x) = a_0 + \dfrac{a_0}{1!} x + \dfrac{a_0}{2!} x^2 + \dfrac{a_0}{3!} x^3 + \ldots$

$f(x) = a_0 \left[1 + \dfrac{x}{1!} + \dfrac{x^2}{2!} + \dfrac{x^3}{3!} + \cdots \right]$

$f(x) = a_0 \left[\sum_{n=0}^{\infty} \dfrac{x^n}{n!} \right]$ Recall: $0! = 1$ and $x^0 = 1$

$f(x) = a_0 e^x$ Recall: $e^x = \sum_{n=0}^{\infty} \dfrac{x^n}{n!}$

Same answer as previous example. $y = Ce^x$

If we want to solve: $P(x)\, y'' + Q(x)\, y' + R(x)\, y = 0$

Near $x = x_0$

We must first, verify that x_0 is a good point.

$$y'' + \frac{Q(x)}{P(x)}\, y' + \frac{R(x)}{P(x)}\, y = 0$$

Evaluate: $P(x_0) \neq 0$ Note: Can't divide by zero!

If $P(x_0) \neq 0$ Then, x_0 is an <u>ordinary point</u>. *

If $P(x_0) = 0$ Then, x_0 is a <u>singular point</u>.

---

* If $P(x_0) \neq 0$ then x_0 is an <u>ordinary point</u> and ∃ (there exists) a solution of DE in the form of a Power Series, centered at x_0

→ The solution is in the form: $y = \sum_{n=0}^{\infty} a_n (x - x_0)^n$

We will focus on the cases when:

$P(x),\ Q(x),\ R(x)$ are polynomials.

Theorem:

A Differential Equation: $P(x) y'' + Q(x) y' + R(x) y = 0$

Has a Power Series (P.S.) solution near x_0

$\qquad$ if x_0 is an ordinary point.

	General Idea: How to solve a Second Order DE with Non-Constant Coefficients $P(x) y'' + Q(x) y' + R(x) y = 0 \qquad$ Near $x = x_0$
1.	Check $P(x_0) \neq 0 \quad \rightarrow \quad x_0$ is an ordinary point.
2.	Assume the DE has solution as P.S. in the form $y = \sum_{n=0}^{\infty} a_n (x - x_0)^n$
3.	Compute y' and y'' to find all the coefficients. Since x_0 is an ordinary point, $\exists$ a Power Series solution in the form: $y = \sum_{n=0}^{\infty} a_n (x - x_0)^n$
4.	Substitute y, y' and y'' into the DE. Get a Recursive Relation, using the DE. Compute the first few coefficients in terms of a_0 and a_1
5.	Write the General Solution (G.S.)

Example: Solve: $y'' + y = 0$ Near $x = x_0 = 0$

Solution:

> NOTE: We solved this problem previously.
>
> $y'' + y = 0$
>
> $r^2 + 1 = 0$ Characteristic Equation.
>
> $r = \pm i$
>
> $y_1 = \cos x$ $y_2 = \sin x$
>
> General Solution: $y = C_1 \cos x + C_2 \sin x$

$P(x) = 1$ $R(x) = 1$ $Q(x) = 0$

$P(x_0) = P(0) = 1 \neq 0$ → $x_0 = 0$ is an ordinary point.

→ ∃ a Power Series (P.S.) solution in the form:
$$y = \sum_{n=0}^{\infty} a_n (x - x_0)^n$$

Compute a_n

Find the 1st and 2nd derivatives:

$y' = \sum_{n=1}^{\infty} n\, a_n (x)^{n-1}$

$y'' = \sum_{n=2}^{\infty} n(n-1)\, a_n (x)^{n-2}$

Substitute:

$y'' + y = 0$

$\sum_{n=2}^{\infty} n(n-1)\, a_n (x)^{n-2} + \sum_{n=0}^{\infty} a_n (x)^n = 0$

$$\sum_{n=2}^{\infty} n(n-1) a_n (x)^{n-2} + \sum_{n=0}^{\infty} a_n (x)^n = 0$$

$$\sum_{n=2}^{\infty} (n+2)(n+1) a_{n+2} (x)^n + \sum_{n=0}^{\infty} a_n (x)^n = \sum_{n=0}^{\infty} 0 \cdot x^n$$

$$\sum_{n=2}^{\infty} [(n+2)(n+1) a_{n+2} + a_n] (x)^n = \sum_{n=0}^{\infty} 0 \cdot x^n$$

$$(n+2)(n+1) a_{n+2} + a_n = 0$$

$$a_{n+2} = \frac{-a_n}{(n+2)(n+1)} \qquad \underline{\text{Recursive Relation}}$$

Find a_n in terms of a_0 and a_1

$$a_{n+2} = \frac{-a_n}{(n+2)(n+1)}$$

$n = 0 \rightarrow a_2 = \dfrac{-a_0}{(2)(1)}$

$n = 1 \rightarrow a_3 = \dfrac{-a_1}{(3)(2)}$

$n = 2 \rightarrow a_4 = \dfrac{-a_2}{(4)(3)} = \dfrac{-(-a_0)}{(4)(3)(2)(1)} = \dfrac{a_0}{4!}$

$n = 3 \rightarrow a_5 = \dfrac{-a_3}{(5)(4)} = \dfrac{-(-a_1)}{(5)(4)(3)(2)} = \dfrac{a_1}{5!}$

$n = 4 \rightarrow a_6 = \dfrac{-a_4}{(6)(5)} = \dfrac{-(a_0)}{(6)(5)4!} = \dfrac{-a_0}{6!}$

$n = 5 \rightarrow a_7 = \dfrac{-a_5}{(7)(6)} = \dfrac{-(a_1)}{(7)(6)5!} = \dfrac{-a_1}{7!}$

(Continued...)

Solution:

$$y = \sum_{n=0}^{\infty} a_n (x)^n = a_0 + a_1 x + a_2 x^2 + a_3 x^3 + \cdots$$

$$y = a_0 + a_1 x - \frac{a_0}{2!} x^2 - \frac{a_1}{3!} x^3 + \frac{a_0}{4!} x^4 + \frac{a_1}{5!} x^5 - \cdots$$

$$y = a_0 \left[1 - \frac{x^2}{2!} + \frac{x^4}{4!} \cdots \right] + a_1 \left[x - \frac{x^3}{3!} + \frac{x^5}{5!} \cdots \right]$$

$$y = a_0 \left[\sum_{n=1}^{\infty} \frac{(-1)^n x^{2n}}{(2n)!} \right] + a_1 \left[\sum_{n=1}^{\infty} \frac{(-1)^n x^{2n+1}}{(2n+1)!} \right]$$

$y = a_0 \cos x + a_1 \sin x$ ➜ Same answer:

$$y = C_1 \cos x + C_2 \sin x$$

Review -- Taylor Series Expansions: (Special Cases)

Expression	Taylor Series
$\dfrac{1}{1-x}$	$\sum_{n=0}^{\infty} x^n$
e^x	$\sum_{n=0}^{\infty} \dfrac{x^n}{n!}$
$\cos x$	$\sum_{n=0}^{\infty} (-1)^n \dfrac{x^{2n}}{(2n)!}$
$\sin x$	$\sum_{n=0}^{\infty} (-1)^{(n-1)} \dfrac{x^{2n-1}}{(2n-1)!} \quad = \quad \sum_{n=0}^{\infty} (-1)^{(n)} \dfrac{x^{2n+1}}{(2n+1)!}$
$\ln(1+x)$	$\sum_{n=0}^{\infty} (-1)^{(n-1)} \dfrac{x^n}{n} \quad = \quad \sum_{n=0}^{\infty} (-1)^{(n+1)} \dfrac{x^n}{n}$
$\tan^{-1} x$	$\sum_{n=0}^{\infty} (-1)^{(n-1)} \dfrac{x^{2n-1}}{(2n-1)} \quad = \quad \sum_{n=0}^{\infty} (-1)^{(n)} \dfrac{x^{2n+1}}{(2n+1)}$

Example: DE near an Ordinary Point

Solve: $y'' - x^2 y = 0$, Near $x = 0$

Solution:
$$P(x) = 1 \qquad Q(x) = 0 \qquad R(x) = -x^2$$

Note: $Q(x)$ is the coefficient of y'

$P(0) = 1 \neq 0 \qquad \rightarrow \qquad x_0 = 0$ is an ordinary point.

∃ a Power Series Solution in the form of: $y = \sum_{n=0}^{\infty} a_n x^n$

$$y = \sum_{n=0}^{\infty} a_n x^n$$
$$y' = \sum_{n=1}^{\infty} n\, a_n x^{n-1}$$
$$y'' = \sum_{n=2}^{\infty} n(n-1)\, a_n x^{n-2}$$

Substitute:

$y'' - x^2 y = 0$ This is the original DE.

$$\sum_{n=2}^{\infty} n(n-1)\, a_n x^{n-2} - x^2 \sum_{n=0}^{\infty} a_n x^n = 0$$

$$\sum_{n=2}^{\infty} n(n-1)\, a_n x^{n-2} - \sum_{n=0}^{\infty} a_n x^{n+2} = 0$$

Change each series to x^n

$$\sum_{n=0}^{\infty} (n+2)(n+1) a_{n+2} x^n - x^2 \sum_{n=2}^{\infty} a_{n-2} x^n = 0$$

Change each series to start at highest starting point. ($n = 2$)

$$(2)(1) a_2 x^0 + (3)(2) a_3 x^1 + \ldots$$
$$+ \sum_{n=2}^{\infty} (n+2)(n+1) a_{n+2} x^n - x^2 \sum_{n=2}^{\infty} a_{n-2} x^n = 0$$

$$2a_2 + 6a_3 x + \ldots$$
$$+ \sum_{n=2}^{\infty} (n+2)(n+1)a_{n+2} x^n - x^2 \sum_{n=2}^{\infty} a_{n-2} x^n = 0$$

Compare "like" terms.

Left Side $= 0 + 0x + \sum_{n=2}^{\infty} 0 \, x^n$

→ $a_2 = 0 \quad a_3 = 0$

$$\sum_{n=2}^{\infty} (n+2)(n+1)a_{n+2} x^n - x^2 \sum_{n=2}^{\infty} a_{n-2} x^n = 0$$
$$(n+2)(n+1)a_{n+2} x^n = a_{n-2} x^n$$
$$(n+2)(n+1)a_{n+2} = a_{n-2}$$
$$a_{n+2} = \frac{a_{n-2}}{(n+2)(n+1)} \qquad n = 2, 3, 4, \ldots \qquad \text{Recursive Eqn.}$$

Find the first few terms:

$n = 2$ → $a_4 = \dfrac{a_0}{(4)(3)}$ Coefficient of x^4

$n = 3$ → $a_5 = \dfrac{a_1}{(5)(4)}$ Coefficient of x^5

$n = 4$ → $a_6 = \dfrac{a_2}{(6)(5)} = 0$ Because $a_2 = 0$

$n = 5$ → $a_7 = \dfrac{a_3}{(7)(6)} = 0$ Because $a_3 = 0$

$n = 6$ → $a_8 = \dfrac{a_4}{(8)(7)} = \dfrac{a_0}{(8)(7)(4)(3)}$

$n = 7$ → $a_9 = \dfrac{a_5}{(9)(8)} = \dfrac{a_1}{(9)(8)(5)(4)}$

$n = 8$ → $a_{10} = \dfrac{a_6}{(10)(9)} = 0$ Because $a_6 = 0$

> Note: $y = C_1 y_1 + C_2 y_2$
>
> Here, we are looking for two coefficients (a_0 and a_1)
>
> So, no need to check too many a_n terms!
>
> For $n > 2$ → $a_n = f(a_0, a_1)$

This is the solution:

$$y = \sum_{n=0}^{\infty} a_n x^n = a_0 + a_1 x + a_2 x^2 + a_3 x^3 + \ldots$$

$$y = a_0 + a_1 x + 0 + 0 + \frac{a_0}{(4)(3)} x^4 + \frac{a_1}{(5)(4)} x^5 + \ldots$$

Separate the a_0 and a_1 terms.

$$y = a_0 (\ ) + a_1 (\ )$$

$$y = a_0 \left(1 + \frac{x^4}{4 \cdot 3} + \frac{x^6}{8 \cdot 7 \cdot 4 \cdot 3} + \cdots \right) + a_1 \left(x + \frac{x^5}{5 \cdot 4} + \cdots \right)$$

DONE.

Extra Practice:

Compute the G.S. of $y'' - xy = 0$, Near $x = 0$ Called "Airy's DE"

Hint: Every third term will be zero.

DE near Ordinary Point Example

Find the G.S. of: $(1 + x^2)y'' - 4xy' + 6y = 0$, Near $x = 0$

Solution:

$$P(x) = 1 + x^2 \qquad Q(x) = -4x \qquad R(x) = 6$$

Check: $P(0) = 1 \neq 0 \rightarrow x = 0$ is an ordinary point.

∃ P.S. solution in the form of: $y = \sum_{n=0}^{\infty} a_n x^n$

$$y = \sum_{n=0}^{\infty} a_n x^n$$
$$y' = \sum_{n=1}^{\infty} n\, a_n x^{n-1}$$
$$y'' = \sum_{n=2}^{\infty} n(n-1)\, a_n x^{n-2}$$

Substitute: (Rearrange first)

$$(1 + x^2)y'' - 4xy' + 6y = 0$$
$$y'' + x^2 y'' - 4xy' + 6y = 0$$
$$\sum_{n=2}^{\infty} n(n-1)\, a_n x^{n-2} + x^2 \sum_{n=2}^{\infty} n(n-1)\, a_n x^{n-2} \quad \ldots$$
$$-\, 4x \sum_{n=1}^{\infty} n\, a_n x^{n-1} + 6 \sum_{n=0}^{\infty} a_n x^n = 0$$

Adjust so all series are in terms of x^n

(Also, bring in the leading x terms)

$$\sum_{n=0}^{\infty} (n+2)(n+1)\, a_{n+2} x^n + \sum_{n=2}^{\infty} n(n-1)\, a_n x^n \quad \ldots$$
$$-\, 4 \sum_{n=1}^{\infty} n\, a_n x^n + 6 \sum_{n=0}^{\infty} a_n x^n = 0$$

Adjust so all series start at highest starting point ($n = 2$)

$(2)(1)a_2 + (3)(2)a_3 x + \sum_{n=2}^{\infty} (n+2)(n+1) a_{n+2} x^n +$
$+ \sum_{n=2}^{\infty} n(n-1) a_n x^n - 4 a_1 x - 4 \sum_{n=2}^{\infty} n a_n x^n +$
$+ 6a_0 + 6 a_1 x + 6 \sum_{n=2}^{\infty} a_n x^n = 0$

$2a_2 + 6a_3 x - 4a_1 x + 6a_0 + 6a_1 x + \ldots$
$\sum_{n=2}^{\infty} [(n+2)(n+1)a_{n+2} + n(n-1)a_n - 4n a_n + 6 a_n] x^n = 0$

Simplify:

$2a_2 + 6a_3 x - 4a_1 x + 6a_0 + 6a_1 x + \cdots$
$\sum_{n=2}^{\infty} [(n+2)(n+1)a_{n+2} + n(n-1)a_n - 4n a_n + 6 a_n] x^n = 0$

$() + ()x + \cdots$
$\quad + \sum_{n=2}^{\infty} [\, () a_n + () a_{n+2} \,] x^n = 0$

$(6a_0 + 2a_2) + (2a_1 + 6a_3)x + \ldots$
$+ \sum_{n=2}^{\infty} [(n(n-1) - 4n + 6) a_n + ((n+2)(n+1))a_{n+2}] x^n = 0$

$(6a_0 + 2a_2) + (2a_1 + 6a_3)x + \ldots$
$+ \sum_{n=2}^{\infty} [(n^2 - 5n + 6) a_n + (n+2)(n+1) a_{n+2}] x^n = 0$

$$(6a_0 + 2a_2) + (2a_1 + 6a_3)x + \ldots$$
$$+ \sum_{n=2}^{\infty} [(n-2)(n-3)a_n + (n+2)(n+1)a_{n+2}]x^n = 0$$

Solve for the first few constants.

$(6a_0 + 2a_2) = 0$

$(3a_0 + a_2) = 0 \qquad \rightarrow \quad a_2 = -3a_0$

$(2a_1 + 6a_3) = 0$

$(a_1 + 3a_3) = 0 \qquad \rightarrow \quad a_3 = \dfrac{-a_1}{3}$

NOTE: Both a_2 and a_3 are in terms of a_0 or a_1

$(n-2)(n-3)a_n + (n+2)(n+1)a_{n+2} = 0$

$$a_{n+2} = -\dfrac{(n-2)(n-3)}{(n+1)(n+2)} a_n \qquad \text{Recursive Equation}$$

Now, we know:

$$a_2 = -3a_0 \qquad \text{and} \qquad a_3 = -\dfrac{a_1}{3}$$

$$a_{n+2} = -\dfrac{(n-2)(n-3)}{(n+1)(n+2)} a_n$$

Solve for more a_n terms:

$$n = 2 \rightarrow a_4 = -\frac{(0)(-1)}{(3)(4)} a_2 = 0$$

$$n = 3 \rightarrow a_5 = -\frac{(1)(0)}{(4)(5)} a_3 = 0$$

$$n = 4 \rightarrow a_6 = -\frac{(2)(1)}{(5)(6)} a_4 = -\frac{(2)(1)}{(5)(6)}(0) = 0$$

$$n = 5 \rightarrow a_7 = -\frac{(3)(2)}{(6)(7)} a_5 = -\frac{(3)(2)}{(6)(7)}(0) = 0$$

$$n = 6 \rightarrow a_8 = -\frac{(4)(3)}{(7)(8)} a_6 = -\frac{(4)(3)}{(7)(8)}(0) = 0$$

→ ALL coefficients in the series ($a_2, a_3, a_4, \ldots$) are zero.

Previously, we found: $a_2 = -3a_0$ and $a_3 = \frac{-a_1}{3}$

General form of the solution is:
$$y = a_0 x^0 + a_1 x^1 + a_2 x^2 + a_3 x^3 + 0 + 0 + 0 + 0$$

Rewrite the solution in terms of just two constants (a_0 and a_1)

$$y = a_0 + a_1 x^1 + (-3a_0) x^2 + \left(-\frac{a_1}{3}\right) x^3$$

$$y = a_0 + a_1 x - 3a_0 x^2 - \left(\frac{1}{3}\right) a_1 x^3$$

$$y = a_0(1 - 3x^2) + a_1 \left(x - \frac{x^3}{3}\right)$$

Notes: Only coefficients: a_0, a_1, a_2, a_3 are NOT zero.
All other coefficients are zero.
Both y_1 and y_2 are polynomials.

TYPES OF SINGULAR POINTS

SINGULAR POINTS

Given: $P(x)y'' + Q(x)y' + R(x)y = 0$ Near $x = x_0$

Check: $P(x_0) \neq 0$ ➔ x_0 is an ordinary point.
BUT IF: $P(x_0) = 0$ ➔ x_0 is a singular point. (BAD Point.)

Singular Points: There are two types of singular points.

First, rewrite the DE $y'' + \dfrac{Q(x)}{P(x)} y' + \dfrac{R(x)}{P(x)} y = 0$

Then, find these limits:

$$p_0 = \lim_{x \to x_0} (x - x_0) \cdot \dfrac{Q(x)}{P(x)}$$

$$q_0 = \lim_{x \to x_0} (x - x_0)^2 \cdot \dfrac{R(x)}{P(x)}$$

Singular Points:

- Regular Singular Point: Both limits exist as real. (not so Bad)
- Irregular Singular Point: One limit does not exist. (Very Bad)

SINGULAR POINTS – Summary

$P(x)y'' + Q(x)y' + R(x)y = 0$		Near $x = x_0$
$y'' + \frac{Q(x)}{P(X)} y' + \frac{R(x)}{P(X)} y = 0$		
$P(x_0) \neq 0$	---	Ordinary Point
$P(x_0) = 0$	BOTH p_0 and q_0 exist	Regular Singular Point
	Either p_0 or q_0 DNE	Irregular Singular Point

Where:

$$p_0 = \lim_{x \to x_0} (x - x_0) \cdot \frac{Q(x)}{P(X)}$$

$$q_0 = \lim_{x \to x_0} (x - x_0)^2 \cdot \frac{R(x)}{P(X)}$$

For Regular Singular Points, assume the solution is in the form:

$y = \sum_{n=0}^{\infty} a_n (x - x_0)^{n+r}$ (Frobenius Series)

Note: There is no constant term in the Frobenius Series.

SINGULAR POINTS - EXAMPLES

Three Examples: Find all singular points of given DE.

And classify them as regular singular or irregular singular points.

Example (1): $x^2 y'' + 2x y' + 7y = 0$

Solution:

$$P(x) = x^2 \qquad Q(x) = 2x \qquad R(x) = 7$$

Find all zeros of $P(x)$

$$P(x) = x^2 = 0 \quad \Rightarrow \quad x = 0$$

Check to see if $x = 0$ is regular or irregular.

Find the two limits:

$$\lim_{x \to x_0} (x - x_0) \cdot \frac{Q(x)}{P(x)} = \lim_{x \to 0} x \cdot \frac{2x}{x^2} = \lim_{x \to 0} 2 = 2$$

$$\lim_{x \to x_0} (x - x_0)^2 \cdot \frac{R(x)}{P(x)} = \lim_{x \to 0} x^2 \cdot \frac{7}{x^2} = \lim_{x \to 0} 7 = 7$$

Both limits exist. $\quad \Rightarrow \quad x = 0$ is a Regular Singular Point.

SINGULAR POINTS – EXAMPLES

Example (2): $(1 - x^2)y'' - 4x\,y' + (x^2 - 2x + 1)y = 0$

Solution:

$$P(x) = (1 - x^2) \qquad Q(x) = 4x \qquad R(x) = (x^2 - 2x + 1)$$

Find all zeros of $P(x)$

$$P(x) = (1 - x^2) = 0 \quad \rightarrow \quad x^2 = 1 \quad \rightarrow \quad x = \pm 1$$

Check $x = 1$ Is it regular or irregular? Find the two limits.

$$\lim_{x \to x_0}(x - x_0) \cdot \frac{Q(x)}{P(x)} = \lim_{x \to 1}(x - 1) \cdot \frac{4x}{1 - x^2}$$

$$= \lim_{x \to 1} \frac{4x(x-1)}{(1-x)(1+x)}$$

$$= \lim_{x \to 1} \frac{-4x}{(1+x)} = -\frac{4}{2} = -2$$

$$\lim_{x \to x_0}(x - x_0)^2 \cdot \frac{R(x)}{P(x)} = \lim_{x \to 1}(x - 1)^2 \cdot \frac{x^2 - 2x + 1}{(1 - x^2)}$$

$$= \lim_{x \to 1} \frac{(x-1)^2 (x-1)(x-1)}{(1-x)(1+x)}$$

$$= \lim_{x \to 1} \frac{-(x-1)^2}{(1+x)} = 0$$

Both limits exist. → $x = 1$ is a Regular Singular Point.

Now, Check $x = -1$ (Continued …)

Check $x = 1$ Is it regular or irregular?   Find the two limits.

$$\lim_{x \to x_0} (x - x_0) \cdot \frac{Q(x)}{P(x)} = \lim_{x \to -1} (x + 1) \cdot \frac{4x}{1 - x^2}$$

$$= \lim_{x \to -1} \frac{4x \, (x+1)}{(1-x)(1+x)}$$

$$= \lim_{x \to -1} \frac{4x}{(1-x)} = \frac{4}{2} = 2$$

$$\lim_{x \to x_0} (x - x_0)^2 \cdot \frac{R(x)}{P(x)} = \lim_{x \to -1} (x + 1)^2 \cdot \frac{x^2 - 2x + 1}{(1 - x^2)}$$

$$= \lim_{x \to -1} \frac{(x+1)^2 \, (x-1)(x-1)}{(1-x)(1+x)}$$

$$= \lim_{x \to -1} -(x + 1)(x - 1) = 0$$

Both limits exist.   ➔ $x = -1$ is a Regular Singular Point.

SINGULAR POINTS – EXAMPLES

Example (3): $x y'' + \frac{\sin x}{x} y' + \cos x \cdot y = 0$

Solution:

$$P(x) = x \qquad Q(x) = \frac{\sin x}{x} \qquad R(x) = \cos x$$

Find all zeros of $P(x)$

$$P(x) = x = 0 \quad \rightarrow \quad x = 0$$

Check $x = 0$. Is it regular or irregular? Find the two limits.

$$\lim_{x \to x_0}(x - x_0) \cdot \frac{Q(x)}{P(x)} = \lim_{x \to 0}(x - 0) \cdot \frac{\sin x}{x^2}$$

$$= \lim_{x \to 0} \frac{\sin x}{x}$$

(Use L'H Rule) $\quad = \lim_{x \to 0} \frac{\cos x}{1} = \cos x = 1$

$$\lim_{x \to x_0}(x - x_0)^2 \cdot \frac{R(x)}{P(x)} = \lim_{x \to 0}(x - 0)^2 \cdot \frac{\cos x}{x}$$

$$= \lim_{x \to 0} x \cdot \cos x = 0$$

$$= \lim_{x \to 1} \frac{-(x-1)^2}{(1+x)} = 0$$

Both limits exist. $\quad \rightarrow \quad x = 0$ is a Regular Singular Point.

Singular Point Example: Compute all singular points of the DE
$$x^3(x^2-1)y'' + \sin x \, y' + \cos x \, y = 0$$

Solution:

$$P(x) = x^3(x^2-1) \qquad Q(x) = \sin x \qquad R(x) = \cos x$$

Singular Points → Find want x-values that make $P(x) = 0$

$x^3(x^2-1) = 0$ → $x = 0, \pm 1$

For: $x = 0$

$$\lim_{x \to x_0}(x-x_0)\frac{Q(x)}{P(x)} = \lim_{x \to 0}(x)\frac{\sin x}{x^3(x^2-1)} = \frac{0}{0}$$

Indeterminate so use L.H.

$$= \lim_{x \to 0} \frac{\cos x}{4x^3 - 2x} = \frac{1}{0} = \infty = DNE$$

Therefore: $x = 0$ is an Irregular Singular Point.

For: $x = 1$

$$\lim_{x \to x_0}(x-x_0)\frac{Q(x)}{P(x)} = \lim_{x \to 1}(x-1)\frac{\sin x}{x^3(x-1)(x+1)} = \frac{\sin 1}{1 \cdot (2)} \quad \text{OK}$$

$$\lim_{x \to x_0}(x-x_0)^2 \frac{R(x)}{P(x)} = \lim_{x \to 1}(x-1)^2 \frac{\cos x}{x^3(x-1)(x+1)}$$

$$= \frac{0 \cdot \cos 1}{2} = \frac{0}{2} = 0 \quad \text{OK}$$

Therefore: $x = 1$ is a Regular Singular Point.

**For: $x = -1$**

$$\lim_{x \to x_0} (x - x_0) \frac{Q(x)}{P(x)} = \lim_{x \to -1} (x+1) \frac{\sin x}{x^3(x-1)(x+1)}$$

$$= \frac{\sin(-1)}{1 \cdot (-2)} \quad \text{OK}$$

$$\lim_{x \to x_0} (x - x_0)^2 \frac{R(x)}{P(x)} = \lim_{x \to -1} (x+1)^2 \frac{\cos x}{x^3(x-1)(x+1)}$$

$$= \frac{0 \cdot \cos(-1)}{(-2)} = \frac{0}{-2} = 0 \quad \text{OK}$$

Therefore: $x = -1$ is a Regular Singular Point.

SUMMARY – Ordinary and Singular Points

$$P(x) \cdot y'' + Q(x) \cdot y' + R(x) \cdot y = 0 \qquad \text{Near } x = x_0$$

$P(x_0)$	Limits: p_0 & q_0	Type	Form of Solution
$\neq 0$	No need to check.	Ordinary	$y = \sum_{n=0}^{\infty} a_n x^n$
$= 0$	Both limits exist.	Regular Singular Point	$y = x^r \cdot \sum_{n=0}^{\infty} a_n x^n = \sum_{n=0}^{\infty} a_n x^{n+r}$
	One or Both limits DNE.	Irregular Singular Point	No solution around this point..

Where: $\quad p_0 = \lim\limits_{x \to x_0} (x - x_0) \dfrac{Q(x)}{P(x)}$

$\qquad\qquad q_0 = \lim\limits_{x \to x_0} (x - x_0)^2 \dfrac{R(x)}{P(x)}$

SERIES SOL'N NEAR A REGULAR SINGULAR POINT

SERIES SOLUTION – Near Regular Singular Point (General Idea)

How to solve a Second Order DE near a Regular Singular Point.

$$P(x) \cdot y'' + Q(x) \cdot y' + R(x) \cdot y = 0 \qquad \text{Near } x = x_0$$

STEP 1:
- Check if $P(x_0) = 0$ (Singular) or $P(x_0) \neq 0$ (Regular)

STEP 2:
- Verify it is a <u>Regular Singular Point</u>.
- Compute the two limits. Verify they both exist.

$$p_0 = \lim_{x \to x_0} (x - x_0) \frac{Q(x)}{P(x)}$$

$$q_0 = \lim_{x \to x_0} (x - x_0)^2 \frac{R(x)}{P(x)}$$

STEP 3:
- Assume the solution is in the form of Frobeneous Series:

$$y = x^r \cdot \sum_{n=0}^{\infty} a_n x^n = \sum_{n=0}^{\infty} a_n x^{n+r}$$

- Note: There are two unknowns: a_n and r.
  Both must be computed.
- Compute y' and y'' (based on y) and substitute into DE.

(Continued ...)

STEP 4:

- Rewrite all P.S. in form of generic term with power of: x^{r+n}

STEP 5:

- Using lowest x-power term, compute the <u>Indicial Equation</u> (eqn. in "r")
- Using the P.S., obtain the recursive relation.
- The quadratic equation in "r" has two roots (r_1 and r_2)
- $r_1 - r_2$ ➔ Three cases   (where: $r_1 > r_2$)
    1. Difference is non-integer (e.g. fraction)
    2. Difference is zero (e.g. repeating roots)
    3. Difference is an integer.

Solving a Second-Order DE near a Regular Singular Point.
More Details about the two roots and 3 Cases:

Case	$r_1 - r_2$	Notes
1	= Non-Integer (e.g. Fraction)	Each value of "r" gives linearly independent (L.I.) solutions. Compute coefficients and write y_1 and y_2
2	= 0 Repeated Roots	Using the largest root, compute coefficient and write first solution. $y_1 = x^r \sum_{n-0}^{\infty} a_n x^n$ Computation of the second solution is tedious. The reduction of order technique can be used to compute the second solution. (Note: In this course, we will NOT have to compute it!)
3	= N = Integer	First solution is based on the larger root of the Indicial Equation. Just plug it into the Recursive equation. Use reduction of order to compute the second solution. (Note: In this course, we will NOT have to compute it!) $y = x^{r_1} \sum_{n=0}^{\infty} a_n(r_1) \cdot x^n$

Solving a Second-Order DE near a Regular Singular Point.

Example: Solve: $4xy'' + 2y' + y = 0$ Near $x_0 = 0$

Solution (Step 1)

$$P(x) = 4x \qquad Q(x) = 2 \qquad R(x) = 1$$

$P(0) = 0$ ➔ Singular point (Bad point)

Solution (Step 2)

Check for Regular or Irregular Singularity.

$$p_0 = \lim_{x \to 0} x \frac{Q(x)}{P(x)} = \lim_{x \to 0} x \frac{2}{4x} = \frac{1}{2} \quad \text{OK}$$

$$q_0 = \lim_{x \to 0} x^2 \frac{R(x)}{P(x)} = \lim_{x \to 0} x^2 \frac{1}{4x} = \frac{x}{4} = \frac{0}{4} = 0 \quad \text{OK}$$

Both finite limits ➔ $x_0 = 0$ is a Regular Singular Point.

Solution (Step 3)

Assume the solution is in the form of Frobeneous Series.

$$y = x^r \cdot \sum_{n=0}^{\infty} a_n x^n = \sum_{n=0}^{\infty} a_n x^{n+r}$$

$$y' = \sum_{n=0}^{\infty} (n+r) a_n x^{n+r-1}$$

$$y'' = \sum_{n=0}^{\infty} (n+r)(n+r-1) a_n x^{n+r-2}$$

Substitute for each term:

$$4xy'' = \sum_{n=0}^{\infty} 4(n+r)(n+r-1) a_n x^{n+r-1}$$

$$2y' = \sum_{n=0}^{\infty} 2(n+r) a_n x^{n+r-1}$$

$$y = \sum_{n=0}^{\infty} a_n x^{n+r}$$

Rewrite in terms of lowest x power (e.g. x^{n+r})

$4xy'' = \sum_{n=-1}^{\infty} 4(n+1+r)(n+r) a_{n+1} x^{n+r}$ $(n \to n+1)$

$2y' = \sum_{n=-1}^{\infty} 2(n+1+r) a_{n+1} x^{n+r}$ $(n \to n+1)$

$y = \sum_{n=0}^{\infty} a_n x^{n+r}$

Rewrite to start all series at highest n (e.g. $n = 0$)

$4xy'' = 4(r)(r-1)a_0 x^{r-1} + \sum_{n=0}^{\infty} 4(n+1+r)(n+r) a_{n+1} x^{n+r}$

$2y' = 2(r)a_0 x^{r-1} + \sum_{n=0}^{\infty} 2(n+1+r) a_{n+1} x^{n+r}$

$y = \sum_{n=0}^{\infty} a_n x^{n+r}$

Rewrite D.E. as one series.

$4(r)(r-1)a_0 x^{r-1} + 2(r)a_0 x^{r-1} + \sum_{n=0}^{\infty} [\ ] x^{n+r} = 0$

$4(r)(r-1)a_0 x^{r-1} + 2(r)a_0 x^{r-1} + \sum_{n=0}^{\infty} \ldots$

$[\, 4(n+r+1)(n+r)a_{n+1} + 2(n+r+1)a_{n+1} + a_n \,] x^{n+r} = 0$

Note: $a_0 x^{r-1} \neq 0$

$\quad 4(r)(r-1)a_0 x^{r-1} + 2(r)a_0 x^{r-1} = 0$

$\quad 4r^2 - 4r + 2r = 0$

$\quad 4r^2 - 2r = 0$

$\quad 2r^2 - r = 0$

$\quad r(2r - 1) = 0 \;\Rightarrow\; r = 0, \tfrac{1}{2}$

With: $r = 0, \frac{1}{2}$

$r_1 - r_2 = \frac{1}{2} - 0 = \frac{1}{2}$

Fraction ➔ Each r value will give one linearally independent solution.

Find Recursive Equation with two unknowns (n and r)

$$4(n+r+1)(n+r)a_{n+1} + 2(n+r+1)a_{n+1} + a_n = 0$$
$$[\,4(n+r+1)(n+r) + 2(n+r+1)\,]a_{n+1} + a_n = 0$$
$$[\,(2n+2r+2)(2n+2r) + (2n+2r+2)\,]a_{n+1} + a_n = 0$$
$$[\,(2n+2r+2)(2n+2r+1)\,]a_{n+1} + a_n = 0$$

$$a_{n+1} = \frac{-a_n}{[\,(2n+2r+2)\cdot(2n+2r+1)\,]} \qquad \text{Recursive Relation}$$

Recall, $r = 0, \frac{1}{2}$ Evaluate for each r-value.

We will get one solution for each r-value.

For $r = \frac{1}{2}$

$$a_{n+1} = \frac{-a_n}{[(2n+2r+2)\cdot(2n+2r+1)]} \quad \text{Recursive Relation}$$

$$a_{n+1} = \frac{-a_n}{[(2n+1+2)\cdot(2n+1+1)]}$$

$$a_{n+1} = \frac{-a_n}{[(2n+3)\cdot(2n+2)]}$$

Compute a few coefficients:

$n = 0 \rightarrow a_1 = \frac{-a_0}{(3)(2)} = \frac{-a_0}{3!}$

$n = 1 \rightarrow a_2 = \frac{-a_1}{(5)(4)} = \frac{-(-a_0)}{5!} = \frac{a_0}{5!}$

$n = 2 \rightarrow a_3 = \frac{-a_2}{(7)(6)} = \frac{-(a_0)}{7!} = \frac{-a_0}{7!}$

$n = 3 \rightarrow a_4 = \qquad\qquad\qquad\qquad = \frac{a_0}{9!}$

First Solution:

$y_1 = x^{r_1} \sum_{n=0}^{\infty} a_n x^n$

$y_1 = x^{\frac{1}{2}} [a_0 + a_1 x + a_2 x^2 + a_3 x^3 + \dots]$

$y_1 = x^{\frac{1}{2}} [a_0 - \frac{a_0}{3!} x + \frac{a_0}{5!} x^2 - \frac{a_0}{7!} x^3 + \frac{a_0}{9!} x^4 + \dots]$

$y_1 = x^{\frac{1}{2}} a_0 [1 - \frac{x}{3!} + \frac{x^2}{5!} - \frac{x^3}{7!} + \frac{x^4}{9!} + \dots]$

$y_1 = x^{\frac{1}{2}} a_0 \sum_{n=0}^{\infty} \frac{(-1)^n x^n}{(2n+1)!} \quad$ First Solution of the DE

To find the Second Solution, use $r_2 = 0$

For $r = 0$

$$a_{n+1} = \frac{-a_n}{[(2n+2r+2)(2n+2r+1)]} = \frac{-a_n}{[(2n+2)(2n+1)]} \quad \text{Recursive Eqn.}$$

Compute a few coefficients:

$$n = 0 \rightarrow a_1 = \frac{-a_0}{(2)(1)} = \frac{-a_0}{2!}$$

$$n = 1 \rightarrow a_2 = \frac{-a_1}{(4)(3)} = \frac{-(-a_0)}{4!} = \frac{a_0}{4!}$$

$$n = 2 \rightarrow a_3 = \frac{-a_2}{(6)(5)} = \frac{-(a_0)}{6!} = \frac{-a_0}{6!}$$

$$n = 3 \rightarrow a_4 = \qquad\qquad\qquad = \frac{a_0}{8!}$$

Second Solution:

$$y_2 = x^{r_2} \sum_{n=0}^{\infty} a_n x^n$$

$$y_2 = x^{r_2} [a_0 + a_1 x + a_2 x^2 + a_3 x^3 + \ldots]$$

$$y_2 = x^0 \left[a_0 - \frac{a_0}{2!} x + \frac{a_0}{4!} x^2 - \frac{a_0}{6!} x^3 + \frac{a_0}{8!} x^4 + \ldots \right]$$

$$y_2 = a_0 \left[1 - \frac{x}{2!} + \frac{x^2}{4!} - \frac{x^3}{6!} + \frac{x^4}{8!} + \ldots \right]$$

$$y_2 = a_0 \sum_{n=0}^{\infty} \frac{(-1)^n x^n}{(2n)!} \quad \text{Second Solution of the D.E.}$$

General Solution for the given DE is: $y = C_1 y_1 + C_2 y_2$

$$y = x^{\frac{1}{2}} a_0 \sum_{n=0}^{\infty} \frac{(-1)^n x^n}{(2n+1)!} + x^0 a_0 \sum_{n=0}^{\infty} \frac{(-1)^n x^n}{(2n)!}$$

$$y = C_1 \sum_{n=0}^{\infty} \frac{(-1)^n x^{n+\frac{1}{2}}}{(2n+1)!} + C_2 \sum_{n=0}^{\infty} \frac{(-1)^n x^n}{(2n)!}$$

SOLVING SECOND-ORDER DE NEAR REGULAR SINULAR POINTS

REVIEW: Solution Guidelines: When x_0 is a Regular Singular Point

To solve: $P(x)y'' + Q(x)y' + R(x)y = 0$ Near x_0 When x_0 is a Regular Singular Point
Verify x_0 is a Singular point. $P(x_0) = 0$ → Singular Point
Verify x_0 is a Regular Singular point. Both limits (p_0 and q_0) exist → Regular
Assume solution in form of Frobenius. $y = x^r \sum_{n=0}^{\infty} a_n x^n = \sum_{n=0}^{\infty} a_n x^{n+r}$
Also calculate/assume y' and y'' (Based on y)
Get Indicial Equation. (Equation in r)
Compute r_1, r_2 Where: $r_1 > r_2$
Values of r_1 and r_2 → One of three cases. • $r_1 - r_2 \neq Integer$ → Two L.I. solutions. • $r_1 = r_2$ (Repeated Root.) → Compute only 1st sol'n. (tricky) • $r_1 - r_2 = Integer$ → Use r_1 to find first solution. (tricky)
Get Recursive Relation. It's a function of n and r.
Obtain Particular Recursive Relation (RR) for r_1 (and r_2)
Compute the coefficients, using the Particular RR (for each r).
Write Solution for y_1 (and y_2)
If you have both y_1 and y_2 write General Sol'n: $C_1 y_1 + C_2 y_2 = 0$

DE Near Regular Singular Points

Example: Solve $3x^2 y'' + 2x y' + x^2 y = 0$ Near: $x = 0$

Solution:

$P(x) = 3x^2 \qquad Q(x) = 2x \qquad R(x) = x^2$

$P(0) = 0 \rightarrow$ Singular Point   (Bad)

To determine if Regular or Irregular Singularity, compute two limits.

$p_0 = \lim\limits_{x \to 0} x \frac{Q(x)}{P(x)} = \lim\limits_{x \to 0} x \frac{2x}{3x^2} = \frac{2}{3}$ OK

$q_0 = \lim\limits_{x \to 0} x^2 \frac{R(x)}{P(x)} = \lim\limits_{x \to 0} x^2 \frac{x^2}{3x^2} = \lim\limits_{x \to 0} \frac{x^2}{3} = \frac{0}{3} = 0$ OK

Both limits exist $\rightarrow$ Regular Singular Point at $x = 0$

Regular Singular Point

$\rightarrow \exists$ a Frobenius Series solution of the D.E. in the form:

$y = x^r \sum_{n=0}^{\infty} a_n x^n = \sum_{n=0}^{\infty} a_n x^{n+r}$

$y' = \sum_{n=0}^{\infty} (n+r) a_n x^{n+r-1}$

$y'' = \sum_{n=0}^{\infty} (n+r)(n+r-1) a_n x^{n+r-2}$

Rewrite each term in terms as Series.

$3x^2 y'' = \sum_{n=0}^{\infty} 3(n+r)(n+r-1) a_n x^{n+r}$

$2x y' = \sum_{n=0}^{\infty} 2(n+r) a_n x^{n+r}$

$x^2 y = \sum_{n=0}^{\infty} a_n x^{n+r+2} = \sum_{n=2}^{\infty} a_{n-2} x^{n+r}$

After x-exponents match, adjust index. Use highest index. ($n = 2$)

[*Terms with* $n = 0$] + [*Terms with* $n = 1$]
$$+ \sum_{n=2}^{\infty} [\text{ Three parts }] x^{n+r} = 0$$

$[\,3(r)(r-1)a_0 x^r + 2(r)a_0 x^r\,]$
$+ [\,3(1+r)(r)a_1 x^{r+1} + 2(1+r)a_1 x^{r+1}\,]$
$+ \sum_{n=2}^{\infty}[\,3(n+r)(n+r-1)a_n + 2(n+r)a_n + a_{n-2}\,]x^{n+r} = 0$

Note: RHS = $0 + 0 + \sum_{n=0}^{\infty}[\,0\,]x^{n+r}$

Compare First Individual Term: Check a_0 Note: $a_0 \neq 0$

$[\,3(r)(r-1)a_0 x^r + 2(r)a_0 x^r\,] = 0$

$[3r(r-1) + 2r\,]a_0 x^r = 0$

$3r^2 - r = 0$

$r(3r - 1) = 0 \;\rightarrow\; r_1 = \dfrac{1}{3}$ and $r_2 = 0$

Compare Roots:

$r_1 - r_2 = \dfrac{1}{3} - 0 = \dfrac{1}{3} = $ *Fraction* $\neq$ *Integer*

→ Each root gives one linearly independent solution.

Compare Second Individual Term:

Check a_1 Note: $a_0 \neq 0$ But a_1 may be zero.

$$[\,3(1+r)(r)a_1 x^{r+1} + 2(1+r)a_1 x^{r+1}\,] = 0$$

$$[\,3r + 3r^2 + 2 + 2r\,]a_1\, x^{r+1} = 0$$

$$[3r^2 + 5r + 2\,]\, a_1\, x^{r+1} = 0$$

But, $r = \frac{1}{3}$ or 0 $\rightarrow$ $a_1 = 0$

Compare Summation Term to find the Recursive Relationship

$$\sum_{n=2}^{\infty} [\,3(n+r)(n+r-1)\,a_n + 2(n+r)a_n + a_{n-2}\,]x^{n+r} = 0$$

$$[\,3(n+r)(n+r-1)\,a_n + 2(n+r)a_n + a_{n-2}\,] = 0$$

$$(n+r)[3n + 3r - 3 + 2]a_n + a_{n-2} = 0$$

$$a_n = \frac{-a_{n-2}}{(n+r)(3n+3r-1)} \qquad \text{General Recursive Relation}$$

Note: a is a function of a previous a.

Previously, we found:

$$r_1 = \frac{1}{3} \quad \text{and} \quad r_2 = 0$$

$$a_n = \frac{-a_{n-2}}{(n+r)(3n+3r-1)} \qquad \text{General Recursive Relation}$$

For $r_1 = \frac{1}{3}$

$$a_n = \frac{-a_{n-2}}{\left(n+\frac{1}{3}\right)(3n)} = \frac{-a_{n-2}}{(3n+1)(n)}$$

$n = 2 \rightarrow a_2 = \dfrac{-a_0}{(7)(2)}$

$n = 3 \rightarrow a_3 = \dfrac{-a_1}{(10)(3)} = 0 \qquad$ Recall: $\quad a_1 = 0$

$n = 4 \rightarrow a_4 = \dfrac{-a_2}{(13)(4)} = \dfrac{a_0}{(2)(4)(7)(13)}$

$n = 5 \rightarrow a_5 = \dfrac{-a_3}{(16)(5)} = 0$

First Solution: $\quad y_1 = x^r \sum_{n=0}^{\infty} a_n x^n$

$y_1 = x^r [a_0 + a_1 x + a_2 x^2 + a_3 x^3 + \ldots]$

$y_1 = x^{\frac{1}{3}} \left[a_0 + 0 - \dfrac{a_0}{2 \cdot 7} x^2 + 0 + \dfrac{a_0}{2 \cdot 4 \cdot 7 \cdot 14} x^4 + \ldots \right]$

$y_1 = a_0 x^{\frac{1}{3}} \left[1 - \dfrac{x^2}{2 \cdot 7} + \dfrac{x^4}{2 \cdot 4 \cdot 7 \cdot 14} + \ldots \right]$

Now, find the Second Solution (For $r = 0$)

Compute the Second Solution:

Previously, we found:

$$r_1 = \frac{1}{3} \quad \text{and} \quad r_2 = 0$$

$$a_n = \frac{-a_{n-2}}{(n+r)(3n+3r-1)} \qquad \text{General Recursive Relation}$$

For $r_2 = 0$

$$a_n = \frac{-a_{n-2}}{(n)(3n-1)}$$

$n = 2 \rightarrow a_2 = \dfrac{-a_0}{(2)(5)}$

$n = 3 \rightarrow a_3 = \dfrac{-a_1}{(3)(8)} = 0 \qquad \text{Recall:} \quad a_1 = 0$

$n = 4 \rightarrow a_4 = \dfrac{-a_2}{(4)(11)} = \dfrac{a_0}{(2)(4)(5)(11)}$

$n = 5 \rightarrow a_5 = \dfrac{-a_3}{(5)(14)} = 0$

Second Solution: $y_2 = x^r \sum_{n=0}^{\infty} a_n x^n$

$$y_2 = x^r [a_0 + a_1 x + a_2 x^2 + a_3 x^3 + \ldots]$$

$$y_2 = x^0 \left[a_0 + 0 - \frac{a_0}{2\cdot 5} x^2 + 0 + \frac{a_0}{2\cdot 4\cdot 5\cdot 11} x^4 + \ldots \right]$$

$$y_2 = a_0 \left[1 - \frac{x^2}{2\cdot 5} + \frac{x^4}{2\cdot 4\cdot 5\cdot 11} + \ldots \right]$$

General Solution: $y = C_1 y_1 + C_2 y_2$

$$y = a_0 x^{\frac{1}{3}} \left[1 - \frac{x^2}{2\cdot 7} + \frac{x^4}{2\cdot 4\cdot 7\cdot 14} + \cdots \right] + a_0 \left[1 - \frac{x^2}{2\cdot 5} + \frac{x^4}{2\cdot 4\cdot 5\cdot 11} + \cdots \right]$$

Example: Use Frobenius Series to solve:

$$x^2 y'' + x\left(\frac{1}{2} + \frac{1}{2}x\right) y' + \frac{1}{2}xy = 0 \quad , \quad \text{Near } x = 0$$

$$2x^2 y'' + xy' + x^2 y' + \frac{1}{2}xy = 0$$

Solution:

$x = 0$ is a regular singular point.

Verify $\lim\limits_{x \to 0} x \dfrac{Q(x)}{P(x)}$ and $\lim\limits_{x \to 0} x^2 \dfrac{R(x)}{P(x)}$ exist.

$y = \sum_{n=0}^{\infty} a_n x^{n+r}$

$y' = \sum_{n=0}^{\infty} (n+r) a_n x^{n+r-1}$

$y'' = \sum_{n=0}^{\infty} (n+r)(n+r-1) a_n x^{n+r-2}$

$2x^2 y'' = \sum_{n=0}^{\infty} 2(n+r)(n+r-1) a_n x^{n+r}$

$xy' = \sum_{n=0}^{\infty} (n+r) a_n x^{n+r}$

$x^2 y' = \sum_{n=0}^{\infty} (n+r) a_n x^{n+r+1} = \sum_{n=1}^{\infty} (n+r) a_n x^{n+r}$

$xy = \sum_{n=0}^{\infty} a_n x^{n+r+1} = \sum_{n=1}^{\infty} a_n x^{n+r}$

$r a_0 x^r + 2r(r-1) a_0 x^r + \sum_{n=1}^{\infty} [\ ] x^{n+r} = 0 + \sum_{n=1}^{\infty} [\,0\,] x^{n+r}$

$r a_0 x^r + 2r(r-1) a_0 x^r$
$+ \sum_{n=1}^{\infty} [\, 2(n+r)(n+r-1) a_n + (n+r-1) a_{n-1} + a_{n-1} \,] x^{n+r}$
$= 0 + \sum_{n=1}^{\infty} [\,0\,] x^{n+r}$

$r + 2r^2 - 2r = 0$

$2r^2 - r = 0$ ➔ $r_2 = 0$, $r_1 = \frac{1}{2}$

Notice: $r_1 - r_2 \neq$ integer

So each "r" value generates one linearly independent (L.I.) solution.

$2(n+r)(n+r-1)a_n + (n+r-1)a_{n-1} + a_{n-1} = 0$

$a_n = \frac{-a_{n-1}}{2n+2r-1}$

For: $r = 0$ ➔ $a_n = \frac{-a_{n-1}}{2n-1}$

$n = 1$ ➔ $a_1 = \frac{-a_0}{1}$

$n = 2$ ➔ $a_2 = \frac{-a_1}{3} = \frac{a_0}{3} = \frac{a_0}{1 \cdot 3}$

$n = 3$ ➔ $a_3 = \frac{-a_2}{5} = \frac{-a_0}{5 \cdot 3 \cdot 1}$

...

$y_1(x) = x^r(a_0 + a_1 x + a_2 x^2 + \ldots)$

$y_1(x) = x^0 \left(a_0 - \frac{a_0}{1}x + \frac{a_0}{1 \cdot 3}x^2 - \frac{a_0}{1 \cdot 3 \cdot 5}x^3 + \cdots \right)$

$y_1(x) = a_0 \left(1 - x + \frac{x^2}{1 \cdot 3} - \frac{x^3}{1 \cdot 3 \cdot 5} + \cdots \right)$

For: $r = 1/2$ ➔ $a_n = \frac{-a_{n-1}}{2n}$

$n = 1$ ➔ $a_1 = \frac{-a_0}{2}$

(Continued ...)

For: $r = 1/2$ $\rightarrow$ $a_n = \dfrac{-a_{n-1}}{2n}$

$n = 1 \rightarrow a_1 = \dfrac{-a_0}{2}$

$n = 2 \rightarrow a_2 = \dfrac{-a_1}{4} = \dfrac{a_0}{4 \cdot 2}$

$n = 3 \rightarrow a_3 = \dfrac{-a_2}{6} = \dfrac{-a_0}{6 \cdot 4 \cdot 2}$

...

$y_2(x) = x^r(a_0 + a_1 x + a_2 x^2 + \ldots)$

$y_2(x) = x^{\frac{1}{2}}\left(a_0 - \dfrac{a_0}{2}x + \dfrac{a_0}{2 \cdot 4}x^2 - \dfrac{a_0}{2 \cdot 4 \cdot 6}x^3 + \cdots\right)$

$y_2(x) = a_0\, x^{\frac{1}{2}}\left(1 - x + \dfrac{x^2}{2 \cdot 4} - \dfrac{x^3}{2 \cdot 4 \cdot 6} + \cdots\right)$

Example: Solve: $xy'' + y = 0$ Near $x = 0$

Also compute radius of convergence for the first solution (y_1).

Write solution in sigma notation.

Solution:

$$P(x) = x \qquad Q(x) = 0 \qquad R(x) = 1$$

$$P(0) = 0 \rightarrow \text{Singular Point (Bad)}$$

Regular or Irregular? → Check Two Limits

$$p_0 = \lim_{x \to 0} x \frac{Q(x)}{P(x)} = \lim_{x \to 0} x \frac{0}{x} = 0 \qquad \text{OK}$$

$$q_0 = \lim_{x \to 0} x^2 \frac{R(x)}{P(x)} = \lim_{x \to 0} x^2 \frac{1}{x} = \lim_{x \to 0} x = 0 \qquad \text{OK}$$

Both limits exist → $x = 0$ is a Regular Singular Point.

Regular Singular Point

→ Solution is in the form of Frobenius Series

Assume: ∃ a Frobenious Series solution in the form:

$$y = x^r \sum_{n=0}^{\infty} a_n x^n = \sum_{n=0}^{\infty} a_n x^{n+r}$$

$$y' = \sum_{n=0}^{\infty} (n+r) a_n x^{n+r-1}$$

$$y'' = \sum_{n=0}^{\infty} (n+r)(n+r-1) a_n x^{n+r-2}$$

Find each term of the original DE: $xy'' + y = 0$

$$xy'' = \sum_{n=0}^{\infty}(n+r)(n+r-1)a_n x^{n+r-1}$$

$$xy'' = \sum_{n=-1}^{\infty}(n+r+1)(n+r)a_{n+1} x^{n+r}$$

$$y = \sum_{n=0}^{\infty} a_n x^{n+r}$$

$xy'' + y = 0$

$$\sum_{n=-1}^{\infty}(n+r+1)(n+r)a_{n+1} x^{n+r} + \sum_{n=0}^{\infty} a_n x^{n+r} = 0$$

Separate first term and add series.

$$(r)(r-1)a_0 x^{r-1} +$$
$$+ \sum_{n=0}^{\infty}[(n+r+1)(n+r)a_{n+1} + a_n] x^{n+r} = 0$$

Set individual term = 0 to solve for "r"   Note: $a_0 \neq 0$

$(r)(r-1)a_0 x^{r-1} = 0$

$(r)(r-1) = 0$ → $r_1 = 1$ and $r_2 = 0$

$r_1 - r_2 = 1 - 0 = 1 = Integer$

Just compute the first solution based on the larger root.

The second solution is too tricky. (Do not try to solve.)

Use Summation term to find Recursive Equation.

$$\sum_{n=0}^{\infty} [(n+r+1)(n+r)a_{n+1} + a_n] x^{n+r} = 0$$

$$(n+r+1)(n+r)a_{n+1} + a_n = 0$$

$$a_{n+1} = \frac{-a_n}{(n+r+1)\cdot(n+r)}$$

For $r_1 = 1$

$$a_{n+1} = \frac{-a_n}{(n+2)(n+1)}$$

Compute first few coefficients.

$n = 0 \;\rightarrow\; a_1 = \frac{-a_0}{2\cdot 1} = \frac{-a_0}{2!}$

$n = 1 \;\rightarrow\; a_2 = \frac{-a_1}{3\cdot 2} = \frac{a_0}{3!\,2!}$

$n = 2 \;\rightarrow\; a_3 = \frac{-a_2}{4\cdot 3} = \frac{-a_0}{4\cdot 3\cdot 3!\,2!} = \frac{-a_0}{4!\,3!}$

$n = 3 \;\rightarrow\; a_4 = \frac{-a_3}{5\cdot 4} = \frac{a_0}{5\cdot 4\cdot 4!\,3!} = \frac{a_0}{5!\,4!}$

Compute first solution:

$$y_1 = x^r [\,a_0 + a_1 x + a_2 x^2 + a_3 x^3 + \ldots\,]$$

$$y_1 = x^1 \left[a_0 - \frac{a_0}{2!\,1!} x + \frac{a_0}{3!\,2!} x^2 - \frac{a_0}{4!\,3!} x^3 + \ldots \right]$$

$$y_1 = a_0\, x \left[1 - \frac{x}{2!\,1!} + \frac{x^2}{3!\,2!} - \frac{x^3}{4!\,3!} + \ldots \right]$$

$$y_1 = a_0\, x \sum_{n=0}^{\infty} \frac{(-1)^n x^n}{(n+1)!\,(n)!} = a_0 \sum_{n=0}^{\infty} \frac{(-1)^n x^{n+1}}{(n+1)!\,(n)!}$$

First Solution:

$$y_1 = a_0 \sum_{n=0}^{\infty} \frac{(-1)^n x^{n+1}}{(n+1)!\,(n)!}$$

Find Radius of Convergence for 1st solution:

Use Ratio Test. Replace n by $n+1$

$$\lim_{n \to \infty} \left| \frac{a_{n+1}}{a_n} \right|$$

$$\lim_{n \to \infty} \left| \frac{(-1)^{n+1} x^{n+2}}{(n+2)!\,(n+1)!} \cdot \frac{(n+1)!\,n!}{(-1)^n x^{n+1}} \right|$$

$$\lim_{n \to \infty} \left| \frac{x}{(n+2)(n+1)} \right| = |x| \lim_{n \to \infty} \left| \frac{1}{(n+2)(n+1)} \right| = |x| \cdot \frac{1}{\infty}$$

$$= |x| \cdot 0 = 0 < 1$$

The limit $= 0 < 1$ for any value of x

→ Radius of Convergence $= \infty$

Example: Solve: $xy'' + y' - y = 0$ Near $x = 0$

And compute the radius of convergence for the first solution.

Solution:

$$P(x) = x \qquad Q(x) = 1 \qquad R(x) = -1$$

$P(0) = 0$ ➔ $x = 0$ is a Singular Point.

Check the two limits to determine if it is a regular or irregular.

$$p_0 = \lim_{x \to 0} x \frac{Q(x)}{P(x)} = \lim_{x \to 0} x \frac{1}{x} = 1 \qquad \text{OK}$$

$$q_0 = \lim_{x \to 0} x^2 \frac{R(x)}{P(x)} = \lim_{x \to 0} x^2 \frac{-1}{x} = \lim_{x \to 0} -x = 0 \qquad \text{OK}$$

Both limits exist ➔ $x = 0$ is a Regular Singular Point.

Regular Singular Point

➔ Solution is in the form of Frobenius Series

Assume: ∃ a Frobenius Series solution in the form:

$$y = x^r \sum_{n=0}^{\infty} a_n x^n = \sum_{n=0}^{\infty} a_n x^{n+r}$$

$$y' = \sum_{n=0}^{\infty} (n+r) a_n x^{n+r-1}$$

$$y'' = \sum_{n=0}^{\infty} (n+r)(n+r-1) a_n x^{n+r-2}$$

Find each term as a series. Adjust to highest power of x

$$xy'' = \sum_{n=0}^{\infty}(n+r)(n+r-1)\,a_n\,x^{n+r-1}$$
$$= \sum_{n=-1}^{\infty}(n+r+1)(n+r)\,a_{n+1}\,x^{n+r}$$

$$y' = \sum_{n=0}^{\infty}(n+r)a_n\,x^{n+r-1}$$
$$= \sum_{n=-1}^{\infty}(n+r+1)a_{n+1}\,x^{n+r}$$

$$-y = -\sum_{n=0}^{\infty}a_n\,x^{n+r}$$

Adjust all series to start at same index (highest). Also add them.

$$(r)(r-1)\,a_0\,x^{r-1} + (r)a_0 x^{r-1} + \ldots$$
$$+ \sum_{n=0}^{\infty}[(n+r+1)(n+r)\,a_{n+1} + (n+r+1)a_{n+1} + a_n]x^{n+r} = 0$$

$$[r^2]\,a_0 x^{r-1} + \sum_{n=0}^{\infty}[(n+r+1)^2\,a_{n+1} + a_n]\,x^{n+r} = 0$$

Use first term to find values of "r"

$$[r^2]\,a_0 x^{r-1} = 0$$
$$r^2 = 0 \;\;\rightarrow\;\; r_1 = r_2 = 0 \;\;\rightarrow\;\; \text{Repeating roots.}$$

We can compute one solution. The 2nd solution is too tricky.

Use the series to find the Recursive equation.

$$(n+r+1)^2\,a_{n+1} + a_n = 0$$
$$a_{n+1} = \frac{-a_n}{(n+r+1)^2}$$

For $r = 0$ Find the Recursive equation.

$$a_{n+1} = \frac{-a_n}{(n+r+1)^2}$$

$$a_{n+1} = \frac{-a_n}{(n+1)^2}$$

Compute the first few coefficients:

$$n = 0 \rightarrow a_1 = \frac{-a_0}{(1)^2}$$

$$n = 1 \rightarrow a_2 = \frac{-a_1}{(2)^2} = \frac{a_0}{(2)^2 (1)^2}$$

$$n = 2 \rightarrow a_3 = \frac{-a_2}{(3)^2} = \frac{-a_0}{(3)^2 (2)^2 (1)^2}$$

$$n = 3 \rightarrow a_4 = \frac{a_0}{(4)!\,(3)^2 (2)^2 (1)^2}$$

Compute first solution:

$$y_1 = x^r [\, a_0 + a_1 x + a_2 x^2 + a_3 x^3 + \ldots \,]$$

$$y_1 = x^0 \left[a_0 - \frac{a_0}{(1)^2} x + \frac{a_0}{(2)^2 (1)^2} x^2 - \frac{a_0}{(3)^2 (2)^2 (1)^2} x^3 + \ldots \right]$$

$$y_1 = a_0 \left[1 - \frac{x}{(1)^2} + \frac{x^2}{(2)^2 (1)^2} - \frac{x^3}{(3)^2 (2)^2 (1)^2} + \ldots \right]$$

$$y_1 = a_0 \sum_{n=0}^{\infty} \frac{(-1)^n x^n}{(n!)^2}$$

The First Solution is: $y_1 = a_0 \sum_{n=0}^{\infty} \frac{(-1)^n x^n}{(n!)^2}$

Find Radius of Convergence for First Solution:

Use Ratio Test.

$$\lim_{n \to \infty} \left| \frac{a_{n+1}}{a_n} \right|$$

$$\lim_{n \to \infty} \left| \frac{(-1)^{n+1} x^{n+1}}{((n+1)!)^2} \cdot \frac{(n!)^2}{(-1)^n x^n} \right|$$

$$\lim_{n \to \infty} \left| \frac{x}{(n+1)^2} \right| = |x| \lim_{n \to \infty} \left| \frac{1}{(n+1)^2} \right| = |x| \cdot \frac{1}{\infty}$$

$$= |x| \cdot 0 = 0 < 1$$

The limit $= 0 < 1$ for any value of x

→ Radius of Convergence $= \infty$

Done!

BESSEL'S DE

BESSEL'S DIFFERENTIAL EQUATION:

We know how to solve: $P(x)y'' + Q(x)y' + R(x)y = 0$ Near x_0
When x_0 is a regular singular point.

A D.E. of the form: $x^2 y'' + xy' + (x^2 - v^2)y = 0$
Is called <u>Bessel's D.E.</u> of order " v ".

Examples of Bessel's D.E. of various order:
- $x^2 y'' + xy' + x^2 y = 0$ → Order = 0
- $x^2 y'' + xy' + (x^2 - 1)y = 0$ → Order = 1
- $x^2 y'' + xy' + \left(s^2 - \frac{1}{4}\right)y = 0$ → Order = ½

Note that there are few cases possible, based on the value of "v".
We will discuss them later. First, let us try to solve an equation.

Note that there are few cases possible, based on the value of "v". We will discuss them later. First, let us try to solve this equation:

Solve: $x^2 y'' + x y' + (x^2 - v^2)y = 0$ Near $x_0 = 0$

Solution:

$$P(x) = x^2 \qquad Q(x) = x \qquad R(x) = x^2 - v^2$$

$P(0) = (0)^2 = 0 \;\rightarrow\; x_0 = 0$ is a singular point.

∃ a solution in terms of Frobenius Series.

Assume this form:

$$y = \sum_{n=0}^{\infty} a_n x^{n+r}$$
$$y' = \sum_{n=0}^{\infty} (n+r) a_n x^{n+r-1}$$
$$y'' = \sum_{n=0}^{\infty} (n+r)(n+r-1) a_n x^{n+r-2}$$

Substitute into: $x^2 y'' + x y' + (x^2 - v^2)y = 0$

$$x^2 y'' = \sum_{n=0}^{\infty} (n+r)(n+r-1) a_n x^{n+r}$$
$$+\; xy' = \sum_{n=0}^{\infty} (n+r) a_n x^{n+r}$$
$$+\; x^2 y = \sum_{n=0}^{\infty} a_n x^{n+r+2} \;=\; \sum_{n=2}^{\infty} a_{n-2} x^{n+r}$$
$$-\; v^2 y = -\sum_{n=0}^{\infty} a_n v^2 x^{n+r}$$

Write equation with just one infinite sum, starting at $n = 2$

Write first few terms separately. (For $n = 0, 1$)

$[r(r-1)a_0 x^r + r a_0 x^r - a_0 v^2 x^r]$
$+ [(r+1)r a_1 x^{r+1} + (r+1)a_1 x^{r+1} - a_1 v^2 x^{r+1}]$
$+ \sum_{n=0}^{\infty}[(n+r)(n+r-1)a_n + (n+r)a_n + a_{n-2} - v^2 a_n] x^{n+r} = 0$

The a_0 terms $= 0$ Note: $a_0 \neq 0$ and $x^r \neq 0$

$\quad [r(r-1) + r - v^2] a_0 x^r = 0$

$\quad r(r-1) + r - v^2 = 0 \quad\quad \leftarrow$ Indicial Equation

$\quad r^2 - v^2 = 0$

$\quad r = \pm v \quad\quad\quad\quad\quad\quad \leftarrow$ Roots of Indicial Equation

The a_1 terms $= 0$

$\quad [(r+1)r + (r+1) - v^2] a_1 x^{r+1} = 0$

$\quad [r^2 + 2r + 1 - v^2] a_1 x^{r+1} = 0$

$\quad$ At $r = \pm v$

$\quad [r^2 + 2r + 1 - v^2] a_1 x^{r+1} = 0$

$\quad$ Since $[r^2 + 2r + 1 - v^2] \neq 0 \quad$ and $x^{r+1} \neq 0 \rightarrow a_1 = 0$

Every term, depending on a_1 will be zero.

Find the Recursive Equation: Note: $\sum [\ ] = 0$

$$\sum_{n=0}^{\infty} [(n+r)(n+r-1)a_n + (n+r)a_n + a_{n-2} - v^2 a_n] x^{n+r} = 0$$
$$[(n+r)(n+r-1)a_n + (n+r)a_n + a_{n-2} - v^2 a_n] = 0$$
$$[(n+r)(n+r-1) + (n+r) - v^2] a_n + a_{n-2} = 0$$
$$[(n+r)(n+r-1+1) + -v^2] a_n + a_{n-2} = 0$$
$$[(n+r)(n+r) - v^2] a_n + a_{n-2} = 0$$
$$[(n+r)^2 - v^2] a_n + a_{n-2} = 0$$

$$a_n = \frac{-a_{n-2}}{(n+r)^2 - v^2} \qquad n = 2, 3, 4 \ldots$$

Recall: $a^2 - b^2 = (a+b)(a-b)$

$$a_n = \frac{-a_{n-2}}{(n+r+v)\cdot(n+r-v)}$$

$$a_n = \frac{-a_{n-2}}{(n+r+v)\cdot(n+r-v)} \qquad n = 2, 3, 4 \ldots$$

Compute first few coefficients: (Recall, the a_1 terms are zero.)

$n = 2 \rightarrow a_2 = \dfrac{-a_0}{(r+2+v)\cdot(r+2-v)}$

$n = 4 \rightarrow a_4 = \dfrac{-a_2}{(r+4+v)\cdot(r+4-v)}$

$ = \dfrac{a_0}{(r+4+v)\cdot(r+4-v)\cdot(r+2+v)\cdot(r+2-v)}$

$ = \dfrac{-a_4}{(r+6+v)\cdot(r+6-v)}$

$ = \dfrac{-a_0}{(r+6+v)\cdot(r+6-v)\cdot(r+4+v)\cdot(r+4-v)\cdot(r+2+v)\cdot(r+2-v)}$

...

$n = 3 \rightarrow a_3 = \dfrac{-a_1}{(r+3+v)\cdot(r+3-v)} = 0$

$n = 5 \rightarrow a_5 = \dfrac{-a_3}{(r+5+v)\cdot(r+5-v)} = 0$

...

Substitute the coefficients in Frobenius Series

$$y = x^r[a_0 + a_1 x + a_3 x^2 + a_4 x^4 + a_5 x^5 + \ldots]$$

$$y = x^r \left[a_0 - \frac{-a_0 x^2}{(r+2+v)\cdot(r+2-v)} + \frac{a_0 x^4}{(r+4+v)\cdot(r+4-v)\cdot(r+2+v)\cdot(r+2-v)} - \ldots \right]$$

Recall: $r_1 = v$ and $r_2 = -v$

For $r_1 = v$

$$y = a_0 x^r \left[1 - \frac{-x^2}{(r+2+v)\cdot(r+2-v)} + \frac{x^4}{(r+4+v)\cdot(r+4-v)\cdot(r+2+v)\cdot(r+2-v)} - \cdots \right]$$

$$y = a_0 x^v \left[1 - \frac{-x^2}{(2v+2)\cdot(2)} + \frac{x^4}{(2v+4)\cdot(4)\cdot(2v+2)\cdot(2)} - \cdots \right]$$

$$y = a_0 x^v \left[1 - \frac{-x^2}{4(v+1)} + \frac{x^4}{8(v+2)\cdot(v+1)\cdot(4)} - \cdots \right]$$

$$y = a_0 x^v \left[1 - \frac{-x^2}{4(v+1)} + \frac{x^4}{8\cdot 4\cdot(v+2)\cdot(v+1)} - \cdots \right] = J_v(x)$$

$y = J_v(x)$ = Bessel's function of the first kind.

For $r_1 = -v$

$$y = a_0 x^r \left[1 - \frac{-x^2}{(r+2+v)\cdot(r+2-v)} + \frac{x^4}{(r+4+v)\cdot(r+4-v)\cdot(r+2+v)\cdot(r+2-v)} - \cdots \right]$$

$$y = a_0 x^{-v} \left[1 - \frac{-x^2}{(2)\cdot(2-2v)} + \frac{x^4}{(4)\cdot(4-2v)\cdot(2-2v)\cdot(2)} - \cdots \right]$$

$$y = a_0 x^{-v} \left[1 - \frac{-x^2}{4(1-v)} + \frac{x^4}{8(2-v)\cdot(1-v)\cdot(4)} - \cdots \right]$$

$$y = a_0 x^{-v} \left[1 - \frac{-x^2}{4(1-v)} + \frac{x^4}{8\cdot 4\cdot(2-v)\cdot(1-v)} - \cdots \right] = J_{-v}(x)$$

$y = J_{-v}(x)$ = Bessel's function of the second kind.

> **Bessel's Solution of the First Kind.**
>
> $$J_v(x) = a_0 x^v \left[1 - \frac{-x^2}{4(v+1)} + \frac{x^4}{8 \cdot 4 \cdot (v+2) \cdot (v+1)} - \ldots \right]$$

For Bessel's D.E. of order 0 and 1, if we put $v = 0$ and $v = 1$ in Bessel's Solution of the First kind, we get:

For $v = 0$

$$J_0(x) = a_0 x^0 \left[1 - \frac{-x^2}{4(1)} + \frac{x^4}{8 \cdot 4 \cdot (2) \cdot (1)} - \ldots \right] \quad \text{Let } a_0 = 1$$

$$J_0(x) = 1 - \frac{-x^2}{4} + \frac{x^4}{8 \cdot 4 \cdot 2} - \ldots$$

$$J_0(x) = \sum_{n=0}^{\infty} \frac{(-1)^n}{(n!)^2 \, 2^{2n}} \cdot x^{2n}$$

For $v = 1$

$$J_1(x) = a_0 x^1 \left[1 - \frac{-x^2}{4(2)} + \frac{x^4}{8 \cdot 4 \cdot (3) \cdot (2)} - \ldots \right] \quad \text{Let } a_0 = 1$$

$$J_1(x) = 1 - \frac{-x^3}{4 \cdot 2} + \frac{x^5}{8 \cdot 4 \cdot 3 \cdot 2} - \ldots$$

$$J_1(x) = \sum_{n=0}^{\infty} \frac{(-1)^n}{n! \, (n+1)! \, 2^{2n+1}} \cdot x^{2n+1}$$

Remarks: (Verify this.)

- The Radius of convergence for $J_0(x)$ and $J_1(x)$ is ∞
- In other words, $J_0(x)$ and $J_1(x)$ are absolutely convergent.
- $J_0'(x) = -J_1(x)$

LAPLACE TRANSFORM

Is there another way to solve a non-homogeneous DE with initial conditions? Many practical engineering problems involve mechanical or electrical systems acted on by the discontinuous or impulse forcing functions. These DEs are slightly more interesting but the solution methods, used previously, may be a too tedious to use.

In this chapter, transformation methods are introduced to solve DEs involving discontinuous forcing and impulse functions. A few Laplace Transforms are derived and tables of Laplace Transforms, for frequently used functions, are provided. Several initial value problems, similar to those in Chapter 2, are solved using Laplace Transforms.

Piecewise continuous functions and discontinuous forcing functions are converted to a form with linear combinations of step functions. After the functions have been converted, Laplace Transforms are used to solve the DE. Several detailed examples are provided.

This chapter concludes with idea of convolution integrals. The emphasis is on problems that typically arise in engineering due to changing forces and impulse forces.

IMPROPER INTEGRATION (REVIEW)

IMPROPER INTEGRATION – Review

Type 1 Improper Integration:

$$\int_a^\infty f(x)\,dx = \lim_{t\to\infty} \int_a^t f(x)\,dx$$

Note: Upper bound is ∞ which is a symbol, not a number.

Example (1) $\int_1^\infty \frac{1}{x}\,dx$ Determine if the integral is Convergent (CGT) or Divergent (DGT).

Solution:

$$\int_1^\infty \frac{1}{x}\,dx = \lim_{t\to\infty} \int_a^t \frac{1}{x}\,dx = \lim_{t\to\infty} [\,\ln|x|\,]_1^t$$

$$= \lim_{t\to\infty} [\ln|t| - \ln|1|] = \lim_{t\to\infty} \ln|t| = \ln|\infty| = \infty$$

→ The Integral is Divergent (DGT).

Example (2) $\int_1^\infty e^{3-2x}\,dx$ Determine if the integral is Convergent or Divergent.

Solution:

$$\int_1^\infty e^{3-2x}\,dx = \lim_{t\to\infty} \int_1^t e^{3-2x}\,dx = \lim_{t\to\infty} \left[\frac{e^{3-2x}}{-2}\right]_1^t$$

$$= \lim_{t\to\infty} \left[\frac{e^{3-2t}}{-2} + \frac{e^3}{2}\right] = \left[\frac{e^{-\infty}}{-2} + \frac{e^3}{2}\right]$$

$$= \left[\frac{-1}{2\,e^\infty} + \frac{e^3}{2}\right] = \left[0 + \frac{e^3}{2}\right] = \frac{e^3}{2}$$

→ The Integral is Convergent (CGT).

Example (3) $\int_e^\infty \dfrac{1}{x \cdot (\ln x)}\, dx$ Determine if the integral is Convergent or Divergent.

Solution:

$$\int_e^\infty \dfrac{1}{x \cdot (\ln x)}\, dx = \lim_{t \to \infty} \int_e^t \left(\dfrac{1}{x}\right) \cdot (\ln x)^{-1}\, dx$$

> **Use U-Substitution.**
>
> Let $u = \ln x$ and $du = \dfrac{1}{x} dx$
>
> $= \lim_{t \to \infty} \int u^{-1}\, du = \lim_{t \to \infty} [\ln u]$

$$\int_e^\infty \dfrac{1}{x \cdot (\ln x)}\, dx = \lim_{t \to \infty} [\ln|\ln|x||]_e^t$$

$$= \lim_{t \to \infty} [\ln|\ln t| - \ln|e|]$$

$$= [\ln|\infty| - \ln|e|] = \infty - 1 = \infty$$

→ The integral is Divergent (DGT).

PIECEWISE CONTINUOUS FUNCTIONS

Let f be any function defined on $[a, b]$, where $[a, b]$ is any partition, containing n equal-length sub-intervals. And let:

$$a = x_0 < x_1 < x_2 \ldots < x_n = b$$

The function f is <u>Piecewise Continuous</u> on $[a, b]$
If two conditions are met:

1) f is continuous on each sub-interval

   $$[x_i, x_{i+1}] \qquad i = 0, 1, 2 \ldots n$$

2) The limit of f exists when approached to the end points WITHIN the sub-interval. Note that the limit approaching each sub-boundary may not match when approached from left and right. But, the limits, approaching the boundaries from within each sub-interval must exist.

PIECEWISE CONTINUOUS FUNCTION – Graphical Example

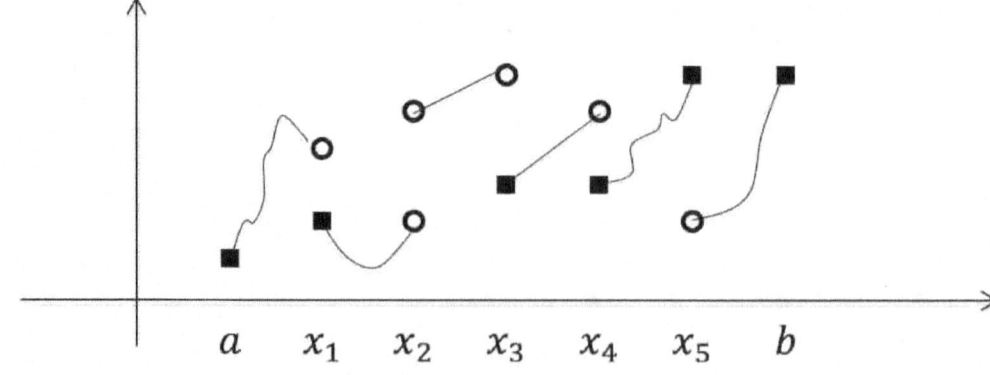

Example: Determine if the functions are continuous, piecewise continuous, or neither.

(1.) $f(x) = \begin{cases} |x| & x < 0 \\ x^2 & 0 \leq x < 2 \\ 4 & x \geq 2 \end{cases}$

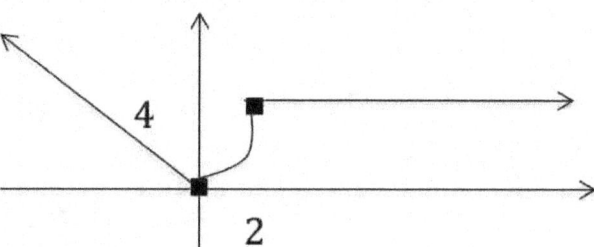

Sketch the graph.
This is a continuous function.

(2.) $f(x) = \begin{cases} -x & x < 0 \\ x^2 + 1 & 0 \leq x < 3 \\ 7 & x \geq 3 \end{cases}$

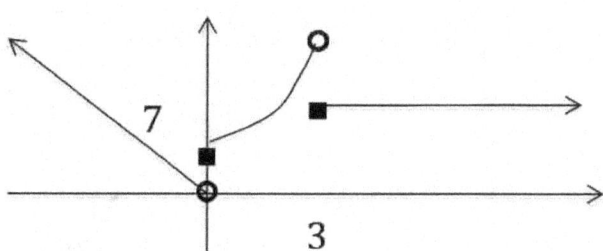

Sketch the graph.
Piecewise Continuous because:
- Continuous in each sub-interval
- Limits exist within each interval.

(3.) $f(x) = \begin{cases} x & x \leq 0 \\ \left(\dfrac{1}{x}\right) & 0 < x \leq 1 \\ 3 & x > 1 \end{cases}$

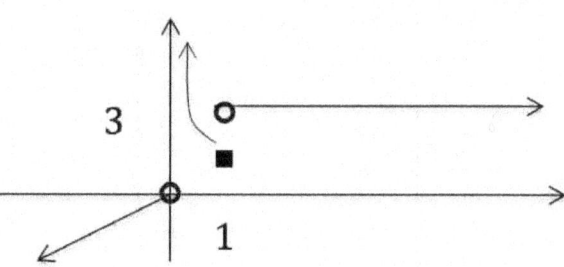

Sketch the graph.
NOT Continuous or
And NOT Piecewise Continuous.
Neither! Because: $\lim\limits_{x \to 0+} f(x) = \infty$

PIECEWISE CONTINUOUS FUNCTIONS can be integrated.

Example: Integrate given function on indicated interval.

$$f(x) = \begin{cases} -2x & -1 \leq x < 0 \\ 3x^2 + 1 & 0 \leq x < 3 \\ 5 & 3 \leq x \leq 4 \end{cases}$$

Note: Limits on all endpoints exist within the sub-intervals.

Solution:

$$\int_{-1}^{4} f(x) \, dx = \int_{-1}^{0} -2x \, dx + \int_{0}^{3} 3x^2 + 1 \, dx + \int_{3}^{4} 5 \, dx$$

$$= [-x^2]_{-1}^{0} + [x^3 + x]_{0}^{3} + [5x]_{3}^{4}$$

$$= [0^2 + (-1)^2] + [3^3 + 3 - 0^3 - 0] + [20 - 15]$$

$$= [\,1\,] + [30] + [5] = 36 = \text{AREA under curve.}$$

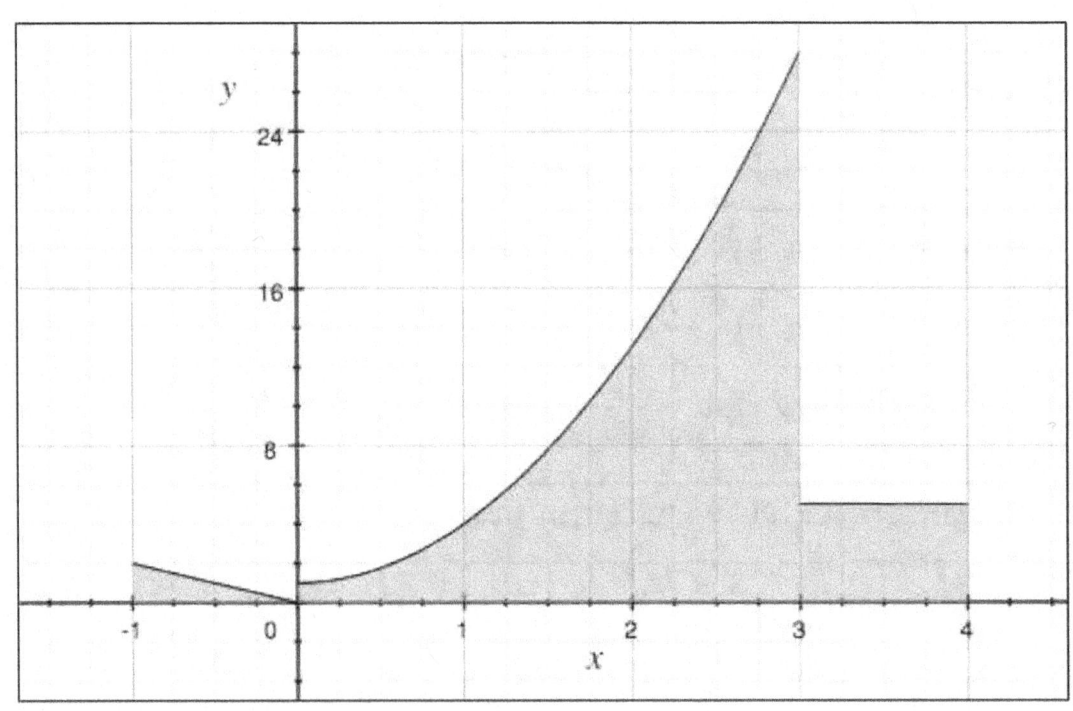

LAPLACE TRANSFORMS

Remarks:
- It is possible to solve DEs by applying Integral Transforms.
- Improper Integration and Piecewise Continuous Functions should be reviewed as needed.

INTEGRAL TRANSFORM of $f(x)$

is the conversion of $f(x)$ to the form: $\int_\alpha^\beta K(s,x) \cdot f(x)\, dx$

Examples of Integral Transforms.
1. Laplace Transform
2. Two-Sided Transform
3. Fourier Transform
4. Fourier Sin Transform
5. Fourier Cos Transform
6. Abel's Transform
7. Hilbert's Transform

Unknowns:

$K(s, x)$ = Kernel of transform

$\alpha = -\infty$ or any real number.

$\beta = \infty$ or any real number.

DEFINITION OF LAPLACE TRANSFORM

LAPLACE TRANSFORM:

➔ $K(s, x) = e^{-sx}$ With: $\alpha = 0$, $\beta = \infty$

The Laplace Transform of $f(x) = L[f(x)]$

$$
\begin{aligned}
L[f(x)] &= \int_0^\infty e^{-sx} \cdot f(x)\, dx \\
&= \lim_{t \to \infty} \int_0^t e^{-sx} \cdot f(x)\, dx \\
&= F(s)
\end{aligned}
$$

Simple Examples: See next few pages.

Simple Examples: Compute the Laplace Transform of each of the following functions.

Example (1)   Find the Laplace Transform of: $f(x) = 1$

Solution: This is a constant function. $s > 0$

$$L[1] = \int_0^\infty e^{-sx} \cdot (1) \, dx$$

$$L[1] = \lim_{t \to \infty} \int_0^t e^{-sx} \, dx = \lim_{t \to \infty} \left[\frac{e^{-sx}}{-s} \right]_0^t$$

$$L[1] = \lim_{t \to \infty} \left[\frac{-e^{-st}}{s} - \frac{-e^0}{s} \right] = \lim_{t \to \infty} \left[\frac{-e^{-st}}{s} + \frac{1}{s} \right]$$

$$L[1] = \frac{-e^{-\infty}}{s} + \frac{1}{s} = 0 + \frac{1}{s} = \frac{1}{s}$$

In General: $L[n] = \frac{n}{s}$ Where n is a positive integer.

Similar examples:

- $L[7] = \frac{7}{s}$
- $L\left[\frac{3}{2}\right] = \frac{3}{2s}$
- $L[\pi] = \frac{\pi}{s}$

Example (2)

Find the Laplace Transform of: $f(x) = e^{ax}$ With $s > a$

Solution:

$$L[e^{ax}] = \int_0^\infty e^{-sx} \cdot e^{ax} \, dx$$

$$L[e^{ax}] = \lim_{t \to \infty} \int_0^t e^{(a-s)x} \, dx = \lim_{t \to \infty} \left[\frac{e^{(a-s)x}}{a-s} \right]_0^t$$

$$L[e^{ax}] = \lim_{t \to \infty} \left[\frac{e^{(a-s)t}}{a-s} - \frac{e^0}{a-s} \right] = \lim_{t \to \infty} \left[\frac{e^{-\infty}}{a-s} - \frac{e^0}{a-s} \right]$$

NOTE: $a - s < 0$

$$L[e^{ax}] = 0 - \frac{1}{a-s} = \frac{-1}{a-s} = \frac{1}{s-a}$$

Similar Example:

$$L[e^{2x+3}] = L[e^{2x}e^3] = e^3 \cdot L[e^{2x}] = e^3 \cdot \frac{1}{s-2} = \frac{e^3}{s-2}$$

Example (3) Find the Laplace Transform of: $f(x) = x$

Solution:

$$L[x] = \int_0^\infty x \cdot e^{-sx} \cdot dx = \lim_{t \to \infty} \int_0^t x \cdot e^{-sx}\, dx$$

Integrate by Parts

Du		$\int dv$
x	+	e^{-sx}
1	−	$-\dfrac{e^{-sx}}{s}$
0	+	$\dfrac{e^{-sx}}{s^2}$

$$L[x] = \lim_{t \to \infty} \left[\frac{-x\, e^{-sx}}{s} - \frac{e^{-sx}}{s^2} \right]_0^t$$

$$L[x] = \lim_{t \to \infty} \left[\frac{-t\, e^{-st}}{s} - \frac{e^{-st}}{s^2} - \frac{0}{s} + \frac{e^0}{s^2} \right]$$

$$L[x] = \lim_{t \to \infty} \left[\frac{-t\, e^{-st}}{s} - \frac{e^{-st}}{s^2} - 0 + \frac{1}{s^2} \right]$$

$$L[x] = \lim_{t \to \infty} \left[\frac{-t}{s\, e^{st}} - \frac{1}{s^2\, e^{st}} - 0 + \frac{1}{s^2} \right]$$

$$L[x] = \lim_{t \to \infty} \left[\frac{-\infty}{\infty} - 0 - 0 + \frac{1}{s^2} \right]$$

(Continued...)

$$L[x] = \lim_{t \to \infty} \left[\frac{-\infty}{\infty} - 0 - 0 + \frac{1}{s^2} \right]$$

Note: $\frac{-\infty}{\infty}$ is Indeterminate

So, use L.H. on that term.

$$L[x] = \lim_{t \to \infty} \left[\frac{-1}{s^2 e^{st}} - 0 - 0 + \frac{1}{s^2} \right]$$

$$L[x] = \lim_{t \to \infty} \left[\frac{-1}{\infty} - 0 - 0 + \frac{1}{s^2} \right]$$

$$L[x] = 0 - 0 - 0 + \frac{1}{s^2} = \frac{1}{s^2}$$

$$L[x] = \frac{1}{s^2}$$

Extra Practice: Verify $L[x^2] = \frac{2}{s^3}$

Hint: Use integration by parts and L.H. Rule two times.

Note: We will soon learn a short way to compute Laplace Transforms of x^n where n is a positive integer.

Example (4)   Find the Laplace Transform of: $f(x) = \cos ax$

Solution:

$$L[\cos ax] = \int_0^\infty \cos ax \cdot e^{-sx} \cdot dx$$

$$L[\cos ax] = \lim_{t \to \infty} \int_0^t \cos ax \cdot e^{-sx} \, dx$$

Integrate by Parts

Du	$\int dv$
$\cos ax$ $\quad +$	e^{-sx}
$-a \sin ax$ $\quad -$	$-\dfrac{e^{-sx}}{s}$
$-a^2 \cos ax$ $\quad +$	$\dfrac{e^{-sx}}{s^2}$

Stop when you get back to the original function.

$$L[\cos ax] = \lim_{t \to \infty} \left[\frac{-\cos ax \, e^{-sx}}{s} + \frac{a \sin ax \, e^{-sx}}{s^2} \right]_0^t$$

$$- \frac{a^2}{s^2} \lim_{t \to \infty} \int_0^t \cos ax \, e^{-sx} \, dx$$

$$L[\cos ax] = \lim_{t \to \infty} \left[\frac{-\cos ax \, e^{-sx}}{s} + \frac{a \sin ax \, e^{-sx}}{s^2} \right]_0^t$$

$$- \frac{a^2}{s^2} \lim_{t \to \infty} \int_0^t \cos ax \, e^{-sx} \, dx$$

$$L[\cos ax] = \lim_{t \to \infty} \left[\frac{-\cos ax \, e^{-sx}}{s} + \frac{a \sin ax \, e^{-sx}}{s^2} \right]_0^t - \frac{a^2}{s^2} L[\cos ax]$$

$$L[\cos ax] \left(1 + \frac{a^2}{s^2}\right) = \lim_{t \to \infty} \left[\frac{-\cos ax \, e^{-sx}}{s} + \frac{a \sin ax \, e^{-sx}}{s^2} \right]_0^t$$

$$L[\cos ax] \cdot \left(\frac{s^2 + a^2}{s^2}\right) = \lim_{t \to \infty} \left[\frac{-\cos ax}{s \, e^{sx}} + \frac{a \sin ax}{s^2 \, e^{sx}} \right]_0^t$$

$$L[\cos ax] \cdot \left(\frac{s^2 + a^2}{s^2}\right) = \lim_{t \to \infty} \left[\frac{-\cos at}{s \, e^{st}} + \frac{a \sin at}{s^2 \, e^{st}} + \frac{\cos 0}{s \, e^0} - \frac{a \sin 0}{s^2 \, e^0} \right]$$

$$L[\cos ax] \cdot \left(\frac{s^2 + a^2}{s^2}\right) = \lim_{t \to \infty} \left[\frac{-\cos at}{\infty} + \frac{a \sin at}{\infty} + \frac{1}{s} - \frac{0}{s^2} \right]$$

$$L[\cos ax] \cdot \left(\frac{s^2 + a^2}{s^2}\right) = \lim_{t \to \infty} \left[0 + 0 + \frac{1}{s} - 0 \right]$$

$$L[\cos ax] \cdot \left(\frac{s^2 + a^2}{s^2}\right) = \frac{1}{s}$$

$$L[\cos ax] \cdot \left(\frac{s^2 + a^2}{s^2}\right) = \frac{1}{s}\left(\frac{s^2}{s^2 + a^2}\right) = \frac{s}{s^2 + a^2}$$

$$L[\cos ax] = \frac{s}{s^2 + a^2}$$

Extra Practice: Verify this: $L[\sin ax] = \dfrac{a}{a^2 + s^2}$

LAPLACE TRANSFORMS – Some we have seen so far:

$L[f(x)]$	$F(s)$
$L[1]$	$\dfrac{1}{s}$
$L[x]$	$\dfrac{1}{s^2}$
$L[e^{ax}]$	$\dfrac{1}{s-a}$
$L[\cos ax]$	$\dfrac{s}{s^2+a^2}$
$L[\sin ax]$	$\dfrac{a}{s^2+a^2}$

Verify: $L[\sin ax] = \dfrac{a}{x^2+a^2}$

Solution:

$L[\sin ax] = \int_0^\infty \sin ax \, e^{-sx} \, dx$

$L[\sin ax] = \lim_{t\to\infty} \int_0^t \sin ax \, e^{-sx} \, dx$

Use the Integration by Parts Tabular Method.

Du		∫ dv
$\sin ax$	+	e^{-sx}
$a\cos ax$	−	$-\dfrac{e^{-sx}}{s}$
$-a^2 \sin ax$	+	$\dfrac{e^{-sx}}{s^2}$

$L[\sin ax] =$

$= \lim_{t\to\infty} \left[\dfrac{-\sin ax \, e^{-sx}}{s} - \dfrac{a\cos ax \, e^{-sx}}{s^2} \right]_0^t - \dfrac{a^2}{s^2} \lim_{t\to\infty} \int_0^t \sin ax \, e^{-sx} \, dx$

$L[\sin ax] = \lim_{t\to\infty} \left[\dfrac{-\sin ax \, e^{-sx}}{s} - \dfrac{a\cos ax \, e^{-sx}}{s^2} \right]_0^t - \dfrac{a^2}{s^2} L[\sin ax]$

$L[\sin ax]\left(1 + \dfrac{a^2}{s^2}\right) = \lim_{t\to\infty} \left[\dfrac{-\sin ax \, e^{-sx}}{s} - \dfrac{a\cos ax \, e^{-sx}}{s^2} \right]_0^t$

Continued...

$$L[\sin ax]\left(1+\frac{a^2}{s^2}\right) = \lim_{t\to\infty}\left[\frac{-\sin ax\, e^{-sx}}{s} - \frac{a\cos ax\, e^{-sx}}{s^2}\right]_0^t$$

$$L[\sin ax]\left(1+\frac{a^2}{s^2}\right)$$
$$= \lim_{t\to\infty}\left[\frac{-\sin at\, e^{-st}}{s} - \frac{a\cos at\, e^{-st}}{s^2} + \frac{\sin 0\, e^0}{s} + \frac{a\cos 0\, e^0}{s^2}\right]$$

$$L[\sin ax]\left(1+\frac{a^2}{s^2}\right) = \lim_{t\to\infty}\left[\frac{-\sin at}{s\, e^{st}} - \frac{a\cos at}{s^2\, e^{st}} + \frac{0}{s} + \frac{a}{s^2}\right]$$

$$L[\sin ax]\left(1+\frac{a^2}{s^2}\right) = \frac{-\sin\infty}{\infty} - \frac{a\cos\infty}{\infty} + \frac{a}{s^2}$$

$$L[\sin ax]\left(1+\frac{a^2}{s^2}\right) = 0 - 0 + \frac{a}{s^2}$$

$$L[\sin ax]\left(\frac{s^2+a^2}{s^2}\right) = \frac{a}{s^2}$$

$$L[\sin ax] = \frac{a}{s^2}\left(\frac{s^2}{s^2+a^2}\right) = \frac{a}{s^2+a^2}$$

$$L[\sin ax] = \frac{a}{s^2+a^2}$$

Example: Compute the Laplace Transform of:

$$e^{ax} \sin bx \quad \text{and} \quad e^{ax} \cos bx$$

Solution:

Recall: $e^{(a+ib)x} = e^{ax} \cdot e^{ibx} = e^{ax}(\cos bx + i \sin bx)$

$$= e^{ax} \cos bx + i e^{ax} \sin bx \quad (1)$$

So,

$$L\left[e^{(a+ib)x}\right] = \frac{1}{s-(a+ib)} = \frac{1}{(s-a)-ib}$$

$$= \frac{1}{(s-a)-ib} \cdot \frac{(s-a)+ib}{(s-a)+ib} = \frac{(s-a)+ib}{(s-a)^2-(ib)^2} = \frac{(s-a)+ib}{(s-a)^2+b^2}$$

$$= \frac{(s-a)}{(s-a)^2+b^2} + i \frac{b}{(s-a)^2+b^2} \quad (2)$$

From (1)

$$L\left[e^{(a+ib)x}\right] = L[e^{ax} \cos bx] + i L[e^{ax} \sin bx] \quad (3)$$

Equate (2) and (3)

$$L[e^{ax} \cos bx] = \frac{(s-a)}{(s-a)^2+b^2} \quad \text{and} \quad L[e^{ax} \sin bx] = \frac{b}{(s-a)^2+b^2}$$

LAPLACE TRANSFORMS – More: (and even more online.)

$L[f(x)]$	$F(s)$
$L[1]$	$\dfrac{1}{s}$
$L[x]$	$\dfrac{1}{s^2}$
$L[e^{ax}]$	$\dfrac{1}{s-a}$
$L[\cos ax]$	$\dfrac{s}{s^2+a^2}$
$L[\sin ax]$	$\dfrac{a}{s^2+a^2}$
$L[e^{ax}\cos bx]$	$\dfrac{s-a}{(s-a)^2+b^2}$
$L[e^{ax}\sin bx]$	$\dfrac{b}{(s-a)^2+b^2}$
$L[x^2]$	$\dfrac{2}{s^3}$
$L[x^3]$	$\dfrac{3!}{s^4}$
$L[x^n]$	$\dfrac{n!}{s^{n+1}}$ *

* Note: $L[x^n] = \dfrac{n!}{s^{n+1}}$ Where n is a positive integer.

Question: What if "n" is not an integer? (more general case.)

Answer: $L[x^p] = \dfrac{\Gamma(p+1)}{s^{p+1}}$ Γ = Gamma Function, $p \in \mathbb{R}$

GAMMA FUNCTION

> **GAMMA FUNCTION:**
>
> $$\Gamma(p+1) = \int_0^\infty x^p \cdot e^{-x}\, dx \qquad \text{For } p > -1$$

GAMMA FUNCTION PROPERTIES (EXAMPLES)

Example (1) $\Gamma(1) = ???$ ➔ $p = 0$

Solution: (Proof)

$\Gamma(1) = \int_0^\infty x^0 \cdot e^{-x}\, dx = \int_0^\infty e^{-x}\, dx$

$\Gamma(1) = \lim_{t \to \infty} \int_0^t e^{-x}\, dx = \lim_{t \to \infty} [(-1)\, e^{-x}]_0^t$

$\Gamma(1) = \lim_{t \to \infty} \left[\dfrac{-1}{e^x}\right]_0^t$

$\Gamma(1) = \lim_{t \to \infty} \left[\dfrac{-1}{e^t} + \dfrac{1}{e^0}\right] = 0 + 1 = 1$

$\Gamma(1) = 1$

Example (2) $\Gamma(p+1) = p \cdot \Gamma(p)$ Recursive Formula.

Solution: (Proof)

$$\Gamma(p+1) = \int_0^\infty x^p \cdot e^{-x}\, dx = \lim_{t \to \infty} \int_0^t x^p \cdot e^{-x}\, dx$$

Use Integration by Parts

Du		$\int dv$
x^p	+	e^{-x}
$p\, x^{p-1}$	−	$-e^{-x}$
STOP (similar format)		

$$\Gamma(p+1) = \lim_{t \to \infty} [-x^p e^{-x}]_0^t + \lim_{t \to \infty} \int_0^t p\, x^{p-1} e^{-x}\, dx$$

$$\Gamma(p+1) = \lim_{t \to \infty}\left[-\frac{t^p}{e^t} + \frac{0^p}{e^0}\right] + p \cdot \lim_{t \to \infty} \int_0^t x^{p-1} e^{-x}\, dx$$

$$\Gamma(p+1) = \lim_{t \to \infty}\left[-\frac{t^p}{e^t} + 0\right] + p \cdot \lim_{t \to \infty} \int_0^t x^{p-1} e^{-x}\, dx$$

$$\Gamma(p+1) = \lim_{t \to \infty}\left[-\frac{p\, t^{p-1}}{e^t}\right] + p \cdot \lim_{t \to \infty} \int_0^t x^{p-1} e^{-x}\, dx$$

Could apply L.H. many times, but that is not necessary. Denominator grows faster than numerator so at infinity, the first limit is zero.

$$\Gamma(p+1) = p \cdot \lim_{t \to \infty} \int_0^t x^{p-1} e^{-x}\, dx$$

$$\Gamma(p+1) = p \cdot \Gamma(p)$$

Example (3) $L[x^3] = ???$

Solution: (Proof)

Recall: $\Gamma(p+1) = p \cdot \Gamma(p)$

Recall: $L[x^p] = \dfrac{\Gamma(p+1)}{s^{p+1}}$

$$\boxed{\begin{aligned} \Gamma(4) &= 3\,\Gamma(3) \\ \Gamma(4) &= 3 \cdot 2 \cdot \Gamma(2) \\ \Gamma(4) &= 3 \cdot 2 \cdot 1 \cdot \Gamma(1) \\ \Gamma(4) &= 3! \end{aligned}}$$

$L[x^3] = \dfrac{\Gamma(1+3)}{s^{1+3}} = \dfrac{\Gamma(4)}{s^4}$

$L[x^3] = \dfrac{3!}{s^4}$

There is no need to use integration by parts 3 times. ☺

Likewise, one can prove:

$$L[x] = \frac{1!}{s^2} = \frac{1}{s^2}$$

Previously proved using the definition of Laplace Transform and Integration by Parts.

Similarly: $L[x^2] = \dfrac{2!}{s^3} = \dfrac{2}{s^3}$

Example (4)   Important Property: $\Gamma(n+1) = n!$

Where: $n =$ Integer

Proof:

$\Gamma(n+1) = n \cdot \Gamma(n)$

$\Gamma(n+1) = n(n-1) \cdot \Gamma(n-1)$

$\Gamma(n+1) = n(n-1)(n-2) \cdot \Gamma(n-2)$

$\Gamma(n+1) = n(n-1)(n-2) \dots (1) \cdot \Gamma(1)$

$\Gamma(n+1) = n!$

Example (5)   Important Property: $\Gamma\left(\frac{1}{2}\right) = \sqrt{\pi}$

Proof:

Long Proof ➔ $\int_0^\infty e^{x^2} \, dx = \sqrt{\pi}$

Example (6) $\Gamma\left(\frac{3}{2}\right) = \frac{1}{2}\sqrt{\pi}$

Proof:

$\Gamma\left(\frac{3}{2}\right) = \Gamma\left(1 + \frac{1}{2}\right)$

$\Gamma\left(\frac{3}{2}\right) = \frac{1}{2} \Gamma\left(\frac{1}{2}\right)$

$\Gamma\left(\frac{3}{2}\right) = \frac{1}{2} \sqrt{\pi}$

Example (7) $\Gamma\left(\frac{7}{2}\right) = \Gamma\left(1 + \frac{5}{2}\right)$

Proof:

$$\Gamma\left(\frac{7}{2}\right) = \frac{5}{2}\Gamma\left(\frac{5}{2}\right)$$

$$\Gamma\left(\frac{7}{2}\right) = \frac{5}{2}\Gamma\left(1 + \frac{3}{2}\right)$$

$$\Gamma\left(\frac{7}{2}\right) = \frac{5}{2}\cdot\frac{3}{2}\Gamma\left(\frac{3}{2}\right)$$

$$\Gamma\left(\frac{7}{2}\right) = \frac{5}{2}\cdot\frac{3}{2}\Gamma\left(1 + \frac{1}{2}\right)$$

$$\Gamma\left(\frac{7}{2}\right) = \frac{5}{2}\cdot\frac{3}{2}\cdot\frac{1}{2}\sqrt{\pi}$$

Summary:

$$L[x^p] = \frac{\Gamma(p+1)}{s^{p+1}} \quad \text{AND} \quad L[x^n] = \frac{n!}{s^{n+1}}$$

GAMMA FUNCTIONS:

$\Gamma(1)$	1
$\Gamma(p+1)$	$p \cdot \Gamma(p)$
$\Gamma(n+1)$	$n!$
$\Gamma\left(\frac{1}{2}\right)$	$\sqrt{\pi}$
$\Gamma\left(\frac{3}{2}\right)$	$\Gamma\left(1+\frac{1}{2}\right) = \frac{1}{2}\Gamma\left(\frac{1}{2}\right) = \frac{1}{2}\sqrt{\pi}$
$\Gamma\left(\frac{5}{2}\right)$	$\frac{5}{2} \cdot \frac{3}{2} \cdot \frac{1}{2} \sqrt{\pi}$

LAPLACE TRANSFORMS – Updated List

$L[f(x)]$	$F(s)$
$L[1]$	$\dfrac{1}{s}$
$L[x]$	$\dfrac{1}{s^2}$
$L[e^{ax}]$	$\dfrac{1}{s-a}$
$L[\cos ax]$	$\dfrac{s}{s^2+a^2}$
$L[\sin ax]$	$\dfrac{a}{s^2+a^2}$
$L[e^{ax}\cos bx]$	$\dfrac{s-a}{(s-a)^2+b^2}$
$L[e^{ax}\sin bx]$	$\dfrac{b}{(s-a)^2+b^2}$
$L[x^2]$	$\dfrac{2}{s^3}$
$L[x^3]$	$\dfrac{3!}{s^4}$
$L[x^n]$	$\dfrac{n!}{s^{n+1}}\quad n = integer > 0$
$L[x^p]$	$\dfrac{\Gamma(p+1)}{s^{p+1}}\quad p = real\ number$
$L[x^n e^{ax}]$	$\dfrac{n!}{(s-a)^{n+1}}$

LAPLACE TRANSFORM of a Derivative

> **THEOREM:** **LAPLACE TRANSFORM OF** $f'(x)$
>
> For Initial Value Problem (I.V.P.) Only
>
> Let $f(x)$ be a continuous function
> and $f'(x)$ be a piecewise continuous function on $0 \leq x \leq A$
>
> Suppose $\exists$ three numbers: m, a, K s.t. $|f(x)| < Ke^{ax}$
> for all $x > m$
>
> Then:
> $$L[f'(x)] = s \cdot L[f(x)] - f'(0)$$

LAPLACE TRANSFORM OF $f'(x)$ -- A more general theorem.

Let $f(x), f'(x), f''(x), \ldots f^{(n-1)}(x)$ Be continuous functions.

and $f^{(n)}(x)$ is piecewise continuous on $0 < x < A$
(last derivative)

Suppose $\exists$ three numbers: K, a, m s.t.

All derivatives bounded by the same value for $x > m$

$$|f(x)| < Ke^{ax}$$
$$|f'(x)| < Ke^{ax}$$
$$|f^{(n-1)}(x)| < Ke^{ax}$$

Then: $L\left[f^{(n)}(x)\right]$ exists and

$$L\left[f^{(n)}(x)\right]$$
$$= s^n L[f(x)] - s^{n-1} \cdot f(0) - s^{n-2} \cdot f'(0) - s^{n-3} \cdot f''(0) - \ldots$$
$$- \ldots s^1 \cdot f^{(n-2)}(0) - f^{(n-1)}(0)$$

Examples:

$$L[y] = L[y]$$
$$L[y'] = s^1 L[y] - y(0)$$
$$L[y''] = s^2 L[y] - s^1 y(0) - y'(0)$$
$$L[y'''] = s^3 L[y] - s^2 y(0) - s^1 y'(0) - y''(0)$$

SOLVE IVP USING LAPLACE TRANSFORM (GENERAL IDEA)

Steps	General Approach: Using Laplace Transforms to solve a DE with Initial Value Problem (I.V.P.)
1.	Apply Laplace Transforms on both sides of the Differential Equation (DE).
2.	Obtain an algebraic equation by using relations between the Laplace Transform (L.T.) of y and its derivatives.
3.	Solve it.
4.	Setup expression as a Laplace Transform (LT.) of a known function.
5.	Use Partial Fractions.
6.	Apply Inverse Laplace Transforms.
7.	Get Particular Solution (P.S.) of the Initial Value Problem (I.V.P.)

Example – Solving a DE <u>without</u> using Laplace Transforms.

Find the Particular Solution of I.V.P. $y'' - 5y' + 6y = 0$

With: $y(0) = 0$ & $y'(0) = 2$

Solution:

$$y'' - 5y' + 6y = 0$$
$$r^2 - 5r + 6 = 0$$
$$(r-3)(r-2) = 0 \quad \rightarrow \quad r = 2, 3$$
$$\rightarrow \quad y_1 = e^{3x} \quad \text{and} \quad y_2 = e^{2x}$$

General Solution: $y = C_1 e^{3x} + C_2 e^{2x}$

$y(0) = 0 \qquad \rightarrow \quad 0 = C_1 + C_2$
$\qquad\qquad\qquad \rightarrow \quad C_1 = -C_2$

$y'(0) = 2 \qquad \rightarrow \quad 2 = 3C_1 + 2C_2 = 3C_1 - 2C_1 = C_1$
$\qquad\qquad\qquad \rightarrow \quad C_2 = -2$

Particular Solution: $y = 2e^{3x} - 2e^{2x}$

Next, do this problem with Laplace Transforms.

Example – Solving a DE using Laplace Transforms.

Find the Particular Solution of I.V.P. $y'' - 5y' + 6y = 0$

With: $y(0) = 0$ & $y'(0) = 2$

Solution:

$L[y''] - L[5y'] - L[6y] = L[0]$

$L[y''] - 5L[y'] - 6L[y] = L[0]$

$[s^2 L[y] - s\, y(0) - y'(0)] - 5[s\, L[y] - y(0)] + 6 L[y] = 0$

$[s^2 L[y] - 0 - 2] - 5[s\, L[y] - 0] + 6 L[y] = 0$

$s^2 L[y] - 2 - 5s\, L[y] + 6 L[y] = 0$

$s^2 L[y] - 5s\, L[y] + 6 L[y] = 2$

$L[y](s^2 - 5s + 6) = 2$

$L[y] = \dfrac{2}{s^2 - 5s + 6}$

$L[y] = \dfrac{2}{(s-3)(s-2)}$ Use Partial Fraction Decomposition.

$\dfrac{2}{(s-3)(s-2)} = \dfrac{A}{s-3} + \dfrac{B}{s-2}$ Find A and B

$2 = A(s-2) + B(s-3)$

$s = 3 \rightarrow 2 = A(1) \rightarrow A = 2$

$s = 2 \rightarrow 2 = B(-1) \rightarrow B = -2$

$L[y] = \dfrac{A}{s-3} + \dfrac{B}{s-2} = \dfrac{2}{s-3} - \dfrac{2}{s-2}$ (Continued ...)

$$L[y] = \frac{2}{s-3} - \frac{2}{s-2} \qquad \text{Recall: } L[e^{ax}] = \frac{1}{s-a}$$

Apply the Inverse Laplace Transform (I.L.T.)

$$L^{-1}[L(y)] = L^{-1}\left[\frac{2}{s-3}\right] - L^{-1}\left[\frac{2}{s-2}\right]$$

$$y = 2L^{-1}\left[\frac{1}{s-3}\right] - 2L^{-1}\left[\frac{1}{s-2}\right]$$

$$y = 2e^{3x} - 2e^{2x} \qquad \text{Same Answer!}$$

Example: Find the P.S. of $y'' - y = \sin 2x$

With: $y(0) = 0$ and $y'(0) = 1$

Solution: Using Laplace Transforms

(not Method of Undetermined Coeff.)

$L[y''] - L[y] = L[\sin 2x]$

$[s^2 L[y] - s\, y(0) - y'(0)] - L[y] = \dfrac{2}{s^2 + 4}$

$[s^2 L[y] - 0 - 1] - L[y] = \dfrac{2}{s^2 + 4}$

$(s^2 - 1)L[y] - 1 = \dfrac{2}{s^2 + 4}$

$(s^2 - 1)L[y] = \dfrac{2}{s^2 + 4} + \dfrac{s^2 + 4}{s^2 + 4}$

$(s^2 - 1)L[y] = \dfrac{s^2 + 6}{s^2 + 4}$

$L[y] = \dfrac{s^2 + 6}{(s^2 - 1)(s^2 + 4)} = \dfrac{s^2 + 6}{(s + 1)(s - 1)(s^2 + 4)}$

Use Partial Fractions to break down the RHS so it can be used with Laplace Transforms.

$$L[y] = \frac{s^2 + 6}{(s+1)(s-1)(s^2+4)}$$

$$\frac{s^2 + 6}{(s+1)(s-1)(s^2+4)} = \frac{A}{s+1} + \frac{B}{s-1} + \frac{Cs + D}{s^2 + 4}$$

$s^2 + 6$
$= A(s-1)(s^2+4) + B(s+1)(s^2+4) + (Cs+D)(s+1)(s-1)$

$s = 1 \;\rightarrow\; 7 = A(2)(5) \;\rightarrow\; A = \frac{7}{10}$

$s = -1 \;\rightarrow\; 7 = B(-2)(5) \;\rightarrow\; B = -\frac{7}{10}$

$s = 0 \;\rightarrow\; 6 = \frac{7}{10}(4) - \frac{7}{10}(-4) + (0+D)(-1)(1)$

$\;\rightarrow\; D = -\frac{4}{10}$

$s = 2 \;\rightarrow\; 10 = \frac{7}{10}(1)(8) - \frac{7}{10}(3)(8) + \left(2C - \frac{4}{10}\right)(3)(2)$

$10 = \frac{56}{10} - \frac{168}{10} + 12C - \frac{24}{10} \;\rightarrow\; C = \frac{59}{30}$

$$L[y] = \left(\frac{7}{10}\right)\frac{1}{s-1} - \left(\frac{7}{10}\right)\frac{1}{s+1} + \left(\frac{59}{30}\right)\frac{s}{s^2+4} - \left(\frac{2}{10}\right)\frac{2}{s^2+4}$$

Apply the Inverse Laplace Transform (I.L.T.)

$$y = \left(\frac{7}{10}\right)e^x - \left(\frac{7}{10}\right)e^{-x} + \left(\frac{59}{30}\right)\cos 2x - \left(\frac{2}{10}\right)\sin 2x$$

Example: Find the particular solution (P.S.) of the Initial Value Problem (I.V.P.) $y''' - 4y' = 5e^{3x}$

With: $y(0) = 1, \quad y'(0) = 0, \quad y''(0) = 2$

Solution:

Note: We could find the homogeneous solution first, to get y_1, y_2, y_3. Then, apply the initial conditions to the particular solution to get: C_1, C_2, C_3. But, this is very long and tedious! So, instead, use Laplace Transforms (L.P.) of derivatives.

Laplace Transforms of Derivatives		
$L[y']$	=	$sL[y] - y(0)$
$L[y'']$	=	$s^2 L[y] - s y(0) - y'(0)$
$L[y''']$	=	$s^3 L[y] - s^2 y(0) - s y'(0) - y''(0)$

Apply Laplace Transforms of Derivatives:

$y''' - 4y' = 5e^{3x}$

$L[y'''] - 4L[y'] = 5 L[e^{3x}]$

$\{s^3 L[y] - s^2 y(0) - s y'(0) - y''(0)\} - 4\{sL[y] - y(0)\} = 5\left\{\dfrac{1}{s-3}\right\}$

$s^3 L[y] - s^2 y(0) - s y'(0) - y''(0) - 4sL[y] + 4y(0) = \dfrac{5}{s-3}$

$L[y](s^3 - 4s) - s^2(1) - (2) + 4 = \dfrac{5}{s-3}$

$L[y](s^3 - 4s) - s^2 + 2 = \dfrac{5}{s-3}$

$$L[y](s^3 - 4s) = \frac{5}{s-3} + s^2 - 2$$

$$L[y](s^3 - 4s) = \frac{5}{s-3} + \frac{(s^2-2)(s-3)}{s-3}$$

$$L[y] = \frac{5 + (s^2-2)(s-3)}{(s-3)(s^3-4s)} = \frac{5 + s^3 - 3s^2 - 2s + 6}{(s-3)(s^3-4s)}$$

$$L[y] = \frac{s^3 - 3s^2 - 2s + 11}{(s-3)(s^3-4s)} = \frac{s^3 - 3s^2 - 2s + 11}{s(s-3)(s^2-4)}$$

$$L[y] = \frac{s^3 - 3s^2 - 2s + 11}{s(s+2)(s-2)(s-3)}$$

Partial Fraction Decomposition:

$$\frac{s^3 - 3s^2 - 2s + 11}{s(s+2)(s-2)(s-3)} = \frac{A}{s} + \frac{B}{x+2} + \frac{C}{s-2} + \frac{D}{s-3}$$

$s = 0 \quad \rightarrow \quad 11 = A(2)(-2)(-3) = A(12) \quad \rightarrow \quad A = \frac{11}{12}$

$s = -2 \quad \rightarrow \quad -8 - 12 + 4 + 11 = B(-2)(-4)(-5) = B(-40)$

$\quad \rightarrow \quad B = \frac{1}{8}$

$s = 2 \quad \rightarrow \quad C = -\frac{3}{8}$

$s = 3 \quad \rightarrow \quad D = \frac{1}{3}$

$$L[y] = \frac{A}{s} + \frac{B}{x+2} + \frac{C}{s-2} + \frac{D}{s-3}$$

$$L[y] = \left(\frac{11}{12}\right)\frac{1}{s} + \left(\frac{1}{8}\right)\frac{1}{s+2} - \left(\frac{3}{8}\right)\frac{1}{s-2} + \left(\frac{1}{3}\right)\frac{1}{s-3}$$

$$L[y] = \left(\frac{11}{12}\right)\frac{1}{s} + \left(\frac{1}{8}\right)\frac{1}{s+2} - \left(\frac{3}{8}\right)\frac{1}{s-2} + \left(\frac{1}{3}\right)\frac{1}{s-3}$$

$$L[y] = \left(\frac{11}{12}\right)L[1] + \left(\frac{1}{8}\right)L[e^{-2x}] - \left(\frac{3}{8}\right)L[e^{2x}] + \left(\frac{1}{3}\right)L[e^{3x}]$$

Apply Inverse Laplace Transforms (I.L.T.)

$$L^{-1}[L[y]] = y = \left(\frac{11}{12}\right)[1] + \left(\frac{1}{8}\right)[e^{-2x}] - \left(\frac{3}{8}\right)[e^{2x}] + \left(\frac{1}{3}\right)[e^{3x}]$$

$$y = \frac{11}{12} + \frac{e^{-2x}}{8} - \frac{3\,e^{2x}}{8} + \frac{e^{3x}}{3} \qquad\qquad \text{P.S.}$$

Particular Solution of the DE.

STEP FUNCTION

"UNIT STEP" Function = "HEAVYCYCLE" Function = $U_c(x)$

$$U_c(x) = \begin{cases} 1 & x \geq c \\ 0 & x < c \end{cases} \quad \text{Where } c \geq 0 \quad \text{(positive)}$$

Graph:

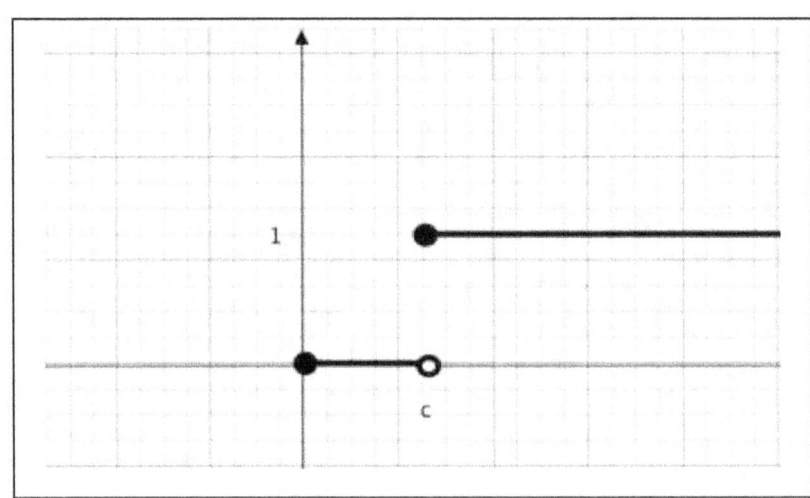

Adding and subtracting the zero part does not affect the result.

$U_2(x) =$ For addition and subtraction		
	$U_2(x) = \begin{cases} 1 & x \geq 2 \\ 0 & x < 1 \end{cases}$	$= U_2(x)$

Note: The RHS of a DE may be a Piecewise Continuous Function!

Examples: Some Heavyside Functions (Step Functions)

Note: $x > 0$ (positive)

#	$U_c(x)$	Graph
1.	$U_0(x) = 1$	
2.	$U_1(x) = \begin{cases} 1 & x \geq 1 \\ 0 & x < 1 \end{cases}$	
3.	$U_\pi(x) = \begin{cases} 1 & x \geq \pi \\ 0 & \text{else} \end{cases}$	

Negative Step Function, $U_c(x)$

Example: $1 - U_2(x) = U_0(x) - U_2(x)$

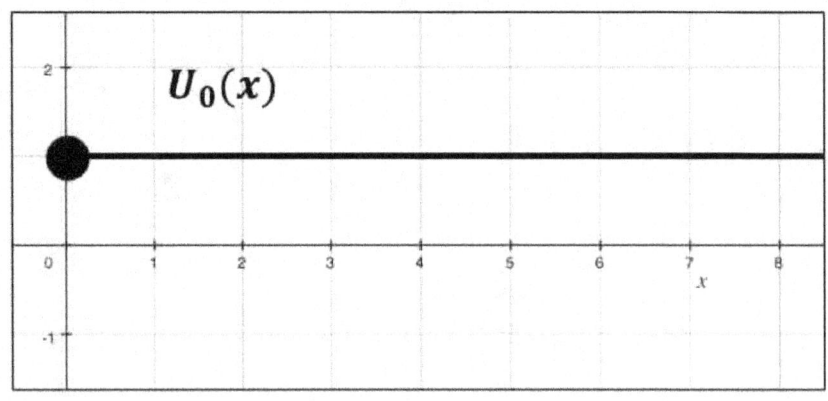

−

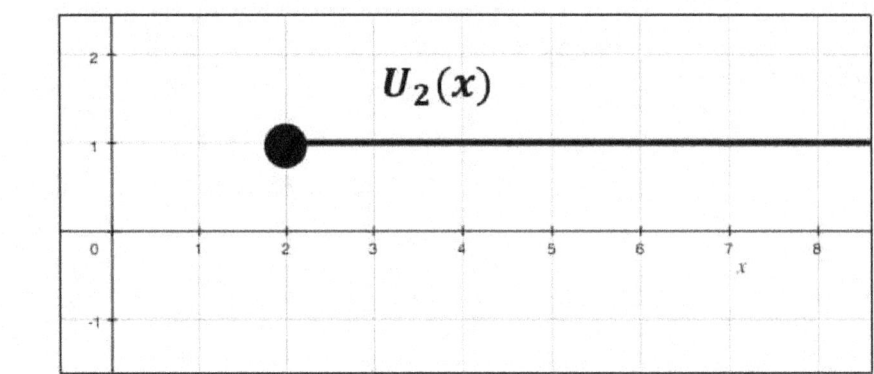

=

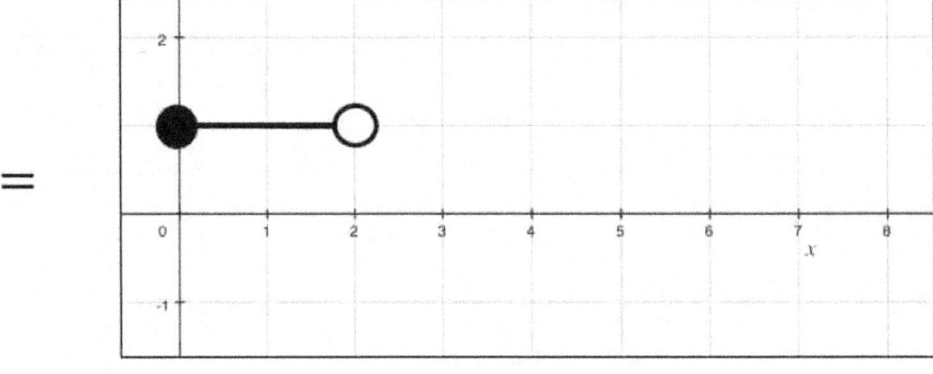

Example: Sketch the graph of: $U_2(x) - U_7(x)$

Solution:

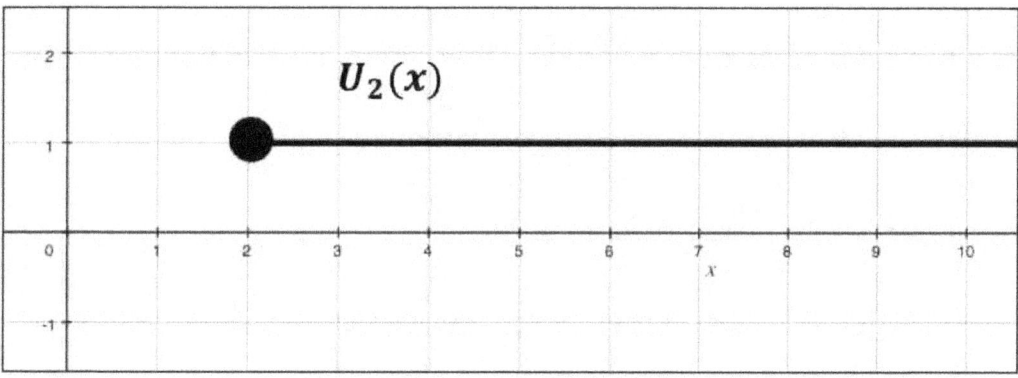

$-$

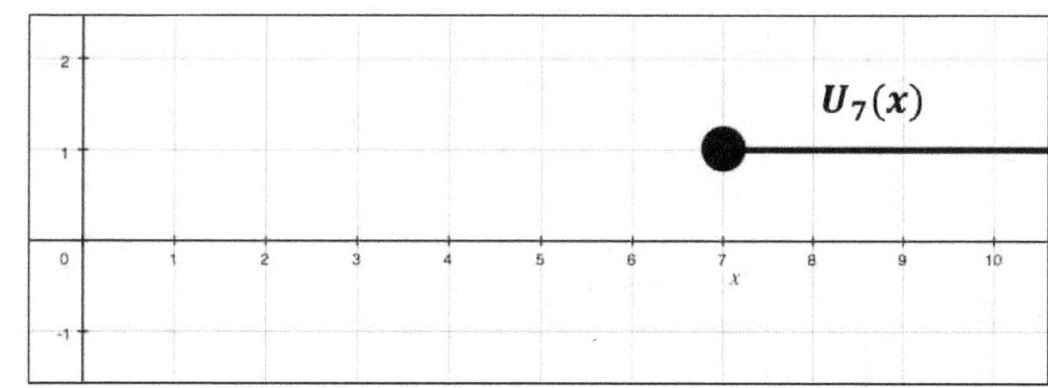

$=$

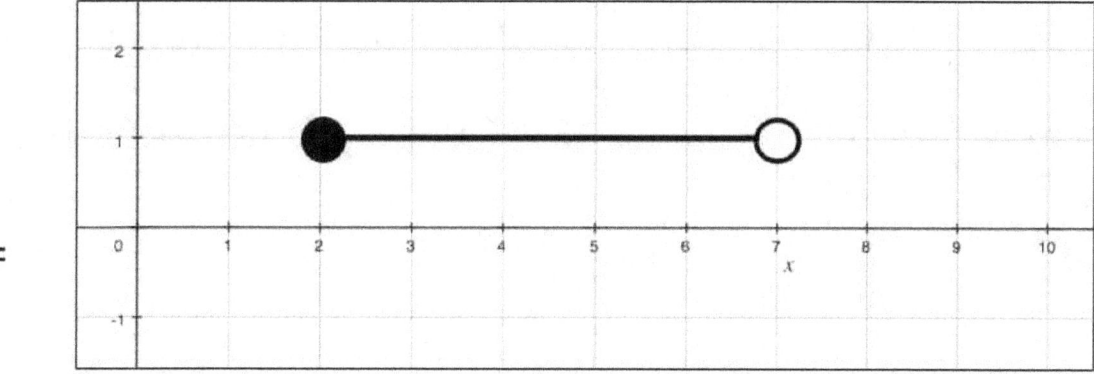

Example: Write the following function as a Linear Combination of Unit Step Functions.

$$f(x) = \begin{cases} \left(\frac{1}{2}\right) & 1 \leq x < 2 \\ 3 & 2 \leq x < 4 \\ 5 & 4 \leq x < 7 \end{cases}$$

Solution: Sketch it first.

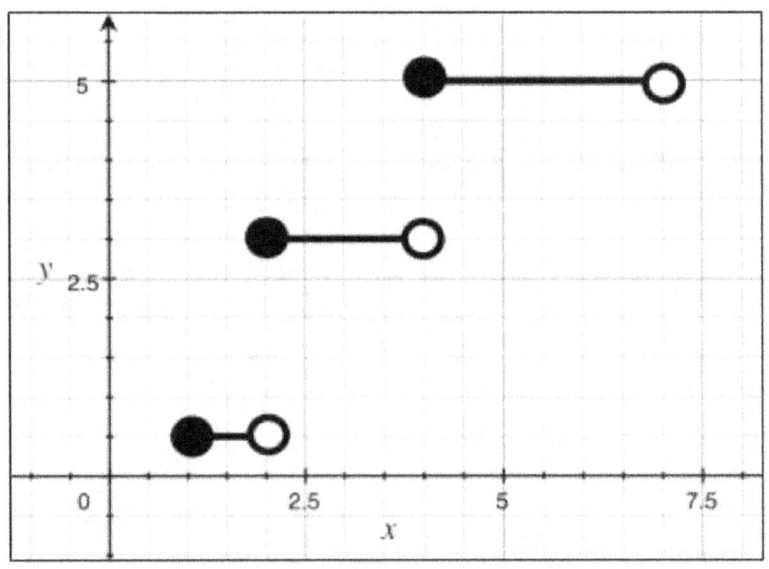

$$f(x) = \frac{1}{2}[U_1(x) - U_2(x)] + 3[U_2(x) - U_4(x)] + 5[U_4(x) - U_7(x)]$$

$$f(x) = \frac{1}{2}U_1(x) - \frac{1}{2}U_2(x) + 3U_2(x) - 3U_4(x) + 5U_4(x) - 5U_7(x)$$

Combine "Like" terms.

$$f(x) = \frac{1}{2}U_1(x) + \frac{5}{2}U_2(x) + 2U_4(x) - 5U_7(x)$$

Laplace Transform of a Unit Step Function:

$$L[\,U_c(x)\,] = \int_0^\infty U_c(x) \cdot e^{-sx}\, dx$$

Recall: $U_c(x) = \begin{cases} 1 & x \geq c \\ 0 & x < c \end{cases}$ Where $c \geq 0$ (positive)

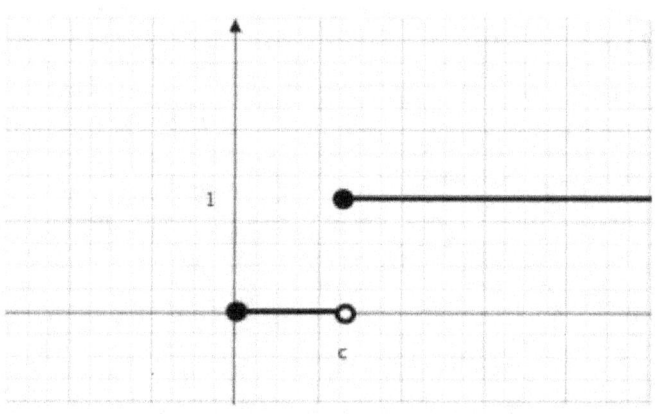

$L[\,U_c(x)\,] = \int_0^c U_c(x) \cdot e^{-sx}\, dx + \int_c^\infty U_c(x) \cdot e^{-sx}\, dx$

$L[\,U_c(x)\,] = \int_0^c 0 \cdot e^{-sx}\, dx + \int_c^\infty 1 \cdot e^{-sx}\, dx$

$L[\,U_c(x)\,] = \int_c^\infty e^{-sx}\, dx = \lim_{t \to \infty} \int_c^t e^{-sx}\, dx$

$L[\,U_c(x)\,] = \lim_{t \to \infty} \left[\dfrac{e^{-sx}}{-s}\right]_c^t = \lim_{t \to \infty} \left[\dfrac{e^{-st}}{-s} - \dfrac{e^{-sc}}{-s}\right]$

$L[\,U_c(x)\,] = \left[0 + \dfrac{e^{-sc}}{s}\right] = \dfrac{e^{-sc}}{s}$

Laplace Transformation of Unit Step Function:

$$L[\,U_c(x)\,] = \dfrac{e^{-sc}}{s}$$

TRANSLATION:

Horizontal Shift to the Right by "C" Units. $f(x) \rightarrow U_c(x) \cdot f(x)$

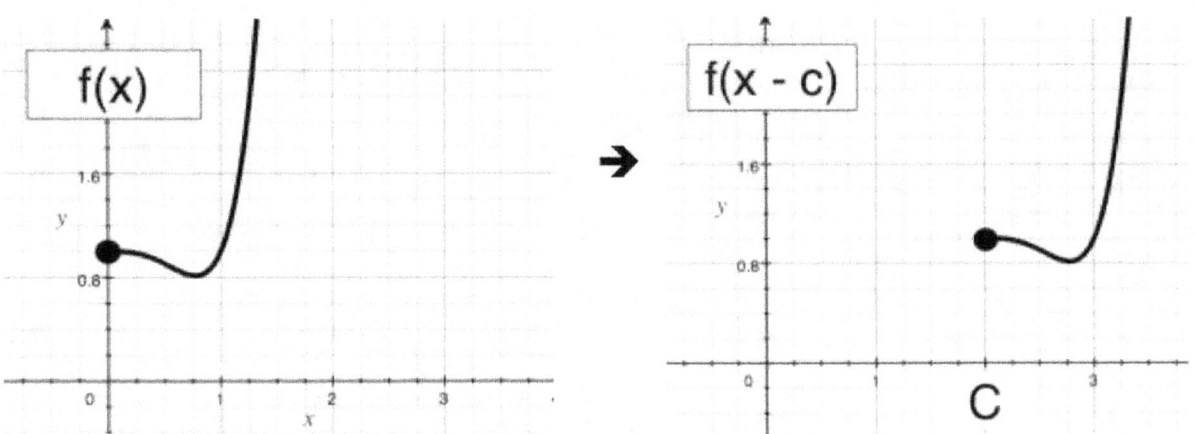

THEOREM:

Let $L[f(x)] = F(s)$ For any $s > a \geq 0$

$$L[U_c(x) \cdot f(x-c)] = e^{-sc} \cdot L[f(x)]$$
$$L[U_c(x) \cdot f(x-c)] = e^{-sc} \cdot F(s)$$

Conversely:

$$L^{-1}[e^{-sc} \cdot L[f(x)]] = U_c(x) \cdot f(x-c)$$

Example (1) Compute the Inverse Laplace Translation (I.L.T) of:

$$F(s) = \frac{2s + 7}{s^2 + 4s + 8}$$

Solution:

Write the right-hand side (RHS) as an Inverse Laplace Translation.

$$F(s) = \frac{2s + 7}{s^2 + 4s + 4 - 4 + 8}$$

$$F(s) = \frac{2s + 7}{(s + 2)^2 + 4} = \frac{2s + 7}{(s + 2)^2 + 2^2}$$

$$F(s) = \frac{2(s+2) - 4 + 7}{(s + 2)^2 + 2^2}$$

$$F(s) = \frac{2(s+2)}{(s + 2)^2 + 2^2} + \frac{3}{(s + 2)^2 + 2^2}$$

$$F(s) = \frac{2(s+2)}{(s + 2)^2 + 2^2} + \frac{1}{2}\left[\frac{3 \cdot 2}{(s + 2)^2 + 2^2}\right]$$

$$F(s) = \frac{2(s+2)}{(s + 2)^2 + 2^2} + \frac{3}{2}\left[\frac{2}{(s + 2)^2 + 2^2}\right] \quad \text{See Tables}$$

$$F(s) = 2\, L[e^{-2x} \cdot \cos 2x] + \frac{3}{2} L[e^{-2x} \cdot \sin 2x]$$

Apply the Inverse Laplace Transform (I.L.T)

$$L^{-1}[F(s)] = 2\, e^{-2x} \cdot \cos 2x + \frac{3}{2} e^{-2x} \cdot \sin 2x$$

$$f(x) = 2\, e^{-2x} \cos 2x + \frac{3}{2} e^{-2x} \sin 2x$$

Example (2) Compute the Inverse Laplace Translation (I.L.T) of:

$$F(s) = \frac{1 - e^{-7s}}{s^4}$$

Solution:

$$F(s) = \frac{1}{s^4} - e^{-7s}\left(\frac{1}{s^4}\right)$$

$$F(s) = \left(\frac{1}{6}\right)\frac{3!}{s^4} - e^{-7s}\left(\frac{1}{6}\right)\left(\frac{3!}{s^4}\right)$$

$$F(s) = \left(\frac{1}{6}\right)L[x^3] - \left(\frac{1}{6}\right)e^{-7s}L[x^3]$$

Apply the Inverse Laplace Translation (I.L.T)

$$L^{-1}[F(s)] = f(x) = \left(\frac{1}{6}\right)x^3 - \left(\frac{1}{6}\right)L^{-1}[e^{-7s}L[x^3]]$$

Recall: Step Function with shift.

$$L^{-1}[e^{-sc} \cdot L[f(x)]] = U_c(x) \cdot f(x - c)$$

$$f(x) = \left(\frac{1}{6}\right)x^3 - \left(\frac{1}{6}\right)L^{-1}[e^{-7s}L[x^3]]$$

$$f(x) = \left(\frac{1}{6}\right)x^3 - \left(\frac{1}{6}\right)U_7(x) \cdot f(x - 7)^3$$

Example (3) Compute the Inverse Laplace Translation (I.L.T) of:

$$F(s) = \frac{4 - e^{-7s}}{s^2 + 9}$$

Solution:

$$F(s) = \frac{4}{s^2 + 9} - e^{-3s} \cdot \frac{1}{s^2 + 9}$$

$$F(s) = \left(\frac{4}{3}\right) \frac{3}{s^2 + 3^2} - \frac{e^{-3s}}{3} \cdot \frac{3}{s^2 + 3^2}$$

$$F(s) = \frac{4}{3} L[\sin 3x] - \frac{e^{-3s}}{3} L[\sin 3x]$$

Apply I.L.T.

>SHIFT

$$L^{-1}[F(s)] = f(x) = \frac{4}{3} \sin 3x - \left(\frac{1}{3}\right) e^{-3s} L[\sin 3x]$$

$$f(x) = \frac{4}{3} \sin 3x - \left(\frac{1}{3}\right) U_3(x) \cdot \sin(3(x-3))$$

$$(x) = \frac{4}{3} \sin 3x - \left(\frac{1}{3}\right) U_3(x) \cdot \sin(3x - 9)$$

Example (4) Compute the Inverse Laplace Translation (I.L.T) of:

$$F(s) = \frac{2s + 3 - e^{-5s}}{s^2 + 2s + 2}$$

Solution:

$$F(s) = \frac{2(s+1) + 1 - e^{-5s}}{(s+1)^2 + (1)^2}$$

$$F(s) = \frac{2(s+1)}{(s+1)^2 + (1)^2} + \frac{1}{(s+1)^2 + (1)^2} - e^{-5s} \cdot \frac{1}{(s+1)^2 + (1)^2}$$

$$F(s) = \left(\frac{1}{2}\right) \frac{(s+1)}{(s+1)^2 + (1)^2} + \frac{1}{(s+1)^2 + (1)^2} - e^{-5s} \cdot \frac{1}{(s+1)^2 + (1)^2}$$

$$F(s) = \left(\frac{1}{2}\right) L[e^{-x} \cos x] + L[e^{-x} \sin x] - e^{-5s} \cdot L[e^{-x} \sin x]$$

Apply I.L.T.

$$L^{-1}[F(s)] = f(x)$$

$$f(x) = \left(\frac{1}{2}\right) e^{-x} \cos x + e^{-x} \sin x - U_5(x) \cdot e^{-(x-5)} \cdot \sin(x - 5)$$

$$f(x) = \left(\frac{1}{2}\right) e^{-x} \cos x + e^{-x} \sin x - U_5(x) \cdot e^{5-x} \cdot \sin(x - 5)$$

Example (1) Express the function as a <u>Linear Combination of a Step Function</u> and compute the <u>Laplace Transform</u>.

$$f(x) = \begin{cases} 1 & 0 \le x < 2 \\ 2 & 2 \le x < 4 \\ 3 & 4 \le x < 6 \end{cases}$$

Solution: Sketch it first.

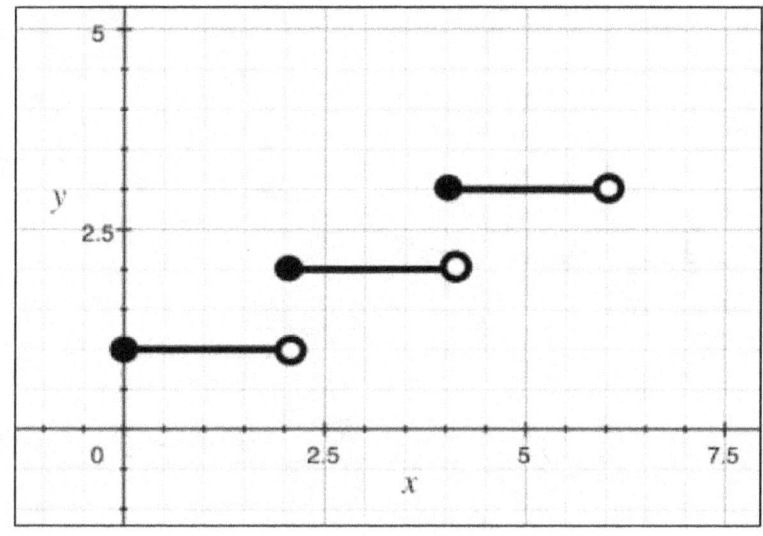

$f(x) = 1\big(U_0(x) - U_2(x)\big) + 2\big(U_2(x) - U_4(x)\big) + 3\big(U_4(x) - U_6(x)\big)$

$f(x) = 1 - U_2(x) + 2U_2(x) - 2U_4(x) + 3U_4(x) - 3U_6(x)$

$f(x) = 1 + U_2(x) + U_4(x) - 3U_6(x)$ Linear Combination
of Step Function

Laplace Transform of the Linear Combination

$L[f(x)] = L[1] + L[U_2(x)] + L[U_4(x)] - 3L[U_6(x)]$

$L[f(x)] = \dfrac{1}{s} + \dfrac{e^{-2s}}{s} + \dfrac{e^{-6s}}{s} + \dfrac{e^{-4s}}{s} - (3)\dfrac{e^{-6s}}{s}$

Example (2) Express the function as a Linear Combination of a Step Function and compute the Laplace Transform.

$$f(x) = \begin{cases} 0 & 0 \leq x < 5 \\ \dfrac{x-5}{5} & 5 \leq x < 10 \\ 1 & 10 \leq x \end{cases}$$

Solution: Sketch it. Note: It is continuous.   Recall: $U_0(x) = 1$

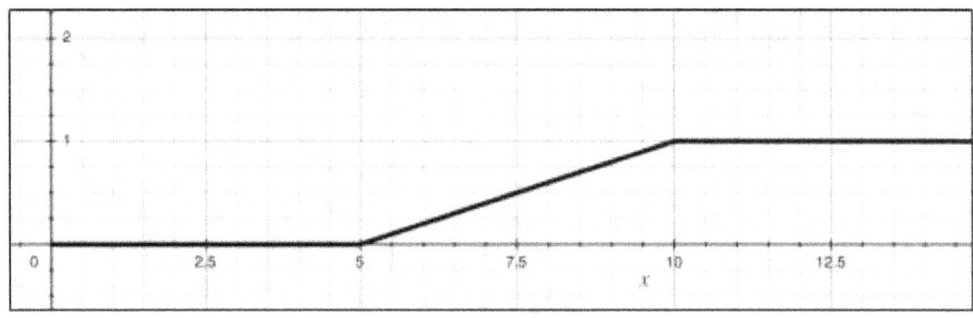

$$f(x) = 0 \cdot [u_0(x) - U_5(x)] + \left(\frac{x-5}{5}\right) \cdot [U_5(x) - U_{10}(x)] + U_{10}(x)$$

$$f(x) = \left(\frac{x-5}{5}\right) U_5(x) - \left(\frac{x-5}{5}\right) U_{10}(x) + \left(\frac{5}{5}\right) U_{10}(x)$$

$$f(x) = \left(\frac{x-5}{5}\right) U_5(x) - \left(\frac{x-10}{5}\right) U_{10}(x)$$

$$L[f(x)] = L\left[\left(\frac{x-5}{5}\right) U_5(x)\right] - L\left[\left(\frac{x-10}{5}\right) U_{10}(x)\right]$$

$$L[f(x)] = \frac{1}{5} \{ L[U_5(x) \cdot (x-5)] - L[U_{10}(x) \cdot (x-10)] \}$$

$$L[f(x)] = \frac{1}{5} \{ e^{-5s} L[x] - e^{-10s} L[x] \}$$

$$L[f(x)] = \frac{1}{5} \left\{ e^{-5s} \left(\frac{1}{s^2}\right) - e^{-10s} \left(\frac{1}{s^2}\right) \right\}$$

$$L[f(x)] = \frac{1}{5} \left\{ \frac{e^{-5s}}{s^2} - \frac{e^{-10s}}{s^2} \right\}$$

Example (3) Express the function as a <u>Linear Combination of a Step Function</u> and compute the <u>Laplace Transform</u>.

$$f(x) = \begin{cases} 0 & 0 \le x < 2 \\ x & 2 \le x < 6 \\ 1 & 6 \le x \end{cases}$$

Solution: Sketch it. Note: It is Piecewise Continuous.

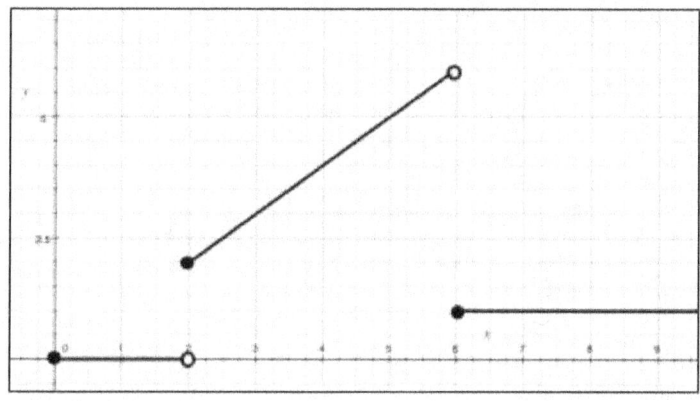

$f(x) = 0[U_0 - U_2] + x[U_2 - U_6] + 1[U_6]$

$f(x) = xU_2 - xU_6 + U_6$

$f(x) = xU_2 - xU_6 + U_6$

$f(x) = U_2(x - 2 + 2) - U_6(x - 6 + 6) + U_6$

$f(x) = U_2(x - 2) + 2U_2 - U_6(x - 6) - 6U_6 + U_6$

$f(x) = U_2(x - 2) + 2U_2 - U_6(x - 6) - 5U_6$

$L[f(x)] = e^{-2s} L[x] + 2 \cdot \dfrac{e^{-2x}}{s} - e^{-6s} L[x] - 5 \cdot \dfrac{e^{-6s}}{s}$

$L[f(x)] = e^{-2s} \left(\dfrac{1}{s^2}\right) + 2 \cdot \dfrac{e^{-2x}}{s} - e^{-6s} \left(\dfrac{1}{s^2}\right) - 5 \cdot \dfrac{e^{-6s}}{s}$

$L[f(x)] = \dfrac{e^{-2s}}{s^2} + \dfrac{2 e^{-2x}}{s} - \dfrac{e^{-6s}}{s^2} - \dfrac{5 e^{-6s}}{s}$

Example (4) Express the function as a <u>Linear Combination of a Step Function</u> and compute the <u>Laplace Transform</u>.

$$g(x) = \begin{cases} e^x & 0 \leq x < 2 \\ 1 & x \geq 2 \end{cases}$$

Solution: Sketch it.

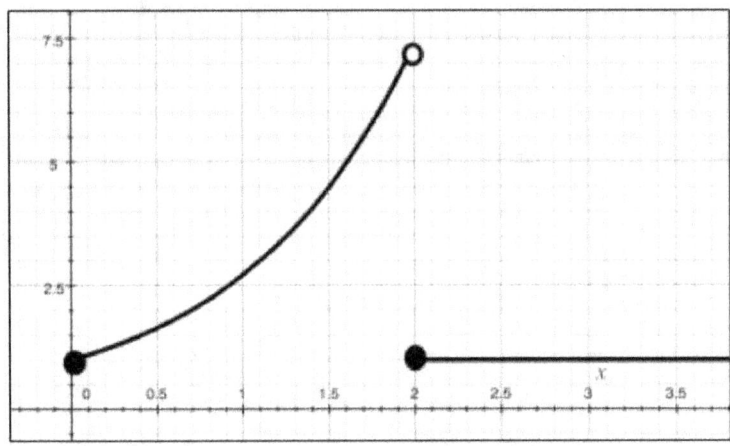

$g(x) = e^x[U_0 - U_2] + U_2$

$g(x) = e^x U_0 - e^x U_2 + U_2$

$g(x) = e^x - e^x U_2 + U_2$

$g(x) = e^x - e^x U_2 \cdot [e^{x-2} \cdot e^{-(x-2)}] + U_2$

$g(x) = e^x - e^{x-x+2} U_2 \cdot e^{x-2} + U_2$

$g(x) = e^x - e^2 U_2 \cdot e^{x-2} + U_2$ Note: e^2 is just a constant.

$L[g(x)] = \dfrac{1}{s-1} - e^2 \cdot e^{-2s} L[e^x] + \dfrac{e^{-2s}}{s}$

$L[g(x)] = \dfrac{1}{s-1} - e^2 \cdot e^{-2s} \left[\dfrac{1}{s-1}\right] + \dfrac{e^{-2s}}{s}$

$L[g(x)] = \dfrac{1}{s-1} - e^2 \cdot \dfrac{e^{-2s}}{s-1} + \dfrac{e^{-2s}}{s}$

Example (5) Express the function as a <u>Linear Combination of a Step Function</u> and compute the <u>Laplace Transform</u>.

$$g(x) = \begin{cases} \sin 2x & 0 \leq x < \frac{\pi}{4} \\ 1 + \cos(2x) & x \geq \frac{\pi}{4} \end{cases}$$

Solution: Sketch it.

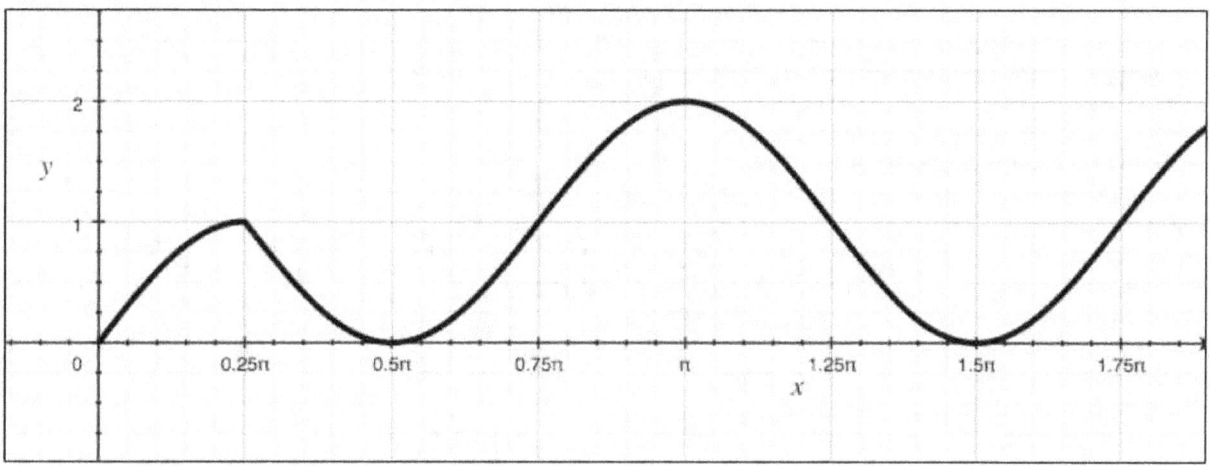

$g(x) = [\sin 2x]\left(U_0 - U_{\frac{\pi}{4}}\right) + [1 + \cos(2x)]U_{\frac{\pi}{4}}$

$g(x) = \sin 2x - \sin 2x \cdot U_{\frac{\pi}{4}} + U_{\frac{\pi}{4}} + \cos(2x) U_{\frac{\pi}{4}}$

$g(x) = \sin 2x + U_{\frac{\pi}{4}} - \cos\left(\frac{\pi}{2} - 2x\right) U_{\frac{\pi}{4}} + \sin\left(\frac{\pi}{2} - 2x\right) U_{\frac{\pi}{4}}$

$g(x) = \sin 2x + U_{\frac{\pi}{4}} \ldots$

$\qquad - \cos\left(-2\left(x - \frac{\pi}{4}\pi\right)\right) U_{\frac{\pi}{4}} + \sin\left(-2\left(x - \frac{\pi}{4}\pi\right)\right) U_{\frac{\pi}{4}}$

$g(x) = \sin 2x + U_{\frac{\pi}{4}} - \cos\left(2\left(x - \frac{\pi}{4}\pi\right)\right) U_{\frac{\pi}{4}} - \sin\left(2\left(x - \frac{\pi}{4}\pi\right)\right) U_{\frac{\pi}{4}}$

Use: $\cos(-\theta) = \cos(\theta)$ and $\sin(-\theta) = -\sin(\theta)$

(Continued ...)

$$g(x) = \sin 2x + U_{\frac{\pi}{4}} - \cos\left(2\left(x - \frac{\pi}{4}\pi\right)\right) U_{\frac{\pi}{4}} - \sin\left(2\left(x - \frac{\pi}{4}\pi\right)\right) U_{\frac{\pi}{4}}$$

$$g(x) = \sin 2x + U_{\frac{\pi}{4}} - U_{\frac{\pi}{4}} \cos\left(2\left(x - \frac{\pi}{4}\pi\right)\right) - U_{\frac{\pi}{4}} \sin\left(2\left(x - \frac{\pi}{4}\pi\right)\right)$$

Apply the Laplace Transform (L.T.)

$$L[g(x)] = L[\sin 2x] + L\left[U_{\frac{\pi}{4}}\right] \ldots$$

$$- L\left[U_{\frac{\pi}{4}} \cos\left(2\left(x - \frac{\pi}{4}\pi\right)\right)\right] - L\left[U_{\frac{\pi}{4}} \sin\left(2\left(x - \frac{\pi}{4}\pi\right)\right)\right]$$

$$L[g(x)] = \frac{2}{s^2+4} + \frac{e^{-\frac{\pi s}{4}}}{s} - e^{-\frac{\pi s}{4}} L[\cos(2x)] - e^{-\frac{\pi s}{4}} L[\sin(2x)]$$

$$L[g(x)] = \frac{2}{s^2+4} + \frac{e^{-\frac{\pi s}{4}}}{s} - e^{-\frac{\pi s}{4}}\left(\frac{s}{s^2+4}\right) - e^{-\frac{\pi s}{4}}\left(\frac{2}{s^2+4}\right)$$

Step	How to Solve an Initial Value Problem (I.V.P) Involving a Step Function (General Idea)
1.	Write the piecewise continuous function as a linear combination of step functions.
2.	Apply the Laplace Transform on both sides of the DE.
3.	Get an algebraic equation function that is multiplied with e^{-cs} term. Label it $H_i(s)$.
4.	Compute the Inverse Laplace Transform (I.L.T.) of each $H_i(s)$ to get $h_i(x)$.
5.	Apply the Inverse Laplace Transform (I.L.T.) on both sides of the algebraic equation and compute the Particular Solution (P.S.).

Example: Find the P.S. of I.V.P. $y'' - 5y' - 6y = f(x)$

Find the Particular Solution of the Initial Value Problem.

$$f(x) = \begin{cases} 1 & 0 \leq x < 2 \\ 2 & 2 \leq x < 4 \\ 3 & 4 \leq x < 6 \end{cases} \quad y(0) = 0 \text{ and } y'(0) = 0$$

Solution:

Write in terms of Linear Combination of Step Function.

This was done previously.

(Step 1)

$f(x) = 1 + U_2(x) + U_4(x) - 3U_6(x)$

(Step 2)

$y'' - 5y' - 6y = 1 + U_2(x) + U_4(x) - 3U_6(x)$

$L[y''] - 5L[y'] - 6L[y] = L[1] + L[U_2] + L[U_4] - 3L[U_6]$

$\{s^2 L[y] - s \cdot y(0) - y'(0)\} - 5\{sL[y] - y(0)\} - 6L[y]$

$$= \frac{1}{s} + \frac{e^{-2s}}{s} + \frac{e^{-4s}}{s} - 3\frac{e^{-6s}}{s}$$

$\{s^2 L[y] - s \cdot 0 - 0\} - 5\{sL[y] - 0\} - 6L[y]$

$$= \frac{1}{s} + \frac{e^{-2s}}{s} + \frac{e^{-4s}}{s} - 3\frac{e^{-6s}}{s}$$

$$s^2 L[y] - 5s L[y] - 6L[y] = \frac{1}{s} + \frac{e^{-2s}}{s} + \frac{e^{-4s}}{s} - 3\frac{e^{-6s}}{s}$$

$$L[y](s^2 - 5s - 6) = \frac{1}{s} + \frac{e^{-2s}}{s} + \frac{e^{-4s}}{s} - 3\frac{e^{-6s}}{s}$$

$$L[y](s+1)(s-6) = \frac{1 + e^{-2s} + e^{-4s} - 3e^{-6s}}{s}$$

$$L[y] = \frac{1 + e^{-2s} + e^{-4s} - 3e^{-6s}}{s(s+1)(s-6)} \quad \rightarrow \text{ Let } H_s = \frac{1}{s(s+1)(s-6)}$$

$$L[y] = (1 + e^{-2s} + e^{-4s} - 3e^{-6s}) \cdot H_s$$

(Step 3) Find the fractional decomposition of H_s

$$H_s = \frac{1}{s(s+1)(s-6)} = \frac{A}{s} + \frac{B}{s+1} + \frac{C}{s-6}$$

$$1 = A(s+1)(s-6) + B(s)(s-6) + C(s)(s+1)$$

$s = 0 \quad \rightarrow 1 = A(1)(-6) \quad \rightarrow A = -\frac{1}{6}$

$s = -1 \quad \rightarrow 1 = B(-1)(-7) \quad \rightarrow B = \frac{1}{7}$

$s = 6 \quad \rightarrow 1 = C(6)(7) \quad \rightarrow C = \frac{1}{42}$

$$H_s = \left(-\frac{1}{6}\right) \cdot \frac{1}{s} + \left(\frac{1}{7}\right) \cdot \frac{1}{s+1} + \left(\frac{1}{42}\right) \cdot \frac{1}{s-6} \quad \text{Apply I.L.T.}$$

$$L^{-1}[H_s] = \left(-\frac{1}{6}\right) L^{-1}\left[\frac{1}{s}\right] + \left(\frac{1}{7}\right) L^{-1}\left[\frac{1}{s+1}\right] + \left(\frac{1}{42}\right) L^{-1}\left[\frac{1}{s-6}\right]$$

$$h(x) = \left(-\frac{1}{6}\right)(1) + \left(\frac{1}{7}\right)(e^{-x}) + \left(\frac{1}{42}\right) e^{-(-6x)}$$

$$h(x) = -\frac{1}{6} + \left(\frac{1}{7}\right)(e^{-x}) + \left(\frac{1}{42}\right) e^{6x}$$

(Step 4) Put H_s back into the equation.

$L[y] = (1 + e^{-2s} + e^{-4s} - 3e^{-6s}) \cdot H_s$

$L[y] = H_s + e^{-2s} H_s + e^{-4s} H_s - 3e^{-6s} H_s$

Apply I.L.T.

$y = L^{-1}[H_s] + L^{-1}[e^{-2s} H_s] + L^{-1}[e^{-4s} H_s] - 3L^{-1}[e^{-6s} H_s]$

$y = h(x) + U_2 \cdot h(x-2) + U_4 \cdot h(x-4) - 3 U_6 \cdot h(x-6)$

Where: $h(x) = -\frac{1}{6} + \left(\frac{1}{7}\right)(e^{-x}) + \left(\frac{1}{42}\right) e^{6x}$

Note: Most of the terms on the RHS are horizontal shifts.

Example: Find the Particular Solution of the Initial Value Problem.

$$y'' + 4y = g(x) \qquad \text{With: } y(0) = y'(0) = 0$$

$$g(x) = \begin{cases} 0 & 0 \le x < 5 \\ \dfrac{x-5}{5} & 5 \le x < 10 \\ 1 & 10 \le x \end{cases}$$

Solution:

Previously, we found the RHS:

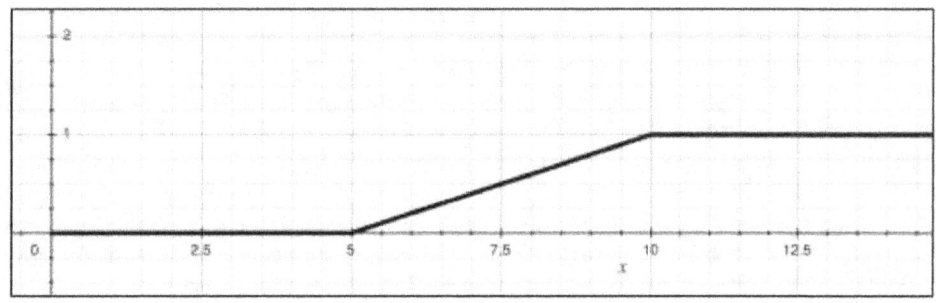

$$g(x) = 0 \cdot [u_0(x) - U_5(x)] + \left(\frac{x-5}{5}\right)[U_5(x) - U_{10}(x)] + U_{10}(x)$$

$$g(x) = \left(\frac{x-5}{5}\right)U_5(x) - \left(\frac{x-5}{5}\right)U_{10}(x) + \left(\frac{5}{5}\right)U_{10}(x)$$

$$g(x) = \left(\frac{x-5}{5}\right)U_5(x) - \left(\frac{x-10}{5}\right)U_{10}(x)$$

$$L[g(x)] = L\left[\left(\frac{x-5}{5}\right)U_5(x)\right] - L\left[\left(\frac{x-10}{5}\right)U_{10}(x)\right]$$

$$L[g(x)] = \frac{1}{5}\{L[U_5(x)\cdot(x-5)] - L[U_{10}(x)\cdot(x-10)]\}$$

$$L[g(x)] = \frac{1}{5}\{e^{-5s}L[x] - e^{-10s}L[x]\}$$

$$L[g(x)] = \frac{1}{5}\left\{e^{-5s}\left(\frac{1}{s^2}\right) - e^{-10s}\left(\frac{1}{s^2}\right)\right\}$$

$$L[g(x)] = \frac{1}{5}\left\{\frac{e^{-5s}}{s^2} - \frac{e^{-10s}}{s^2}\right\}$$

$$y'' - y' - 12y = g(x)$$
$$L[y''] - L[y'] - 12L[y] = L[g(x)]$$
$$\{s^2 L[y] - s \cdot y(0) - y'(0)\} - \{s L[y] - y(0)\} - 12L[y] = L[g(x)]$$
$$s^2 L[y] - s L[y] - 12L[y] = L[g(x)]$$
$$L[y] \cdot (s^2 - s - 12) = L[g(x)]$$
$$L[y] \cdot (s+3)(s-4) = L[g(x)] \qquad L[g(x)] \text{ found previously.}$$

$$L[y] \cdot (s+3)(s-4) = \frac{1}{s-1} - e^2 \cdot \frac{e^{-2s}}{s-1} + \frac{e^{-2s}}{s}$$

$$L[y] = \left(\frac{1}{s-1} - e^2 \cdot \frac{e^{-2s}}{s-1} + \frac{e^{-2s}}{s} \right) \cdot \left(\frac{1}{(s+3)(s-4)} \right)$$

$$L[y] = \left(\frac{1 - e^{-2s+2}}{s-1} + \frac{e^{-2s}}{s} \right) \cdot \left(\frac{1}{(s+3)(s-4)} \right)$$

$$L[y] = (1 - e^{-2s+2}) \frac{1}{(s-1)(s+3)(s-4)} + (e^{-2s}) \frac{1}{(s)(s+3)(s-4)}$$

$$L[y] = (1 - e^{-2s+2}) \cdot H_1 + (e^{-2s}) \cdot H_2$$

$$H_1 = \frac{1}{(s-1)(s+3)(s-4)} = \frac{A}{s-1} + \frac{B}{s+3} + \frac{C}{s-4}$$

$$1 = A(s+3)(s-4) + B(s-1)(s-4) + C(s-1)(s+3)$$

$s = 1 \quad \rightarrow \quad 1 = A(4)(-3) \qquad \rightarrow \quad A = -\frac{1}{12}$

$s = -3 \quad \rightarrow \quad 1 = B(-4)(-8) \qquad \rightarrow \quad B = \frac{1}{32}$

$s = 4 \quad \rightarrow \quad 1 = C(3)(7) \qquad \rightarrow \quad C = \frac{1}{21}$

$$H_s = \frac{A}{s} + \frac{B}{s^2} + \frac{(Cs+D)}{s^2+4}$$

$$H_s = \left(\frac{1}{4}\right)\frac{1}{s^2} - \left(\frac{1}{4}\right)\frac{1}{s^2+4}$$

$$H_s = \left(\frac{1}{4}\right)\frac{1}{s^2} - \left(\frac{1}{8}\right)\frac{2}{s^2+4}$$

Apply Inverse Laplace Transformation (I.L.T.)

$$h(x) = \left(\frac{1}{4}\right) \cdot L^{-1}\left[\frac{1}{s^2}\right] - \left(\frac{1}{8}\right) \cdot L^{-1}\left[\frac{2}{s^2+2^2}\right]$$

$$h(x) = \left(\frac{1}{4}\right) \cdot (x) - \left(\frac{1}{8}\right) \cdot \sin 2x$$

Recall:

$$L[y] = \frac{1}{5}(e^{-5s} - e^{-10s}) \cdot H_s$$

$$L[y] = \frac{1}{5}e^{-5s} \cdot H_s - \frac{1}{5}e^{-10s} \cdot H_s$$

Apply Inverse Laplace Transformation (I.L.T.)

$$y = \frac{1}{5} U_5 \cdot h(x-5) - \frac{1}{5} U_{10} \cdot h(x-10)$$

Where: $h(x) = \frac{x}{4} - \frac{\sin 2x}{8}$ (see above)

Example: Find the P.S. of D.E. $\quad y'' - y' - 12y = g(x)$

With: $y(0) = y'(0) = 0$ and $g(x) = \begin{cases} e^x & 0 \leq x < 2 \\ 1 & x \geq 2 \end{cases}$

Solution: (Previously started.)

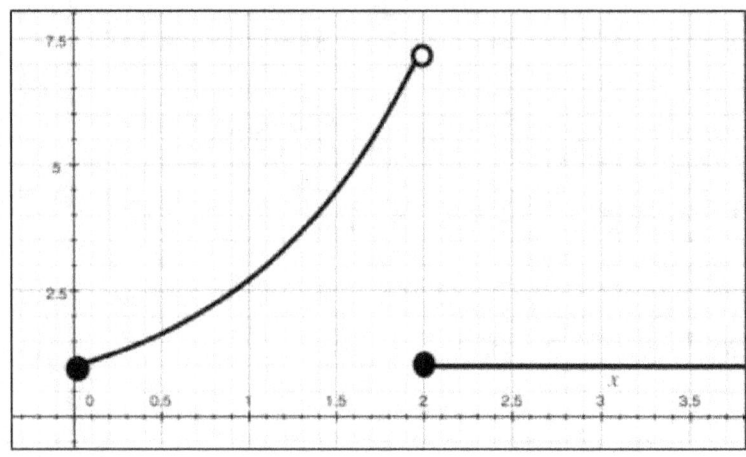

$g(x) = e^x[U_0 - U_2] + U_2$

$g(x) = e^x U_0 - e^x U_2 + U_2$

$g(x) = e^x - e^x U_2 + U_2$

$g(x) = e^x - e^x U_2 \cdot [e^{x-2} \cdot e^{-(x-2)}] + U_2$

$g(x) = e^x - e^{x-x+2} U_2 \cdot e^{x-2} + U_2$

$g(x) = e^x - e^2 U_2 \cdot e^{x-2} + U_2$

$L[g(x)] = \dfrac{1}{s-1} - e^2 \cdot e^{-2s} L[e^x] + \dfrac{e^{-2s}}{s}$

$L[g(x)] = \dfrac{1}{s-1} - e^2 \cdot e^{-2s} \left[\dfrac{1}{s-1}\right] + \dfrac{e^{-2s}}{s}$

$L[g(x)] = \dfrac{1}{s-1} - e^2 \cdot \dfrac{e^{-2s}}{s-1} + \dfrac{e^{-2s}}{s}$

$$y'' - y' - 12y = g(x)$$
$$L[y''] - L[y'] - 12L[y] = L[g(x)]$$
$$\{s^2 L[y] - s \cdot y(0) - y'(0)\} - \{s L[y] - y(0)\} - 12L[y] = L[g(x)]$$
$$s^2 L[y] - s L[y] - 12L[y] = L[g(x)]$$
$$L[y] \cdot (s^2 - s - 12) = L[g(x)]$$
$$L[y] \cdot (s+3)(s-4) = L[g(x)] \qquad L[g(x)] \text{ found previously.}$$

$$L[y] \cdot (s+3)(s-4) = \frac{1}{s-1} - e^2 \cdot \frac{e^{-2s}}{s-1} + \frac{e^{-2s}}{s}$$

$$L[y] = \left(\frac{1}{s-1} - e^2 \cdot \frac{e^{-2s}}{s-1} + \frac{e^{-2s}}{s}\right) \cdot \left(\frac{1}{(s+3)(s-4)}\right)$$

$$L[y] = \left(\frac{1 - e^{-2s+2}}{s-1} + \frac{e^{-2s}}{s}\right) \cdot \left(\frac{1}{(s+3)(s-4)}\right)$$

$$L[y] = (1 - e^{-2s+2}) \frac{1}{(s-1)(s+3)(s-4)} + (e^{-2s}) \frac{1}{(s)(s+3)(s-4)}$$

$$L[y] = (1 - e^{-2s+2}) \cdot H_1 + (e^{-2s}) \cdot H_2$$

$$H_1 = \frac{1}{(s-1)(s+3)(s-4)} = \frac{A}{s-1} + \frac{B}{s+3} + \frac{C}{s-4}$$

$$1 = A(s+3)(s-4) + B(s-1)(s-4) + C(s-1)(s+3)$$

$s = 1$ → $1 = A(4)(-3)$ → $A = -\frac{1}{12}$

$s = -3$ → $1 = B(-4)(-8)$ → $B = \frac{1}{32}$

$s = 4$ → $1 = C(3)(7)$ → $C = \frac{1}{21}$

$$H_1 = \left(-\frac{1}{12}\right)\frac{1}{s-1} + \left(\frac{1}{32}\right)\frac{1}{s+3} + \left(\frac{1}{21}\right)\frac{1}{s-4}$$

$$h_1(x) = \left(-\frac{1}{12}\right)\cdot L^{-1}\left[\frac{1}{s-1}\right] + \left(\frac{1}{32}\right)\cdot L^{-1}\left[\frac{1}{s+3}\right] + \left(\frac{1}{21}\right)\cdot L^{-1}\left[\frac{1}{s-4}\right]$$

$$h_1(x) = \left(-\frac{1}{12}\right)\cdot e^x + \left(\frac{1}{32}\right)\cdot e^{-3x} + \left(\frac{1}{21}\right)\cdot e^{4x}$$

$$H_2 = \frac{1}{(s)(s+3)(s-4)} = \frac{A}{s} + \frac{B}{s+3} + \frac{C}{s-4}$$

$$1 = A(s+3)(s-4) + B(s)(s-4) + C(s)(s+3)$$

$s = 0$ → $1 = A(3)(-4)$ → $A = -\frac{1}{12}$

$s = -3$ → $1 = B(-3)(-7)$ → $B = \frac{1}{21}$

$s = 4$ → $1 = C(4)(7)$ → $C = \frac{1}{28}$

$$H_2 = \left(-\frac{1}{12}\right)\frac{1}{s} + \left(\frac{1}{21}\right)\frac{1}{s+3} + \left(\frac{1}{28}\right)\frac{1}{s-4}$$

$$h_2(x) = \left(-\frac{1}{12}\right)\cdot L^{-1}\left[\frac{1}{s}\right] + \left(\frac{1}{21}\right)\cdot L^{-1}\left[\frac{1}{s+3}\right] + \left(\frac{1}{28}\right)\cdot L^{-1}\left[\frac{1}{s-4}\right]$$

$$h_2(x) = \left(-\frac{1}{12}\right)\cdot (1) + \left(\frac{1}{21}\right)\cdot e^{-3x} + \left(\frac{1}{28}\right)\cdot e^{4x}$$

$L[y] = (1 - e^{-2s+2})\cdot H_1 + (e^{-2s})\cdot H_2$

$L[y] = (1)\cdot H_1 - (e^{-2s+2})\cdot H_1 + (e^{-2s})\cdot H_2$

$L[y] = H_1 - (e^2)(e^{-2s+2})(e^{-2})\cdot H_1 + (e^{-2s})\cdot H_2$

$L[y] = H_1 - (e^2)(e^{-2s})\cdot H_1 + (e^{-2s})\cdot H_2$

$$L[y] = H_1 - (e^2)(e^{-2s}) \cdot H_1 + (e^{-2s}) \cdot H_2$$

$$y = L^{-1}[H_1] - (e^2) \cdot L^{-1}[(e^{-2s}) \cdot H_1] + L^{-1}[(e^{-2s}) \cdot H_2]$$

$$y = h_1(x) - (e^2) \cdot U_2 \cdot h_1(x-2) + U_2 \cdot h_2(x-2)$$

Where:

$$h_1(x) = \left(-\frac{1}{12}\right) \cdot e^x + \left(\frac{1}{32}\right) \cdot e^{-3x} + \left(\frac{1}{21}\right) \cdot e^{4x}$$

$$h_2(x) = \left(-\frac{1}{12}\right) \cdot (1) + \left(\frac{1}{21}\right) \cdot e^{-3x} + \left(\frac{1}{28}\right) \cdot e^{4x}$$

Example: Solve the Initial Value Problem (I.V.P)
using Laplace Transforms (L.T.)

$$y'' + y = g(x)$$

With: $g(x) = \begin{cases} x & 0 \leq x < 2 \\ 4 & x \geq 2 \end{cases}$ And: $\begin{cases} y(0) = 0 \\ y'(0) = 1 \end{cases}$

Solution:

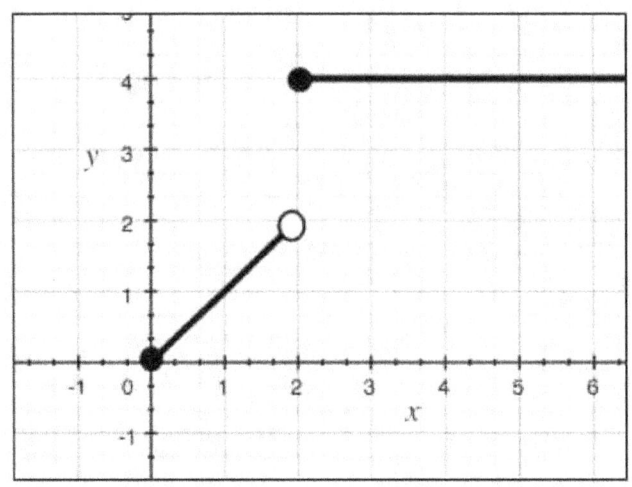

Rewrite equation as a linear combination of a step function.

$$g(x) = x(from\ 0\ to\ 2) + 4(from\ 2)$$
$$g(x) = x(U_0(x) - U_2(x)) + 4U_2(x)$$
$$g(x) = x[1 - U_2(x)] + 4U_2(x)$$
$$g(x) = x - xU_2(x) + 4U_2(x)$$

Equation: $g(x) = x - xU_2(x) + 4U_2(x)$

Get into the form: $f(x - c)U_c(x)$

$$g(x) = x - (x - 2)U_2(x) - 2U_2(x) + 4U_2(x)$$
$$g(x) = x - (x - 2)U_2(x) + 2U_2(x)$$

Now, we have:

$$y'' + y = x - (x - 2)U_2(x) + 2U_2(x) \quad \text{With: } \begin{cases} y(0) = 0 \\ y'(0) = 1 \end{cases}$$

Take Laplace Transformation of both sides of the equation.

$$L[y''] + L[y] = L[x] - L[(x - 2)U_2(x)] + L[2U_2(x)]$$

$$s^2 L[y] - s \cdot y(0) - y'(0) + L[y] = (???)$$
$$s^2 L[y] - 0 - 1 + L[y] = (???)$$
$$L[y] \cdot (s^2 + 1) = 1 + (???)$$

Use: $\boxed{L[f(x - c)U_c(x)] = e^{-sc} \cdot L[f(x)]}$

$$L[y](s^2 + 1) = 1 + L[x] - L[(x - 2)U_2(x)] + L[2U_2(x)]$$

$$L[y](s^2 + 1) = 1 + \frac{1}{s^2} - \frac{1}{s^2}e^{-2s} + 2\frac{e^{-2s}}{s}$$

$$L[y] = \frac{1}{s^2+1} + \frac{1}{s^2(s^2+1)} - \frac{1}{s^2(s^2+1)}e^{-2s} + \frac{2e^{-2s}}{s(s^2+1)}$$

$$L[y] = H_1(s) + H_2(s) - H_2(s) \cdot e^{-2s} + H_3(s) \cdot e^{-2s}$$

Use Partial Fractional Decomposition and ILT to find:
$H_1(s), H_2(s), H_3(s),$ and $h_1(x), h_2(x), h_3(x)$

$$H_1(s) = \frac{1}{s^2+1}$$

$$L^{-1}[H_1(s)] = L^{-1}\left[\frac{1}{s^2+1}\right]$$

$$h_1(x) = \sin x$$

$$H_2(s) = \frac{1}{s^2(s^2+1)} = \frac{A}{s} + \frac{B}{s^2} + \frac{Cs+D}{s^2+1}$$

$1 = As(s^2+1) + B(s^2+1) + (Cs+D)s^2$

$1 = As^3 + As + Bs^2 + B + Cs^3 + Ds^2$

$1 = s^3(A+C) + s^2(B+D) + s(A) + (B)$

➔ $B = 1, A = 0, C = 0, D = -1$

$$H_2(s) = \frac{1}{s^2} - \frac{1}{s^2+1}$$

$$L^{-1}[H_2(s)] = L^{-1}\left[\frac{1}{s^2}\right] - L^{-1}\left[\frac{1}{s^2+1}\right]$$

$$h_2(x) = x - \sin x$$

$$H_3(s) = \frac{2}{s(s^2+1)} = \frac{A}{s} + \frac{Bs+C}{s^2+1}$$

$$2 = A(s^2+1) + (Bs+C)(s)$$

$$2 = As^2 + A + Bs^2 + Cs$$

$$2 = s^2(A+B) + s(C) + (A)$$

$$\rightarrow A = 2, \ C = 0, \ B = -2$$

$$H_3(s) = \frac{2}{s} - \frac{2s}{s^2+1} = 2\left(\frac{1}{s}\right) - 2\left(\frac{s}{s^2+1}\right)$$

$$L^{-1}[H_3(s)] = 2L^{-1}\left[\frac{1}{s}\right] - 2L^{-1}\left[\frac{s}{s^2+1}\right]$$

$$h_3(x) = 2(1) - 2\cos x$$

$$h_3(x) = 2 - 2\cos x$$

Solution:

$$L[y] = H_1(s) + H_2(s) - H_2(s)\cdot e^{-2s} + H_3(s)\cdot e^{-2s}$$

$$y = h_1(x) - h_2(x) - U_2(x)\cdot h_2(x-2) + U_2(x)\cdot h_3(x-2)$$

Where: $h_1(x)$, $h_2(x)$, $h_3(x)$ are known.

IMPULSE FUNCTION

IMPULSE FUNCTION	
A large force, applied for a short period of time. (shock)	
Equation	$g(x) = d_\tau(x) = \begin{cases} \frac{1}{2\tau} & -\tau < x < \tau \\ 0 & \text{else} \end{cases}$
Graph of Impulse Function With Long Interval	*(graph showing a low flat pulse between $-\tau$ and τ)*
Graph of Impulse Function With Short Interval	*(graph showing a tall narrow spike near 0 between $-\tau$ and τ)*
Two Properties of Impulse Function	$\lim_{\tau \to 0} d_\tau(x) = \delta(x) = 0$ NOTE: $\delta(x) = 0$ When $x \neq 0$ $\lim_{\tau \to 0} \int_{-\infty}^{\infty} d_\tau(x) = \lim_{\tau \to 0} \int_{-\infty}^{\infty} \delta(x) = 1$
Total Impulse of Force	$\int_{-\tau}^{\tau} g(x)\, dx = \int_{-\infty}^{\infty} g(x)\, dx$

DIRAE-DELTA FUNCTION	
Unit Impulse at Origin	$\delta(x) = 0 \quad \text{When } x \neq 0$ $\int_{-\infty}^{\infty} \delta(x) = 1$
Unit Impulse near $x = x_0$	$\delta(x - x_0) = 0 \quad \text{When } x \neq x_0$ $\int_{-\infty}^{\infty} \delta(x - x_0) = 1$
Laplace Transformation of $\delta(x)$	$L[\delta(x)] = e^{-0 \cdot s} = e^0 = 1$
Laplace Transformation of $\delta(x - x_0)$	$L[\delta(x - x_0)] = e^{-x_0 \cdot s}$

Updated table of Laplace Transforms of some frequently used functions.

LAPLACE TRANSFORMS	
$L[f(x)]$	$= F(s)$
$L[1]$	$= \dfrac{1}{s}$
$L[x^n]$	$= \dfrac{n!}{s^{n+1}}$, $n = $ positive integer
$L[x^p]$	$= \dfrac{\Gamma(p+1)}{x^{p+1}}$, $p = $ real number
$L[e^{ax}]$	$= \dfrac{1}{s-a}$
$L[\sin ax]$	$= \dfrac{a}{s^2 + a^2}$
$L[\cos ax]$	$= \dfrac{s}{s^2 + a^2}$
$L[e^{ax} \sin bx]$	$= \dfrac{b}{(s-a)^2 + b^2}$
$L[e^{ax} \cos bx]$	$= \dfrac{(s-a)}{(s-a)^2 + b^2}$
$L[x^n e^{ax}]$	$= \dfrac{n!}{(s-a)^{n+1}}$
$L[U_c(x)]$	$= \dfrac{e^{-cs}}{s}$
$L[U_c(x) \cdot f(x-c)]$	$= e^{-cs} \cdot L[f(x)]$
$L[\delta(x - x_0)]$	$= e^{-x_0 \cdot s}$

Example: Use Laplace Transforms (L.T.) to solve the I.V.P.
$$y''' + y'' + 3y' - 5y = \delta(x-1)$$
With: $y(0) = 0$, $y'(0) = 2$, $y''(0) = -2$

Solution:
$$L[y'''] + L[y''] + 3L[y'] - 5L[y] = L[\delta(x-1)]$$

$s^3 L[y] - \cancel{s^2 y(0)} - sy'(0) - y''(0) + s^2 L[y] - \cancel{sy(0)} - y'(0) +$
$\qquad\qquad + 3L[y] - \cancel{3y(0)} - 5L[y] = e^{-s}$

$s^3 L[y] - sy'(0) - y''(0) + s^2 L[y] - y'(0) + 3L[y] - 5L[y] = e^{-s}$

$(s^3 + s^2 + 3s - 5) \cdot L[y] - s(2) + 2 - 2 = e^{-s}$

$(s^3 + s^2 + 3s - 5) L[y] = 2s + e^{-s}$

$L[y] = \dfrac{2s}{s^3+s^2+3s-5} + \dfrac{1}{s^3+s^2+3s-5} \cdot e^{-s}$

$L[y] = H_1 + H_2 \cdot e^{-s}$

$H_1 = \dfrac{2s}{s^3+s^2+3s-5} = \dfrac{2s}{(s-1)(s^2+2s+5)} = \dfrac{A}{s-1} + \dfrac{Bs+C}{s^2+2s+5}$

$2s = A(s^2 + 2s + 5) + (Bs + C)(s - 1)$

(Continued ...)

$$2s = A(s^2 + 2s + 5) + (Bs + C)(s - 1)$$

$s = 1 \rightarrow 2 = 8A$ $\qquad \rightarrow A = \frac{1}{4}$

$s = 0 \rightarrow 0 = 5A + C(-1)$ $\qquad \rightarrow C = 5A = \frac{5}{4}$

$s = -1 \rightarrow -2 = \frac{1}{4}(4) + \left(B(-1) + \frac{5}{4}\right)(-2)$

$\qquad \qquad \frac{3}{2} = -B + \frac{5}{4} \qquad \rightarrow B = \frac{5}{4} - \frac{3}{2} = -\frac{1}{4}$

$$H_1 = \frac{A}{s-1} + \frac{Bs + C}{s^2 + 2s + 5} = \frac{A}{s-1} + \frac{Bs + C}{s^2 + 2s + 1 + 4} = \frac{A}{s-1} + \frac{B(s+1) - B + C}{(s+1)^2 + (2)^2}$$

$$H_1(s) = \frac{1}{4}\left(\frac{1}{s-1}\right) + \frac{-\frac{1}{4}(s+1) + \frac{1}{4} + \frac{5}{4}}{(s+1)^2 + (2)^2}$$

$$H_1(s) = \frac{1}{4}\left(\frac{1}{s-1}\right) + \frac{1}{4} \frac{s+1}{(s+1)^2+(2)^2} + \frac{3}{4} \frac{2}{(s+1)^2+(2)^2}$$

Inverse Laplace Transform (I.L.T.)

$$h_1(x) = \frac{1}{4}e^x + \frac{1}{4}e^{-x}\cos 2x + \frac{3}{4}e^{-x}\sin 2x$$

$$H_2(s) = \frac{1}{s^3 + s^2 + 3s - 5} = \frac{1}{(s-1)(s^2 + 2s + 5)} = \frac{A}{s-1} + \frac{Bs + C}{s^2 + 2s + 5}$$

$$1 = A(s^2 + 2s + 5) + (Bs + C)(s - 1)$$

$$1 = A(s^2 + 2s + 5) + (Bs + C)(s - 1)$$

$$s = 1 \rightarrow A = \frac{1}{8}$$

$$s = 0 \rightarrow 1 = 5A + C(-1) \quad\quad \rightarrow C = \frac{5}{8} - 1 = -\frac{3}{8}$$

$$s = -1 \rightarrow 1 = \frac{1}{8}(4) + \left(-B + \left(-\frac{3}{8}\right)\right)(-2)$$

$$\frac{1}{2} = \left(-B - \frac{3}{8}\right)(-2)$$

$$\frac{1}{4} = -B - \frac{3}{8} \quad\quad \rightarrow B = -\frac{3}{8} + \frac{1}{4} = -\frac{1}{8}$$

$$H_2(s) = \frac{A}{s-1} + \frac{Bs + C}{s^2 + 2s + 5} = \frac{A}{s-1} + \frac{B(s+1) - B + C}{(s+1)^2 + (2)^2}$$

$$H_2(s) = \frac{1}{8}\left(\frac{1}{s-1}\right) + \frac{-\frac{1}{8}(s+1) + \frac{1}{8} - \frac{3}{8}}{(s+1)^2 + (2)^2}$$

$$H_2(s) = \frac{1}{8}\left(\frac{1}{s-1}\right) - \frac{1}{8}\frac{(s+1)}{(s+1)^2 + (2)^2} - \frac{3}{16}\frac{(2)}{(s+1)^2 + (2)^2}$$

Inverse Laplace Transform (I.L.T.)

$$h_2(x) = \frac{1}{8}e^x - \frac{1}{8}e^{-x}\cos 2x - \frac{3}{16}e^{-x}\sin 2x$$

Recall:

$$L[y] = H_1 + H_2 \cdot e^{-s} \quad\quad\quad \text{(I.L.T.)}$$

$$L^{-1}[y] = y = h_1(x) + U_1 \cdot h_2(x - 1)$$

Example: Solve the IVP
$$y'' - 3y' - 10y = \sin 2x + U_5(x) + \delta(x - 10)$$
With: $y(0) = y'(0) = 0$

Solution: Apply L.T. to both sides.
$$L[y''] - 3L[y'] - 10L[y] = L[\sin 2x] + L[U_5(x)] + L[\delta(x - 10)]$$
$$(s^2 - 3s - 10) \cdot L[y] = \frac{2}{s^2+4} + \frac{e^{-5s}}{s} + e^{-10s}$$
$$(s - 5)(s + 2) \cdot L[y] = \frac{2}{s^2+4} + \frac{e^{-5s}}{s} + e^{-10s}$$

$$L[y] = \frac{2}{(s^2+4)(s-5)(s+2)} + \frac{e^{-5s}}{s(s-5)(s+2)} + \frac{e^{-10s}}{(s-5)(s+2)}$$

$$L[y] = (2)\frac{1}{(s^2+4)(s-5)(s+2)} + (e^{-5s})\frac{1}{s(s-5)(s+2)} + (e^{-10s})\frac{1}{(s-5)(s+2)}$$

$$L[y] = (2)H_1 + (e^{-5s})H_2 + (e^{-10s})H_3$$

Use Partial Fractions three times.

$$H_1 = \frac{1}{(s^2+4)(s-5)(s+2)} = \frac{As+B}{s^2+4} + \frac{C}{s-5} + \frac{D}{s+2}$$

$$1 = (As + B)(s - 5)(s + 2) + C(s^2 + 4)(s + 2) + D(s^2 + 4)(s - 5)$$

$s = -2 \quad \rightarrow 1 = D(8)(-7) \rightarrow D = -\frac{1}{56}$

$s = 5 \quad \rightarrow 1 = C(29)(7) \rightarrow C = \frac{1}{203}$

(Continued...)

$s = 0 \quad \rightarrow 1 = (B)(-5)(2) + \left(\frac{1}{203}\right)(4)(2) + \left(-\frac{1}{56}\right)(4)(-5)$

$\rightarrow B = -\frac{7}{116}$

$s = 1 \quad \rightarrow$

$1 = \left(A - \frac{1}{116}\right)(-4)(3) + \left(\frac{1}{203}\right)(5)(3) + \left(-\frac{1}{56}\right)(5)(-4)$

$\rightarrow A = -\frac{9}{232}$

$H_1 = \left(-\frac{1}{232}s - \frac{7}{116}\right)\frac{1}{s^2+4} + \left(\frac{1}{203}\right)\frac{1}{s-5} + \left(-\frac{1}{56}\right)\frac{1}{s+2} H_1 =$

$-\left(\frac{1}{232}\right)\frac{s}{s^2+4} - \left(\frac{7}{116}\right)\frac{1}{s^2+4} + \left(\frac{1}{203}\right)\frac{1}{s-5} - \left(\frac{1}{56}\right)\frac{1}{s+2}$

$H_1 = -\left(\frac{1}{232}\right)\frac{s}{s^2+4} - \left(\frac{7}{116}\right)\frac{1}{s^2+4} + \left(\frac{1}{203}\right)\frac{1}{s-5} - \left(\frac{1}{56}\right)\frac{1}{s+2}$

$h_1(x) = -\left(\frac{1}{232}\right) \cdot L^{-1}\left[\frac{s}{s^2+2^2}\right] - \left(\frac{7}{116}\right) \cdot L^{-1}\left[\frac{1}{s^2+2^2}\right]$

$\qquad + \left(\frac{1}{203}\right) \cdot L^{-1}\left[\frac{1}{s-5}\right] - \left(\frac{1}{56}\right) \cdot L^{-1}\left[\frac{1}{s+2}\right]$

$h_1(x) = -\left(\frac{1}{232}\right)\cos 2x - \left(\frac{7}{116}\right)\sin 2x + \left(\frac{1}{203}\right)e^{5x} - \left(\frac{1}{56}\right)e^{-2x}$

$$H_2 = \frac{1}{s(s-5)(s+2)} = \frac{A}{s} + \frac{B}{s-5} + \frac{C}{s+2}$$

$$1 = A(s-5)(s+2) + B(s)(s+2) + C(s)(s-5)$$

$s = 0 \;\rightarrow\; 1 = A(-5)(2) \;\rightarrow\; A = -\frac{1}{10}$

$s = 5 \;\rightarrow\; 1 = B(5)(7) \;\rightarrow\; B = \frac{1}{35}$

$s = -2 \;\rightarrow\; 1 = C(-2)(-7) \;\rightarrow\; C = \frac{1}{14}$

$$H_2 = \left(-\frac{1}{10}\right)\frac{1}{s} + \left(\frac{1}{35}\right)\frac{1}{s-5} + \left(\frac{1}{14}\right)\frac{1}{s+2}$$

$$h_2(x) = \left(-\frac{1}{10}\right) \cdot L^{-1}\left[\frac{1}{s}\right] + \left(\frac{1}{35}\right) \cdot L^{-1}\left[\frac{1}{s-5}\right] + \left(\frac{1}{14}\right) \cdot L^{-1}\left[\frac{1}{s+2}\right]$$

$$h_2(x) = \left(-\frac{1}{10}\right) \cdot (1) + \left(\frac{1}{35}\right) \cdot e^{5x} + \left(\frac{1}{14}\right) \cdot e^{-2x}$$

$$H_3 = \frac{1}{(s-5)(s+2)} = \frac{A}{s-5} + \frac{B}{s+2}$$

$$1 = A(s+2) + B(s-5)$$

$s = 5 \;\rightarrow\; 1 = A(7) \;\rightarrow\; A = \frac{1}{7}$

$s = -2 \;\rightarrow\; 1 = B(-7) \;\rightarrow\; B = -\frac{1}{7}$

$$H_3 = \left(\frac{1}{7}\right)\frac{1}{s-5} - \left(\frac{1}{7}\right)\frac{1}{s+2}$$

$$h_3(x) = \left(\frac{1}{7}\right) \cdot L^{-1}\left[\frac{1}{s-5}\right] - \left(\frac{1}{7}\right) \cdot L^{-1}\left[\frac{1}{s+2}\right]$$

$$h_3(x) = \left(\frac{1}{7}\right) \cdot e^{5x} - \left(\frac{1}{7}\right) \cdot e^{-2x}$$

$$L[y] = (2)H_1 + (e^{-5s})H_2 + (e^{-10s})H_3$$

$$y = 2 \cdot h_1(x) + U_5 \cdot h_2(x-5) + U_{10} \cdot h_3(x-10)$$

Where:

$$h_1(x) = -\left(\frac{1}{232}\right)\cos 2x - \left(\frac{7}{116}\right)\sin 2x + \left(\frac{1}{203}\right)e^{5x} - \left(\frac{1}{56}\right)e^{-2x}$$

$$h_2(x) = \left(-\frac{1}{10}\right)(1) + \left(\frac{1}{35}\right)e^{5x} + \left(\frac{1}{14}\right)e^{-2x}$$

$$h_3(x) = \left(\frac{1}{7}\right)e^{5x} - \left(\frac{1}{7}\right)e^{-2x}$$

Example: Find the particular solution (P.S.) of
$$y'' - y' - 6y = \sin 2x + \delta(x - 4) + U_6(x)$$
With: $y(0) = y'(0) = 0$

Solution:

$$L[y''] - L[y'] - L[6y] = L[\sin 2x] + L[\delta(x-4)] + L[U_6(x)]$$

$$\{s^2 L[y] - sy(0) - y'(0)\} - \{s L[y] - y(0)\} - 6 L[y]$$
$$= \frac{2}{s^2+4} + e^{-4s} + \frac{e^{-6s}}{s}$$

$$\{s^2 L[y] - 0 - 0\} - \{s L[y] - 0\} - 6 L[y] = \frac{2}{s^2+4} + e^{-4s} + \frac{e^{-6s}}{s}$$

$$s^2 L[y] - s L[y] - 6 L[y] = \frac{2}{s^2+4} + e^{-4s} + \frac{e^{-6s}}{s}$$

$$L[y] \cdot (s^2 - s - 6) = \frac{2}{s^2+4} + e^{-4s} + \frac{e^{-6s}}{s}$$

$$L[y] \cdot (s-3)(s+2) = \frac{2}{s^2+4} + e^{-4s} + \frac{e^{-6s}}{s}$$

$$L[y] = \frac{2}{(s-3)(s+2)(s^2+4)} + e^{-4s}\left(\frac{1}{(s-3)(s+2)}\right) + e^{-6s}\left(\frac{1}{s(s-3)(s+2)}\right)$$

$$L[y] = H_1(s) + e^{-4s} \cdot H_2(s) + e^{-6s} \cdot H_3(s)$$

Use Partial Fraction Decomposition three times.

$$H_1(s) = \frac{2}{(s-3)(s+2)(s^2+4)} = \frac{A}{s-3} + \frac{B}{s+2} + \frac{Cx+D}{s^2+4}$$

$$2 = A(s+2)(s^2+4) + B(s-3)(s^2+4) + (Cx+D)(s-3)(s+2)$$

$s = 3 \quad \rightarrow \quad 2 = A(5)(13) \quad \rightarrow \quad A = \frac{2}{65}$

$s = -2 \quad \rightarrow \quad 2 = B(5)(8) \quad \rightarrow \quad B = -\frac{1}{20}$

$s = 0 \quad \rightarrow \quad 2 = \left(\frac{2}{65}\right)(2)(4) + \left(-\frac{1}{20}\right)(-3)(4) + D(-3)(2)$

$\quad\quad\quad \rightarrow \quad D = -\frac{5}{26}$

$s = 1 \quad \rightarrow \quad C = \frac{1}{52}$

$$H_1(s) = \frac{2}{65}\left(\frac{1}{s-3}\right) - \frac{1}{20}\left(\frac{1}{s+2}\right) + \frac{1}{52}\left(\frac{s}{s^2+4}\right) - \frac{5}{26}\left(\frac{1}{s^2+4}\right)$$

$h_1(x) = L^{-1}[H_1(s)]$

$$h_1(x) = \left(\frac{2}{65}\right)e^{3x} - \left(\frac{1}{20}\right)e^{-2x} + \left(\frac{1}{52}\right)\cos 2x - \left(\frac{5}{52}\right)\sin 2x$$

$$H_2(s) = \frac{1}{(s-3)(s+2)} = \frac{A}{s-3} + \frac{B}{s+2}$$

$1 = A(s+2) + B(s-3)$

$s = 3 \quad \rightarrow \quad 1 = A(5) \quad \rightarrow \quad A = \frac{1}{5}$

$s = -2 \quad \rightarrow \quad 1 = B(-5) \quad \rightarrow \quad B = -\frac{1}{5}$

$$H_2(s) = \frac{1}{5}\left(\frac{1}{s-3}\right) - \frac{1}{5}\left(\frac{1}{s+2}\right)$$

$$L^{-1}[H_2(s)] = h_2(x) = \frac{1}{5} \cdot e^{3x} - \frac{1}{5} \cdot e^{-2x}$$

$$H_3(s) = \frac{1}{s(s-3)(s+2)} = \frac{A}{s} + \frac{B}{s-3} + \frac{C}{s+2}$$

$$1 = A(s-3)(s+2) + B(s)(s+2) + C(s)(s-3)$$

$s = 0$ → $1 = A(-3)(2)$ → $A = -\frac{1}{6}$

$s = 3$ → $1 = B(3)(5)$ → $B = \frac{1}{15}$

$s = -2$ → $1 = C(-2)(-5)$ → $C = \frac{1}{10}$

$$H_3(s) = -\frac{1}{6}\left(\frac{1}{s}\right) + \frac{1}{15}\left(\frac{1}{s-3}\right) + \frac{1}{10}\left(\frac{1}{s+2}\right)$$

$$L^{-1}[H_3(s)] = h_3(x) = -\frac{1}{6}(1) + \frac{1}{15}(e^{3x}) + \frac{1}{10}(e^{-2x})$$

$$L[y] = H_1(s) + e^{-4s} \cdot H_2(s) + e^{-6s} \cdot H_3(s)$$

$$y = L^{-1}[H_1(s)] + L^{-1}[e^{-4s} \cdot H_2(s)] + L^{-1}[e^{-6s} \cdot H_3(s)]$$

$$y = h_1(x) + U_4(x) \cdot h_2(x-4) + U_6(x) \cdot h_3(x-6)$$

This is the Particular Solution (P.S) of the given DE.

With Step Function, Step Force, and Impulse Force.

Note:
- $h_1(x)$, $h_2(x)$, $h_3(x)$ and $U_n(s)$ were found previously.
- They are functions of "x" or "s" but they may be simply written as h_n or as U_n

CONVOLUTION INTEGRAL

CONVOLUTION INTEGRAL

CONVOLUTION INTEGRAL of $f(x)$ and $g(x)$	$h(x) = \int_0^x f(x - \tau) \cdot g(\tau)\, d\tau$ AND $h(x) = \int_0^x f(\tau) \cdot f(x - \tau)\, d\tau$
Laplace Transformation of $h(x)$	$H(s) = L[h(x)] = F(s) \cdot G(s)$
Here,	$L[f(x)] = F(s)$ AND $L[g(x)] = G(s)$ Both exist for all $s > a \geq 0$

Laplace Transform (L.T.) of a Convolution Integral

Example (1) $\quad h(x) = \int_0^x (x - \tau)^5 \cdot \cos(3\tau) \, d\tau$

$f(x) = x^5$	$g(x) = \cos(3x)$
$L[f(x)] = L[x^5]$	$L[g(x)] = L[\cos(3x)]$
$F(s) = \dfrac{5!}{s^6} = \dfrac{120}{s^6}$	$G(s) = \dfrac{s}{s^2 + 9}$

$$L[h(x)] = H(s) = F(s) \cdot G(s)$$

$$H(s) = \left(\frac{120}{s^6}\right) \cdot \left(\frac{s}{s^2+9}\right) = \frac{120}{s^5(s^2+9)}$$

Example (2) $\quad h(x) = \int_0^x e^{5x} \cdot \sin(7(x - \tau)) \, d\tau$

$f(x) = e^{5x}$	$g(x) = \sin(7x)$
$L[f(x)] = L[e^{5x}]$	$L[g(x)] = L[\sin(7x)]$
$F(s) = \dfrac{1}{s - 5}$	$G(s) = \dfrac{7}{s^2 + 49}$

$$L[h(x)] = H(s) = F(s) \cdot G(s)$$

$$H(s) = \left(\frac{1}{s-5}\right) \cdot \left(\frac{7}{s^2+49}\right) = \frac{7}{(s-5) \cdot (s^2+49)}$$

Now, go backwards and express your answer in terms of the Convolution Integral.

Example (1)

Compute the Inverse Laplace Transformation (I.L.T.) of $H_1(s)$ and express your answer in the form of Convolution Integrals.

$$H_1(s) = \frac{1}{(s+2)\cdot(s^2+9)}$$

Solution:

$H_1(s) = F(s)\cdot G(s)$

$F(s) = \frac{1}{s+2}$	$G(s) = \frac{1}{s^2+9} = \frac{1}{3}\cdot\frac{3}{(s^2+9)}$
$f(x) = L^{-1}[F(s)]$	
$f(x) = L^{-1}\left[\frac{1}{s+2}\right]$	$g(x) = L^{-1}[G(s)]$
$f(x) = e^{-2x}$	$g(x) = \left(\frac{1}{3}\right) L^{-1}\left[\frac{3}{(s^2+9)}\right]$
	$g(x) = \left(\frac{1}{3}\right)\sin(3x)$

$$h_1(x) = \int_0^x f(x-\tau)\cdot g(\tau)\, d\tau = \int_0^x f(\tau)\cdot g(x-\tau)\, d\tau$$

$$h_1(x) = \int_0^x e^{-2(x-\tau)}\cdot\left(\frac{1}{3}\right)\sin(3\tau)\, d\tau$$

$$h_1(x) = \int_0^x e^{-2\tau}\cdot\left(\frac{1}{3}\right)\sin(3(x-\tau))\, d\tau$$

Example (2)

Compute the Inverse Laplace Transformation (I.L.T.) of $H_2(s)$ and express your answer in the form of Convolution Integrals.

$$H_2(s) = \frac{1}{(s+3)} \cdot \frac{s}{s^2+4}$$

Solution:

$$H_2(s) = F(s) \cdot G(s)$$

$F(s) = \frac{1}{(s+3)}$	$G(s) = \frac{s}{s^2+4}$
$f(x) = L^{-1}[F(s)]$	$g(x) = L^{-1}[G(s)]$
$f(x) = L^{-1}\left[\frac{1}{(s+3)}\right]$	$g(x) = L^{-1}\left[\frac{s}{s^2+4}\right]$
$f(x) = e^{-3x}$	$g(x) = \cos(2x)$

$$h_2(x) = \int_0^x f(x-\tau) \cdot g(\tau) \, d\tau = \int_0^x f(\tau) \cdot g(x-\tau) \, d\tau$$

$$h_2(x) = \int_0^x e^{-3(x-\tau)} \cdot \cos(2\tau) \, d\tau$$

$$h_2(x) = \int_0^x e^{-3\tau} \cdot \cos(2(x-\tau)) \, d\tau$$

Example (3)

Compute the Inverse Laplace Transformation (I.L.T.) of $H_3(s)$ and express your answer in the form of Convolution Integrals.

$$H_3(s) = \frac{1}{s^4\,(s+3)^5}$$

Solution:

$H_3(s) = F(s) \cdot G(s)$

$F(s) = \frac{1}{s^4} = \frac{1}{6} \cdot \frac{3!}{s^4}$	$G(s) = \frac{1}{(s+3)^5} = \frac{1}{24} \cdot \frac{4!}{(s+3)^5}$
$f(x) = L^{-1}[F(s)]$	$g(x) = L^{-1}[G(s)]$
$f(x) = L^{-1}\left[\frac{1}{6} \cdot \frac{3!}{s^4}\right]$	$g(x) = L^{-1}\left[\frac{1}{24} \cdot \frac{4!}{(s+3)^5}\right]$
$f(x) = \left(\frac{1}{6}\right) \cdot L^{-1}\left[\frac{3!}{s^4}\right]$	$g(x) = \left(\frac{1}{24}\right) \cdot L^{-1}\left[\frac{4!}{(s+3)^5}\right]$
$f(x) = \left(\frac{1}{6}\right) \cdot x^3 = \frac{x^3}{6}$	$g(x) = \left(\frac{1}{24}\right) e^{-3x} \cdot x^4$

$h_3(x) = \int_0^x f(x-\tau) \cdot g(\tau)\, d\tau = \int_0^x f(\tau) \cdot g(x-\tau)\, d\tau$

$h_3(x) = \int_0^x \frac{(x-\tau)^3}{6} \cdot \left(\frac{\tau^4}{24}\right) e^{-3\tau}\, d\tau = \int_0^x \frac{\tau^3}{6} \left(\frac{(x-\tau)^4}{24}\right) e^{-3(x-\tau)}\, d\tau$

$h_3(x) = \left(\frac{1}{144}\right) \int_0^x (x-\tau)^3 \cdot \tau^4 e^{-3\tau}\, d\tau$

$h_3(x) = \left(\frac{1}{144}\right) \int_0^x \tau^3 (x-\tau)^4 e^{-3(x-\tau)}\, d\tau$

Example: Find the Particular Solution (P.S.) of the given DE.

$y'' + 9y = g(x)$ With: $y(0) = 1$ and $y'(0) = 3$

Note: $g(x)$ could be any function.

Solution:

$L[y''] + L[9y] = L[g(x)]$

$\{s^2 L[y] - sy(0) - y'(0)\} + \{9 L[y]\} = G(s)$

$\{s^2 L[y] - s(1) - 3\} + \{9 L[y]\} = G(s)$

$s^2 L[y] - s - 3 + 9 L[y] = G(s)$

$L[y](s^2 + 9) - s - 3 = G(s)$

$L[y](s^2 + 9) = s + 3 + G(s)$

$L[y] = \dfrac{s+3}{(s^2+9)} + \left(\dfrac{s+3}{s^2+9}\right) \cdot G(s)$

$L[y] = \dfrac{s}{(s^2+9)} + \dfrac{3}{(s^2+9)} + \left(\dfrac{1}{s^2+9}\right) \cdot G(s)$

$y = L^{-1}\left[\dfrac{s}{(s^2+9)}\right] + L^{-1}\left[\dfrac{3}{(s^2+9)}\right] + L^{-1}\left[\left(\dfrac{s+3}{s^2+9}\right) \cdot G(s)\right]$

$y = \cos(3x) + \sin(3x) + L^{-1}\left[\left(\dfrac{1}{s^2+9}\right) \cdot G(s)\right]$

Convolution Integral

$$F(s) = \frac{1}{s^2 + 9} = \left(\frac{1}{3}\right)\frac{3}{s^2 + 9}$$

$$f(x) = \left(\frac{1}{3}\right)L^{-1}\left[\frac{3}{s^2 + 9}\right]$$

$$f(x) = \left(\frac{1}{3}\right)\sin(3x)$$

$$G(s) = G(s)$$

$$g(x) = L^{-1}[G(s)]$$

$$g(x) = g(x)$$

$$f(x) = \left(\frac{1}{3}\right)\sin(3x) \quad \text{AND} \quad g(x) = g(x)$$

Convolution Integral $= L^{-1}[F(s) \cdot G(s)] = \int_0^x f(x - \tau) \cdot g(\tau)\, d\tau$

$$\int_0^x f(x - \tau) \cdot g(\tau)\, d\tau = \left(\frac{1}{3}\right)\int_0^x \sin 3(x - \tau) \cdot g(\tau)\, d\tau$$

Therefore:

$$y = \cos(3x) + \sin(3x) + L^{-1}\left[\left(\frac{1}{s^2 + 9}\right) \cdot G(s)\right] \quad \text{(found previously)}$$

$$y = \cos(3x) + \sin(3x) + \left(\frac{1}{3}\right)\int_0^x \sin 3(x - \tau) \cdot g(\tau)\, d\tau$$

Example: Find the Particular Solution (P.S.) of the given DE.

$y'' + 3y' + 2y = \cos(ax)$ With: $y(0) = 0$ and $y'(0) = 1$

Solution:

$s^2 L[y] - sy(0) - y'(0) + esL[y] - 3y(0) + 2L[y] = L[\cos(ax)]$

$L[y](s^2 + 3s + 2) - 1 = \dfrac{s}{a^2 + s^2}$

$L[y](s+2)(s+1) = 1 + \dfrac{s}{a^2 + s^2}$

$L[y] = \dfrac{1}{(s+2)(s+1)} + \left(\dfrac{s}{a^2+s^2}\right) \cdot \left(\dfrac{1}{(s+2)(s+1)}\right)$

$L[y] = F(s) + F(s) \cdot \left(\dfrac{s}{a^2+s^2}\right)$

Partial Fraction Decomposition:

$F(s) = \dfrac{1}{(s+2)(s+1)} = \dfrac{A}{s+2} + \dfrac{B}{s+1}$

$1 = A(s+1) + B(s+2)$

$s = -2 \;\rightarrow\; 1 = A(-1) \;\rightarrow\; A = -1$

$s = -1 \;\rightarrow\; 1 = B(1) \;\rightarrow\; B = 1$

$F(s) = \dfrac{1}{(s+2)(s+1)} = \dfrac{-1}{s+2} + \dfrac{1}{s+1}$

$$F(s) = \frac{-1}{s+2} + \frac{1}{s+1}$$

$$L[y] = F(s) + F(s) \cdot \left(\frac{s}{\alpha^2 + s^2}\right)$$

$$L[y] = \left(\frac{-1}{s+2} + \frac{1}{s+1}\right) + \left(\frac{-1}{s+2} + \frac{1}{s+1}\right) \cdot \left(\frac{s}{\alpha^2 + s^2}\right)$$

$$L[y] = -\frac{1}{s+2} + \frac{1}{s+1} - \left(\frac{1}{s+2}\right) \cdot \left(\frac{s}{\alpha^2 + s^2}\right) + \left(\frac{1}{s+1}\right) \cdot \left(\frac{s}{\alpha^2 + s^2}\right)$$

$$\underbrace{\phantom{\left(\frac{1}{s+2}\right) \cdot \left(\frac{s}{\alpha^2 + s^2}\right)}}_{\text{Convolution Integral \#1}} \quad \underbrace{\phantom{\left(\frac{1}{s+1}\right) \cdot \left(\frac{s}{\alpha^2 + s^2}\right)}}_{\text{Convolution Integral \#2}}$$

Convolution Integral #1:

$F(s) = \frac{1}{s+2}$	$G(s) = \frac{s}{\alpha^2 + s^2}$
$f(x) = L^{-1}\left[\frac{1}{s+2}\right]$	$g(x) = L^{-1}\left[\frac{s}{\alpha^2 + s^2}\right]$
$f(x) = e^{-2x}$	$g(x) = \cos(\alpha x)$

$$L^{-1}[F(s) \cdot G(s)] = \int_0^x f(x-\tau) \cdot g(\tau) \, d\tau$$
$$= \int_0^x e^{-2(x-\tau)} \cdot \cos(\alpha \tau) \, d\tau$$

Convolution Integral #2:

$F(s) = \frac{1}{s+1}$	$G(s) = \frac{s}{\alpha^2 + s^2}$
$f(x) = L^{-1}\left[\frac{1}{s+1}\right]$	$g(x) = L^{-1}\left[\frac{s}{\alpha^2 + s^2}\right]$
$f(x) = e^{-x}$	$g(x) = \cos(\alpha x)$

$$L^{-1}[F(s) \cdot G(s)] = \int_0^x f(x-\tau) \cdot g(\tau)\, d\tau$$
$$= \int_0^x e^{-(x-\tau)} \cdot \cos(\alpha \tau)\, d\tau$$

$$L[y] = -\frac{1}{s+2} + \frac{1}{s+1} - \left(\frac{1}{s+2}\right)\cdot\left(\frac{s}{\alpha^2+s^2}\right) + \left(\frac{1}{s+1}\right)\cdot\left(\frac{s}{\alpha^2+s^2}\right)$$

$$y = -L^{-1}\left[\frac{1}{s+2}\right] + L^{-1}\left[\frac{1}{s+1}\right]$$
$$- L^{-1}\left[\left(\frac{1}{s+2}\right)\cdot\left(\frac{s}{\alpha^2+s^2}\right)\right] + L^{-1}\left[\left(\frac{1}{s+1}\right)\cdot\left(\frac{s}{\alpha^2+s^2}\right)\right]$$

$$y = -e^{-2x} + e^{-x}$$
$$- \int_0^x e^{-2(x-\tau)}\cdot\cos(\alpha\tau)\, d\tau + \int_0^x e^{-(x-\tau)}\cos(\alpha\tau)\, d\tau$$

SYSTEMS OF FIRST-ORDER DE

There are many physical problems that involve a number of separate elements, related to each other in some way (e.g. electrical networks). In such cases, we get systems of DEs.

Systems of first-order linear DEs with constant coefficients, is introduced in this chapter. Some elementary aspects of linear algebra are used to unify the presentation. Reducing systems of higher-order DEs with constant coefficients is discussed. Eigenvectors and eigenvalues are used to construct the fundamental solution set for the system. Three main cases, based on eigenvalues (real and distinct, repeated, or complex) are discussed. The method of variation parameter is used in this setting to solve non-homogeneous DEs.

Phase portraits and phase planes for linear systems are discussed in the second half of this chapter. All possible cases for equilibrium solutions are discussed in detail with supporting examples. There are many DEs, especially nonlinear ones, that are not susceptible to analytical solution in any reasonable manner. An efficient way to handle locally linear systems, using an approximation method, is discussed.

This chapter concludes with two interesting applications involving nonlinear systems: Competing Species and Predator-Prey systems.

SYSTEMS OF FIRST-ORDER LINEAR DEs

SYSTEMS OF FIRST-ORDER DEs (I.V.P Form)

$$x_1' = F_1(t, x_1, x_2, x_3, \ldots x_n)$$
$$x_2' = F_2(t, x_1, x_2, x_3, \ldots x_n)$$
$$x_3' = F_3(t, x_1, x_2, x_3, \ldots x_n)$$

.
.
.

$$x_n' = F_1(t, x_1, x_2, x_3, \ldots x_n)$$

Need one initial condition for each DE.

$$x_1(t_0) = x_1^0$$
$$x_2(t_0) = x_2^0$$
$$x_3(t_0) = x_3^0$$

.
.
.

$$x_n(t_0) = x_n^0$$

Remark: Every higher-order DE can be transformed into a system of First-Order DEs.

Differential Equations (DEs)

- First-Order Linear DE $\quad x' + p(t)x = g(t)$
- Second-Order Linear DE $\quad x'' + p(t)x' + q(t) = g(t)$

SYSTEM OF FIRST-ORDER LINEAR DEs

Just rearrange the above equation.

$$x_1' = P_{11}(t)\,x_1 + P_{12}(t)\,x_2 + \ldots P_{1n}(t)\,x_n + g_1(t)$$
$$x_2' = P_{21}(t)\,x_1 + P_{22}(t)\,x_2 + \ldots P_{2n}(t)\,x_n + g_2(t)$$
$$\ldots$$
$$x_n' = P_{n1}(t)\,x_1 + P_{n2}(t)\,x_2 + \ldots P_{nn}(t)\,x_n + g_n(t)$$

Initial Data: (All are numbers.)

$$x_1(t_0) = x_1^0$$
$$x_2(t_0) = x_2^0$$
$$\ldots$$
$$x_n(t_0) = x_n^0$$

THEOREM: Existence and Uniqueness
If $\quad P_{11}(t),\ P_{12}(t),\ \ldots\ P_{nn}(t) \quad$ are Continuous
And $g_1(t),\ g_2(t),\ \ldots\ g_n(t)$ are Continuous
On some interval (I) containing x_0
Then: There exists a Unique Solution of the System.

3 Examples: For each example, transform one higher-order DE into a system of First-Order linear equations.

Example (1) $x'' - 2x' - 3x = 0$

Solution:

Label them:

$x_1 = x$ Start here, then differentiate.

$x_1' = x' = x_2$ ➔ $x'' = x_2'$

$x_2' - 2x_2 - 3x_1 = 0$ Rewrite the original equation.

Now, we have a system First-Order DE equivalent to the original Second-Order DE

$$\begin{cases} x_1' = x_2 \\ x_2' = 3x_1 + 2x_2 \end{cases}$$

Example (2) $y''' - 2y' - 5 = 0$ (No y'' term.)

Solution:

Highest order = 3 ➜ Three First-Order linear equations.

$\quad y_1 = y$ \hfill Start here, then differentiate.

$\quad y_1' = y' = y_2$

$\quad y_2' = y'' = y_3$

$\quad y_3' = y''' = 2y_2 + 5$ \hfill Use the original equation
\hfill to get the last linear equation.

Now, we have a system of First-Order linear equations equivalent to the original Higher-Order DE

$$\begin{cases} y_1' = y_2 \\ y_2' = y_3 \\ y_3' = 2y_2 + 5 \end{cases}$$

Example (3) $v^{(4)} - 3v''' + 2v' - 5v = \sin t$ (No v'' term.)

Solution:

Transform to linear DE

$$\begin{aligned}
v_1 &= v & &\text{Start here, then differentiate.} \\
v_1' &= v_2 & &= v' \\
v_2' &= v_3 & &= v'' \\
v_3' &= v_4 & &= v''' \\
v_4' - 3v_4 + 2v_2 - 5v_1 &= \sin t & &\text{Use original equation.}
\end{aligned}$$

The system is:

$$\begin{cases} v_1' &= v_2 \\ v_2' &= v_3 \\ v_3' &= v_4 \\ v_4' &= 5v_1 - 2v_2 + 3v_4 + \sin t \end{cases}$$

EIGENVALUES AND EIGENVECTORS

	EIGEN VALUES & EIGEN VECTORS
Background	Consider: $Av = \lambda v$ Where: $A = n \times n$ matrix and $v = n \times 1$ vector. λ = scalar. $$\begin{bmatrix} a_{11} & a_{12} & \dots & a_{1n} \\ a_{21} & a_{22} & \dots & a_{2n} \\ \dots & & & \\ a_{n1} & a_{n2} & \dots & a_{nn} \end{bmatrix} \cdot \begin{bmatrix} v_1 \\ v_2 \\ \dots \\ v_n \end{bmatrix} = \lambda \cdot \begin{bmatrix} v_1 \\ v_2 \\ \dots \\ v_n \end{bmatrix}$$ $\quad\quad\quad n \times n \quad\quad\quad n \times 1 \;\rightarrow\; n \times 1$ $Av - \lambda v\quad\; = 0$ $Av - \lambda I_n v\; = 0$ $(A - \lambda I_n)v = 0$
Characteristic Equation	$\det(A - \lambda I_n) = 0$
Eigen Values	Eigen Values = Roots of the Characteristic Equation.
Eigen Vectors	Eigen Vectors = Non-zero vectors v, satisfying: $Av = \lambda v$

<u>Recommendation</u>: Review matrices and basic operations in the Appendix.

Two EXAMPLES: Compute Eigen Values and Eigen Vectors

Example (1) $A = \begin{pmatrix} 1 & 2 \\ 0 & 5 \end{pmatrix}$ This is a 2×2 matrix.

Solution:

$$\det(A - \lambda \cdot I_2) = 0$$

$$\det\left(\begin{bmatrix} 1 & 2 \\ 0 & 5 \end{bmatrix} - \lambda \begin{bmatrix} 1 & 0 \\ 0 & 1 \end{bmatrix}\right) = 0$$

$$\det\left(\begin{bmatrix} 1 & 2 \\ 0 & 5 \end{bmatrix} + \begin{bmatrix} -\lambda & 0 \\ 0 & -\lambda \end{bmatrix}\right) = 0$$

$$\det\left(\begin{bmatrix} 1-\lambda & 2 \\ 0 & 5-\lambda \end{bmatrix}\right) = 0$$

$(1-\lambda)(5-\lambda) - (2)(0) = 0$ ➔ $\lambda = 1, 5$ Eigen Values

Compute the Eigen Vector corresponding to $\lambda = 1$

$$Av = \lambda v$$

$$\begin{pmatrix} 1 & 2 \\ 0 & 5 \end{pmatrix} \cdot \begin{pmatrix} v_1 \\ v_2 \end{pmatrix} = 1 \cdot \begin{pmatrix} v_1 \\ v_2 \end{pmatrix}$$

$$\begin{pmatrix} v_1 + 2v_2 \\ 5v_2 \end{pmatrix} = \begin{pmatrix} v_1 \\ v_2 \end{pmatrix}$$

$v_1 + 2v_2 = v_1$ ➔ $5v_2 = v_2$ ➔ $v_2 = 0$

$v_1 + 2v_2 = v_1$ ➔ $v_1 = v_1$ ➔ v_1 is Arbitrary

Since v_1 is arbitrary, let $v_1 = 1$.   NOT = $v_2 = 0$

First Eigen Vector is: $V = \begin{pmatrix} v_1 \\ v_2 \end{pmatrix} = \begin{pmatrix} 1 \\ 0 \end{pmatrix}$

(Continued...)

Compute the Eigen Vector corresponding to $\lambda = 5$

$$Av = \lambda v$$

$$\begin{pmatrix} 1 & 2 \\ 0 & 5 \end{pmatrix} \cdot \begin{pmatrix} v_1 \\ v_2 \end{pmatrix} = 5 \cdot \begin{pmatrix} v_1 \\ v_2 \end{pmatrix}$$

$$\begin{pmatrix} v_1 + 2v_2 \\ 5v_2 \end{pmatrix} = \begin{pmatrix} 5v_1 \\ 5v_2 \end{pmatrix}$$

$$v_1 + 2v_2 = 5v_1$$
$$5v_2 = 5v_2 \quad \rightarrow \quad v_2 \text{ is Arbitrary}$$

Since v_2 is arbitrary, let $v_2 = 1$.   (could $= 0$ if $v_1 \neq 0$)

$$v_1 + 2v_2 = 5v_1$$
$$v_1 + 2(1) = 5v_1$$
$$2 = 4v_1 \quad \rightarrow \quad v_1 = \frac{1}{2}$$

Second Eigen Vector is: $V = \begin{pmatrix} v_1 \\ v_2 \end{pmatrix} = \begin{pmatrix} \frac{1}{2} \\ 1 \end{pmatrix}$

NOTE: If $v_2 = 0$ Then: $v_1 + 2(0) = 5v_1 \rightarrow v_1 = 0$

This is NOT permitted because $v_1 = 0$ AND $v_2 = 0$

Example (2) $A = \begin{pmatrix} 3 & -2 \\ 4 & -1 \end{pmatrix}$ This is a 2×2 matrix.

Solution:

$$\det(A - \lambda \cdot I_2) = 0$$

$$\det\left(\begin{bmatrix} 3 & -2 \\ 4 & -1 \end{bmatrix} - \lambda \begin{bmatrix} 1 & 0 \\ 0 & 1 \end{bmatrix}\right) = 0$$

$$\det\left(\begin{bmatrix} 3 & -2 \\ 4 & -1 \end{bmatrix} + \begin{bmatrix} -\lambda & 0 \\ 0 & -\lambda \end{bmatrix}\right) = 0$$

$$\det\left(\begin{bmatrix} 3-\lambda & -2 \\ 4 & -1-\lambda \end{bmatrix}\right) = 0$$

$$(3-\lambda)(-1-\lambda) + (2)(4) = 0$$

$$(3-\lambda)(-1-\lambda) + 8 = 0$$

$$-3 - 3\lambda + \lambda + \lambda^2 + 8 = 0$$

$$\lambda^2 - 2\lambda + 5 = 0$$

$$\lambda = \frac{2 \pm \sqrt{4 - 4(1)(5)}}{2(1)} = \frac{2 \pm \sqrt{-16}}{2} = 1 \pm 2i \quad \rightarrow \quad \text{Two Eigen Values}$$

Compute the Eigen Vector corresponding to $\lambda = 1 + 2i$

$$Av = \lambda v$$

$$\begin{pmatrix} 3 & -2 \\ 4 & -1 \end{pmatrix} \cdot \begin{pmatrix} v_1 \\ v_2 \end{pmatrix} = (1 + 2i) \cdot \begin{pmatrix} v_1 \\ v_2 \end{pmatrix}$$

$$\begin{pmatrix} 3v_1 - 2v_2 \\ 4v_1 - v_2 \end{pmatrix} = \begin{pmatrix} (1 + 2i) \cdot v_1 \\ (1 + 2i) \cdot v_2 \end{pmatrix}$$

(1) $3v_1 - 2v_2 = (1 + 2i) \cdot v_1$

(2) $4v_1 - v_2 = (1 + 2i) \cdot v_2$

These two equations turn out to be the same.

(1) $2v_1 - 2v_2 = (2i) \cdot v_1$

$v_1(2 - 2i) = 2v_2$

$v_1(1 - i) = v_2$

(2) $4v_1 - 2v_2 = (2i) \cdot v_2$

$4v_1 = v_2(2 + 2i)$

$2v_1 = v_2(1 + i)$

$2v_1(1 - i) = v_2(1 + i)(1 - i)$

$2v_1(1 - i) = v_2(1^2 - i^2)$

$2v_1(1 - i) = v_2(2)$

$v_1(1 - i) = v_2$ ➔ The two equations are the same.

Since both equations are the same, both v_1 and v_2 are arbitrary. But, they are related with this equation: $\quad v_1(1-i) = v_2$

Let: $v_1 = 1 \quad \rightarrow \quad (1)(1-i) = v_2 \quad \rightarrow \quad v_2 = 1-i$

$$V = \begin{pmatrix} v_1 \\ v_2 \end{pmatrix} = \begin{pmatrix} 1 \\ 1-i \end{pmatrix} \quad \text{This is the first Eigen Vector}$$

Corresponding to $\lambda = 1 + 2i$

Now, find the Eigen Vector corresponding to $\lambda = 1 - 2i$

$Av = \lambda v$

$$\begin{pmatrix} 3 & -2 \\ 4 & -1 \end{pmatrix} \cdot \begin{pmatrix} v_1 \\ v_2 \end{pmatrix} = (1-2i) \cdot \begin{pmatrix} v_1 \\ v_2 \end{pmatrix}$$

$$\begin{pmatrix} 3v_1 - 2v_2 \\ 4v_1 - v_2 \end{pmatrix} = \begin{pmatrix} (1-2i) \cdot v_1 \\ (1-2i) \cdot v_2 \end{pmatrix}$$

This gives two equations with two unknowns.

Two equations and two unknowns:

$$3v_1 - 2v_2 = (1 - 2i) \cdot v_1$$
$$4v_1 - v_2 = (1 - 2i) \cdot v_2$$

These two equations turn out to be the same.

(1) $\quad 2v_1 - 2v_2 = (-2i) \cdot v_1$

$\quad\quad v_1(2 + 2i) = 2v_2$

$\quad\quad v_1(1 + i) = v_2$

(2) $\quad 4v_1 - 2v_2 = (-2i) \cdot v_2$

$\quad\quad 4v_1 = v_2(2 - 2i)$

$\quad\quad 2v_1 = v_2(1 - i)$

$\quad\quad 2v_1(1 + i) = v_2(1 - i)(1 + i)$

$\quad\quad 2v_1(1 + i) = v_2(1^2 - i^2)$

$\quad\quad 2v_1(1 + i) = v_2(2)$

$\quad\quad v_1(1 + i) = v_2$ $\quad$ → The two equations are the same.

Since both equations are the same, both v_1 and v_2 are arbitrary.
But, they are related with this equation: $\quad v_1(1 + i) = v_2$

$\quad\quad\quad$ Let: $v_1 = 1$ $\quad$→$\quad (1)(1 + i) = v_2$ $\quad$→$\quad v_2 = 1 + i$

$$V = \begin{pmatrix} v_1 \\ v_2 \end{pmatrix} = \begin{pmatrix} 1 \\ 1 + i \end{pmatrix} \quad \text{This is the first Eigen Vector}$$

$\quad\quad\quad\quad\quad\quad\quad\quad\quad\quad\quad\quad$ Corresponding to $\lambda = 1 - 2i$

EIGEN VALUES & EIGEN VECTORS - REVIEW

Example: Compute all Eigen Values & corresponding Eigen Vectors.

$$\text{For:} \quad A = \begin{pmatrix} 3 & 2 & 2 \\ 1 & 4 & 1 \\ -2 & -4 & -1 \end{pmatrix}$$

Solution:

To compute Eigen Values, use: $\det(A - \lambda \cdot I_3) = 0$

$$\det\left(\begin{pmatrix} 3 & 2 & 2 \\ 1 & 4 & 1 \\ -2 & -4 & -1 \end{pmatrix} + \begin{pmatrix} -\lambda & 0 & 0 \\ 0 & -\lambda & 0 \\ 0 & 0 & -\lambda \end{pmatrix}\right) = 0$$

$$\det\begin{pmatrix} (3-\lambda) & 2 & 2 \\ 1 & (4-\lambda) & 1 \\ -2 & -4 & (-1-\lambda) \end{pmatrix} = 0$$

$(3-\lambda)[(4-\lambda)(-1-\lambda) - (-4)] - 2[(-1-\lambda) - (-2)]$
$\qquad\qquad + 2[-4 - (-2)(4-\lambda)] = 0$

$(3-\lambda)(-4 - 4\lambda + \lambda + \lambda^2 + 4) - 2(1-\lambda) + 2(-4 + 8 - 2\lambda) = 0$

$(3-\lambda)(\lambda^2 - 3\lambda) - 2(1-\lambda) + 2(4 - 2\lambda) = 0$

$(3-\lambda)(\lambda^2 - 3\lambda) - 2(1-\lambda) + 4(2 - \lambda) = 0$

$(3-\lambda)(\lambda^2 - 3\lambda) - 2 + 2\lambda + 8 - 4\lambda = 0$

$(3-\lambda)(\lambda^2 - 3\lambda) + 6 - 2\lambda = 0$

$(3-\lambda)(\lambda^2 - 3\lambda) + 2(3 - \lambda) = 0$

$(3-\lambda)(\lambda^2 - 3\lambda + 2) = (3-\lambda)(\lambda - 2)(\lambda - 1) = 0$

$(3-\lambda)(\lambda-2)(\lambda-1) = 0$ → $\lambda = 1, 2, 3$

Three Linearly Independent Eigen Values.

For $\lambda = 1$ Compute Eigen Vector corresponding to $\lambda = 1$

$$\boxed{\text{Definition of Eigen Vector:} \quad Av = \lambda v}$$

$$\begin{pmatrix} 3 & 2 & 2 \\ 1 & 4 & 1 \\ -2 & -4 & -1 \end{pmatrix} \cdot \begin{pmatrix} v_1 \\ v_2 \\ v_3 \end{pmatrix} = 1 \begin{pmatrix} v_1 \\ v_2 \\ v_3 \end{pmatrix}$$

$$\begin{pmatrix} 3v_1 + 2v_2 + 2v_3 \\ v_1 + 4v_2 + v_3 \\ -2v_1 - 4v_2 - v_3 \end{pmatrix} = \begin{pmatrix} v_1 \\ v_2 \\ v_3 \end{pmatrix}$$

(1.) $3v_1 + 2v_2 + 2v_3 = v_1$ → $v_1 + v_2 + v_3 = 0$

(2.) $v_1 + 4v_2 + v_3 = v_2$ → $v_1 + 3v_2 + v_3 = 0$

(3.) $-2v_1 - 4v_2 - v_3 = v_3$ → $-2v_1 - 4v_2 - 2v_3 = 0$

(1.) and (2.) → $2v_2 = 0$ → $v_2 = 0$

With $v_2 = 0$ → (3.) is a linear combination of (1.)
→ No Unique Sol'n.

With $v_2 = 0$ → (2.) $v_1 + v_3 = 0$ → $v_1 = -v_3$

Select: $v_1 = 1$ → $v_3 = -1$

Therefore, the corresponding Eigen Vector is: $V_1 = \begin{pmatrix} 1 \\ 0 \\ -1 \end{pmatrix}$

For $\lambda = 2$ $\quad$ Compute Eigen Vector corresponding to $\lambda = 2$

$$\boxed{\text{Definition of Eigen Vector:} \quad Av = \lambda v}$$

$$\begin{pmatrix} 3 & 2 & 2 \\ 1 & 4 & 1 \\ -2 & -4 & -1 \end{pmatrix} \cdot \begin{pmatrix} v_1 \\ v_2 \\ v_3 \end{pmatrix} = 2 \begin{pmatrix} v_1 \\ v_2 \\ v_3 \end{pmatrix}$$

$$\begin{pmatrix} 3v_1 + 2v_2 + 2v_3 \\ v_1 + 4v_2 + v_3 \\ -2v_1 - 4v_2 - v_3 \end{pmatrix} = \begin{pmatrix} 2v_1 \\ 2v_2 \\ 2v_3 \end{pmatrix}$$

(1.) $3v_1 + 2v_2 + 2v_3 = 2v_1$ → $v_1 + 2v_2 + 2v_3 = 0$

(2.) $v_1 + 4v_2 + v_3 = 2v_2$ → $v_1 + 2v_2 + v_3 = 0$

(3.) $-2v_1 - 4v_2 - v_3 = 2v_3$ → $-2v_1 - 4v_2 - 3v_3 = 0$

(1.) and (3.) → $v_3 = 0$

With $v_3 = 0$ → (3.) is a linear combination of (2.)

$\quad\quad\quad\quad\quad$ → No Unique Sol'n.

With $v_3 = 0$ → (1.) $v_1 + 2v_2 = 0$ → $v_1 = -2v_2$

Select: $v_2 = 1$ → $v_1 = -2$

Therefore, the corresponding Eigen Vector is: $V_2 = \begin{pmatrix} -2 \\ 1 \\ 0 \end{pmatrix}$

For $\lambda = 3$ Compute Eigen Vector corresponding to $\lambda = 3$

$$\boxed{\text{Definition of Eigen Vector:} \quad Av = \lambda v}$$

$$\begin{pmatrix} 3 & 2 & 2 \\ 1 & 4 & 1 \\ -2 & -4 & -1 \end{pmatrix} \cdot \begin{pmatrix} v_1 \\ v_2 \\ v_3 \end{pmatrix} = 3 \begin{pmatrix} v_1 \\ v_2 \\ v_3 \end{pmatrix}$$

$$\begin{pmatrix} 3v_1 + 2v_2 + 2v_3 \\ v_1 + 4v_2 + v_3 \\ -2v_1 - 4v_2 - v_3 \end{pmatrix} = \begin{pmatrix} 3v_1 \\ 3v_2 \\ 3v_3 \end{pmatrix}$$

(1.) $3v_1 + 2v_2 + 2v_3 = 3v_1$ → $2v_2 + 2v_3 = 0$

(2.) $v_1 + 4v_2 + v_3 = 3v_2$ → $v_1 + v_2 + v_3 = 0$

(3.) $-2v_1 - 4v_2 - v_3 = 3v_3$ → $-2v_1 - 4v_2 - 4v_3 = 0$

(1.) and (2.) → $v_1 = 0$

With $v_1 = 0$ → (2.) is a linear combination of (1.)
→ No Unique Sol'n.

With $v_1 = 0$ → (3.) $-4v_2 - 4v_3 = 0$ → $v_2 = -v_3$

Select: $v_3 = 1$ → $v_2 = -1$

Therefore, the corresponding Eigen Vector is: $V_3 = \begin{pmatrix} 0 \\ -1 \\ 1 \end{pmatrix}$

Systems of Differential Equations:

It is possible to rewrite a system of DEs in the form: $x' = Ax$

$$\boxed{\text{System of DEs:} \quad x' = Ax}$$

Examples:

Rewrite each of the following systems in the form: $x' = Ax$

System 1: $\begin{cases} x_1' = 2x_1 + 3x_2 \\ x_2' = 4x_1 - 5x_2 \end{cases} \rightarrow \begin{pmatrix} x_1' \\ x_2' \end{pmatrix} = \begin{pmatrix} 2 & 3 \\ 4 & -5 \end{pmatrix} \cdot \begin{pmatrix} x_1 \\ x_2 \end{pmatrix}$

System 2: $\begin{cases} u_1' = 2u_1 + 3u_2 - u_3 \\ u_2' = u_1 - u_3 \\ u_3' = 3u_2 \end{cases} \rightarrow \begin{pmatrix} u_1' \\ u_2' \\ u_3' \end{pmatrix} = \begin{pmatrix} 2 & 3 & -1 \\ 1 & 0 & -1 \\ 0 & 3 & 0 \end{pmatrix} \cdot \begin{pmatrix} u_1 \\ u_2 \\ u_3 \end{pmatrix}$

HOMOGENEOUS SYSTEM WITH CONST. COEFF.

HOMOGENEOUS SYSTEM WITH CONSTANT COEFFICIENTS

How to solve: $x' = Ax$ (Keep it simple. Let $A = 2 \times 2$ matrix.)	
Step 1	Express as: $x' = Ax$
Step 2	Compute Eigen Values by solving: $\det(A - \lambda \cdot I_2) = 0$ Get λ_1 and λ_2 → 3 Possible Cases (See Below)
Step 3	Get the General Solution.

Case	λ_1 & λ_2 → General Solution of DE $\quad x = \begin{pmatrix} x_1 \\ x_2 \end{pmatrix}$
I	λ_1, λ_2 are Real and Distinct. → $x = C_1 V_1 e^{\lambda_1 t} + C_2 V_2 e^{\lambda_2 t}$ Use: $Av = \lambda v$ to find V_1 & V_2 (Use it twice, once for each λ)
II	$\lambda_1 = p + iq$ and $\lambda_2 = p - iq$ → $x = C_1 V_1 e^{pt} \cos(qt) + C_2 V_2 e^{pt} \sin(qt)$ Use: $Av = \lambda v$ to find V_1 & V_2 (Use it twice, once for each λ)
III	$\lambda_1 = \lambda_2 = \lambda$ (Tricky!) Sometimes, using $Av = \lambda v$ once, we get two, unrelated Eigen vectors (not related and they are free). In this case: $x = C_1 V_1 e^{\lambda t} + C_2 V_2 e^{\lambda t}$
IV	$\lambda_1 = \lambda_2 = \lambda$ (Tricky!) Use: $Av = \lambda v$ $\qquad$ to find V_1 Use: $(A - \lambda \cdot I_2) V_2 = V_1$ $\quad$ to find V_2 Then get: $x_2 = x_1 t + V_2 e^{\lambda t}$ → $x = C_1 V_1 e^{\lambda t} + C_2 (x_1 t + V_2 e^{\lambda t})$

Case	λ_1 & λ_2 ➔ General Solution of DE $x = \begin{pmatrix} x_1 \\ x_2 \end{pmatrix}$
I	λ_1, λ_2 are Real and Distinct. ➔ $x = C_1 V_1 e^{\lambda_1 t} + C_2 V_2 e^{2t}$ • Opposite Signs. Saddle node is unstable. • Both negative. As $t \to \infty$ ➔ $x \to 0$ Stable Node. • Both positive. As $t \to \infty$ ➔ $x \to \infty$ Unstable Node
II	$\lambda_1 = p + iq$ and $\lambda_2 = p - iq$ ➔ $x = C_1 V_1 e^{pt} \cos(qt) + C_2 V_2 e^{pt} \sin(qt)$ • Real part Negative. As $t \to \infty$ ➔ $x \to 0$ • Origin is "Spiral Node." Asymptotic & stable solutions. • Real part Positive. As $t \to \infty$ ➔ $x \to \infty$ Unbounded, unstable solution. • Real part is Zero. Constant Closed Circle. (Oval). Solution will NOT converge. Bounded and closed.
III	$\lambda_1 = \lambda_2$ and we get two L.I. Eigen Vectors. In this case: $x = C_1 V_1 e^{\lambda t} + C_2 V_2 e^{\lambda t}$ • $\lambda =$ Positive ➔ Unstable "Star Node" • $\lambda =$ Negative ➔ Stable "Star Node"
IV	$\lambda_1 = \lambda_2$ Tricky! Use: $Av = \lambda v$ to find V_1 Use: $(A - \lambda \cdot I_2) V_2 = V_1$ to find V_2 Then get: $x_2 = x_1 t + V_2 e^{\lambda t}$ ➔ $x = C_1 V_1 e^{\lambda t} + C_2 (x_1 t + V_2 e^{\lambda t})$ • $\lambda =$ Positive ➔ Unstable "Improper Node" • $\lambda =$ Negative ➔ Stable "Improper Node"

Some Graphs:

Case	λ_1 & λ_2 → General Solution of DE $\quad x = \begin{pmatrix} x_1 \\ x_2 \end{pmatrix}$
I	λ_1, λ_2 are Real and Distinct. → $x = C_1 V_1 e^{\lambda_1 t} + C_2 V_2 e^{2t}$ • Opposite Signs. → Unstable Saddle Node. • Both Negative → Stable Node • Both Positive → Unstable Node
II	$\lambda_1 = p + iq$ and $\lambda_2 = p - iq$ → $x = C_1 V_1 e^{pt} \cdot \cos(qt) + C_2 V_2 e^{pt} \cdot \sin(qt)$ • Spiral if Real part ≠ 0 • Oval if Real part = 0 • Stable (Converges) if Real part is Negative. • Unstable (Diverges) if Real part is Positive.

This will be discussed in more detail, soon.

Example: Solve $\begin{cases} x_1' = x_1 + x_2 \\ x_2' = 4x_1 + x_2 \end{cases}$

Solution:

Find Matrix A

$$\begin{pmatrix} x_1' \\ x_2' \end{pmatrix} = \begin{pmatrix} 1 & 1 \\ 4 & 1 \end{pmatrix} \cdot \begin{pmatrix} x_1 \\ x_2 \end{pmatrix} \quad \rightarrow \quad A = \begin{pmatrix} 1 & 1 \\ 4 & 1 \end{pmatrix}$$

Find Eigen Values:

$$\boxed{\det(A - \lambda \cdot I_3) = 0}$$

$$\det \begin{vmatrix} (1-\lambda) & 1 \\ 4 & (1-\lambda) \end{vmatrix} = 0$$

$(1-\lambda)^2 - 4 = 0$

$\lambda^2 - 2\lambda + 1 - 4 = 0$

$\lambda^2 - 2\lambda - 3 = 0$

$(\lambda - 3)(\lambda + 1) = 0 \qquad \rightarrow \lambda_1 = 3 \quad \lambda_2 = -1$

Both Eigen Vectors are L.I.

For $\lambda_1 = 3$

$$Av = \lambda v$$

$$\begin{pmatrix} 1 & 1 \\ 4 & 1 \end{pmatrix} \begin{pmatrix} v_1 \\ v_2 \end{pmatrix} = 3 \begin{pmatrix} v_1 \\ v_2 \end{pmatrix}$$

$v_1 + v_2 = 3v_1$ ➔ $v_2 = 2v_1$

$4v_1 + v_2 = 3v_2$ ➔ $2v_2 = 4v_1$

These are the same equation ➔ Make an arbitrary choice.

Select: $v_1 = 1$ ➔ $v_2 = 2$

$$V_1 = \begin{pmatrix} 1 \\ 2 \end{pmatrix}$$

For $\lambda_2 = -1$

$$Av = \lambda v$$

$$\begin{pmatrix} 1 & 1 \\ 4 & 1 \end{pmatrix} \begin{pmatrix} v_1 \\ v_2 \end{pmatrix} = -1 \begin{pmatrix} v_1 \\ v_2 \end{pmatrix}$$

$v_1 + v_2 = -v_1$ ➔ $v_2 = -2v_1$

$4v_1 + v_2 = -v_2$ ➔ $-2v_2 = 4v_1$

These are the same equation ➔ Make an arbitrary choice.

Select: $v_1 = 1$ ➔ $v_2 = -2$

$$V_2 = \begin{pmatrix} 1 \\ -2 \end{pmatrix}$$

Optional:

Check to make sure V_1 and V_2 are linearly independent. (L.I.)

$$\det \begin{vmatrix} 1 & 1 \\ 2 & -2 \end{vmatrix} = (1)(-2) - (1)(2) = -2 - 2 = -4 \neq 0 \rightarrow \text{L.I.}$$

General Solution:

$$\boxed{x = \begin{pmatrix} x_1 \\ x_2 \end{pmatrix} = C_1 V_1 e^{\lambda_1 t} + C_2 V_2 e^{\lambda_2 t}}$$

$$x = \begin{pmatrix} x_1 \\ x_2 \end{pmatrix} = C_1 \begin{pmatrix} 1 \\ 2 \end{pmatrix} e^{3t} + C_2 \begin{pmatrix} 1 \\ -2 \end{pmatrix} e^{-t}$$

$$\begin{pmatrix} x_1 \\ x_2 \end{pmatrix} = C_1 \begin{pmatrix} e^{3t} \\ 2e^{3t} \end{pmatrix} + C_2 \begin{pmatrix} e^{-t} \\ -2e^{-t} \end{pmatrix}$$

Example: Solve the following system. $\begin{cases} x_1' = -x_1 - 4x_2 \\ x_2' = x_1 - x_2 \end{cases}$

Solution:

$$x' = A \cdot x$$

$$\begin{pmatrix} x_1' \\ x_2' \end{pmatrix} = \begin{pmatrix} -1 & -4 \\ 1 & -1 \end{pmatrix} \cdot \begin{pmatrix} x_1 \\ x_2 \end{pmatrix}$$

$$A = \begin{pmatrix} -1 & -4 \\ 1 & -1 \end{pmatrix}$$

$\det(A - \lambda_2) = 0$ ← For Eigen Values

$$\begin{vmatrix} (-1-\lambda) & -4 \\ 1 & (-1-\lambda) \end{vmatrix} = 0$$

$(-1-\lambda)(-1-\lambda) - (-4)(1) = 0$

$(-1)(1+\lambda)(-1)(1+\lambda) + 4 = 0$

$(1+\lambda)(1+\lambda) + 4 = 0$

$1 + 2\lambda + \lambda^2 + 4 = 0$

$\lambda^2 + 2\lambda + 5 = 0$

$$\lambda = \frac{-2 \pm \sqrt{4 - 4(1)(5)}}{2} = \frac{-2 \pm \sqrt{-16}}{2} = -1 \pm 2i$$

Eigen Values: $\lambda_1 = -1 + 2i$ and $\lambda_2 = -1 - 2i$

Now, find the Eigen Vector for each Eigen Value.

For: $\lambda_1 = -1 + 2i$

$Av = \lambda v$

$$\begin{pmatrix} -1 & -4 \\ 1 & -1 \end{pmatrix} \cdot \begin{pmatrix} v_1 \\ v_2 \end{pmatrix} = (-1 + 2i) \cdot \begin{pmatrix} v_1 \\ v_2 \end{pmatrix}$$

$$\begin{pmatrix} -v_1 - 4v_2 \\ v_1 - v_2 \end{pmatrix} = \begin{pmatrix} (-1 + 2i) \cdot v_1 \\ (-1 + 2i) \cdot v_2 \end{pmatrix}$$

(1.) $-v_1 - 4v_2 = (-1 + 2i) \cdot v_1$

(2.) $v_1 - v_2 = (-1 + 2i) \cdot v_2$

- These two equations are Linearly Dependent (L.D.) so they are just two copies of the same equation.
- So, you only need to use one equation and assign a non-zero value to one of the variables.

$v_1 - v_2 = (-1 + 2i) \cdot v_2$

$v_1 = v_2 - v_2 + 2i \cdot v_2$

$v_1 = 2i \cdot v_2$ ➔ Select: $v_2 = 1$ ➔ $v_1 = 2i$

The first Eigen Vector: $V_1 = \begin{pmatrix} 2i \\ 1 \end{pmatrix}$

The first Solution: $x_1 = V_1 \, e^{-t} \cos(2t)$

For: $\lambda_n = r + qi$ ➔ $x_n = \begin{cases} V_n \, e^{rt} \cos(qt) & q > 0 \\ V_n \, e^{rt} \sin(qt) & q < 0 \end{cases}$

For: $\lambda_2 = -1 - 2i$

$Av = \lambda v$

$$\begin{pmatrix} -1 & -4 \\ 1 & -1 \end{pmatrix} \cdot \begin{pmatrix} v_1 \\ v_2 \end{pmatrix} = (-1 - 2i) \cdot \begin{pmatrix} v_1 \\ v_2 \end{pmatrix}$$

$$\begin{pmatrix} -v_1 - 4v_2 \\ v_1 - v_2 \end{pmatrix} = \begin{pmatrix} (-1 - 2i) \cdot v_1 \\ (-1 - 2i) \cdot v_2 \end{pmatrix}$$

(1.) $\quad -v_1 - 4v_2 = (-1 - 2i) \cdot v_1$

(2.) $\quad v_1 - v_2 = (-1 - 2i) \cdot v_2$

- These two equations are Linearly Dependent (L.D.) so they are just two copies of the same equation.
- So, you only need to use one equation and assign a non-zero value to one of the variables.

$v_1 - v_2 = (-1 - 2i) \cdot v_2$

$v_1 = v_2 - v_2 - 2i \cdot v_2$

$v_1 = -2i \cdot v_2 \qquad \rightarrow$ Select: $v_2 = 1 \; \rightarrow \; v_1 = -2i$

The first Eigen Vector: $\quad V_2 = \begin{pmatrix} -2i \\ 1 \end{pmatrix}$

The first Solution: $\quad x_2 = V_2 \, e^{-t} \sin(2t)$

For: $\lambda_n = r + qi \quad \rightarrow \quad x_n = \begin{cases} V_n \, e^{rt} \cos(qt) & q > 0 \\ V_n \, e^{rt} \sin(qt) & q < 0 \end{cases}$

337

Previously, we found two solutions: x_1 and x_2

$$x_1 = V_1 e^{-t} \cos(2t) = \begin{pmatrix} 2i \\ 1 \end{pmatrix} e^{-t} \cos(2t)$$

$$x_2 = V_2 e^{-t} \sin(2t) = \begin{pmatrix} -2i \\ 1 \end{pmatrix} e^{-t} \sin(2t)$$

General Solution:

$$x = C_1 x_1 + C_2 x_2$$

$$x = C_1 V_1 e^{-t} \cos(2t) + C_2 V_2 e^{-t} \sin(2t)$$

Where:

$$V_1 = \begin{pmatrix} 2i \\ 1 \end{pmatrix} \quad \text{and} \quad V_2 = \begin{pmatrix} -2i \\ 1 \end{pmatrix}$$

Previously, we found two solutions: x_1 and x_2

$$x_1 = V_1 e^{-t} \cos(2t) = \begin{pmatrix} 2i \\ 1 \end{pmatrix} e^{-t} \cos(2t)$$

$$x_2 = V_2 e^{-t} \sin(2t) = \begin{pmatrix} -2i \\ 1 \end{pmatrix} e^{-t} \sin(2t)$$

General Solution:

$$x = C_1 x_1 + C_2 x_2$$

$$x = C_1 V_1 e^{-t} \cos(2t) + C_2 V_2 e^{-t} \sin(2t)$$

Where:

$$V_1 = \begin{pmatrix} 2i \\ 1 \end{pmatrix} \quad \text{and} \quad V_2 = \begin{pmatrix} -2i \\ 1 \end{pmatrix}$$

Example: With Repeated Root.

Solve this system: $x' = \begin{pmatrix} 4 & -2 \\ 8 & -4 \end{pmatrix} x$

Solution:

$$x' = A \cdot x$$

$$\begin{pmatrix} x_1' \\ x_2' \end{pmatrix} = \begin{pmatrix} 4 & -2 \\ 8 & -4 \end{pmatrix} \cdot \begin{pmatrix} x_1 \\ x_2 \end{pmatrix}$$

$$A = \begin{pmatrix} 4 & -2 \\ 8 & -4 \end{pmatrix}$$

$\det(A - \lambda_2) = 0$ ← For Eigen Values

$$\begin{vmatrix} (4 - \lambda) & -2 \\ 8 & (-4 - \lambda) \end{vmatrix} = 0$$

$(4 - \lambda)(-4 - \lambda) - (-2)(8) = 0$

$-16 - 4\lambda + 4\lambda + \lambda^2 + 16 = 0$

$\lambda^2 = 0$

Eigen Values: $\lambda_1 = \lambda_2 = 0$

For: $\lambda = 0$

$$Av = \lambda v$$

$$\begin{pmatrix} 4 & -2 \\ 8 & -4 \end{pmatrix} \cdot \begin{pmatrix} v_1 \\ v_2 \end{pmatrix} = (0) \cdot \begin{pmatrix} v_1 \\ v_2 \end{pmatrix}$$

$$\begin{pmatrix} 4v_1 - 2v_2 \\ 8v_1 - 4v_2 \end{pmatrix} = \begin{pmatrix} 0 \\ 0 \end{pmatrix}$$

(1.) $4v_1 - 2v_2 = 0$

(2.) $8v_1 - 4v_2 = 0$

These two equations are <u>Linearly Dependent</u> because they are just two copies of the same equation. So, you only need to use one equation and assign a non-zero value to one of the variables.

$4v_1 - 2v_2 = 0$

$2v_1 = v_2$ ➔ Select: $v_1 = 1$ ➔ $v_2 = 2$

The Eigen Vector: $V = V_1 = \begin{pmatrix} 1 \\ 2 \end{pmatrix}$

We know $V_1 = \begin{pmatrix} 1 \\ 2 \end{pmatrix}$ Now, find V_2

$$\boxed{(A - \lambda \cdot I_2) V_2 = V_1 \qquad \text{for Repeated Roots}}$$

$(A - \lambda \cdot I_2) V_2 = V_1$

$\begin{pmatrix} (4-\lambda) & -2 \\ 8 & (-4-\lambda) \end{pmatrix} \begin{bmatrix} v_1 \\ v_2 \end{bmatrix} = \begin{pmatrix} 1 \\ 2 \end{pmatrix}$

$\begin{pmatrix} (4) & -2 \\ 8 & (-4) \end{pmatrix} \begin{bmatrix} v_1 \\ v_2 \end{bmatrix} = \begin{pmatrix} 1 \\ 2 \end{pmatrix}$

$4v_1 - 2v_2 = 1 \quad \rightarrow \quad 2v_1 - v_2 = \frac{1}{2} \quad \rightarrow \quad v_2 = 2v_1 - \frac{1}{2}$

$8v_1 - 4v_2 = 2 \quad \rightarrow \quad 2v_1 - v_2 = \frac{1}{2} \quad$ SAME

Select $v_1 = 1 \quad \rightarrow \quad v_2 = 2(1) - \frac{1}{2} = \frac{3}{4}$

$V_2 = \begin{bmatrix} 1 \\ \begin{pmatrix} 3 \\ 4 \end{pmatrix} \end{bmatrix}$

Recall:	Case III	$\lambda_1 = \lambda_2$	Tricky

Use: $Av = \lambda v$ to find V_1
Use: $(A - \lambda \cdot I_2) V_2 = V_1$ to find V_2
Get: $x_2 = x_1 t + V_2 e^{\lambda t} \quad \rightarrow \quad x = C_1 V_1 e^{\lambda t} + C_2(x_1 t + V_2 e^{\lambda t})$

We found: $\lambda = \lambda_1 = \lambda_2 = 0$ ➔ Repeated Roots with $\lambda = 0$

Used: $Av = \lambda v$ to find V_1 ➔ $V_1 = \begin{bmatrix} 1 \\ 2 \end{bmatrix}$

Used: $(A - \lambda \cdot I_2)V_2 = V_1$ ➔ $V_2 = \begin{bmatrix} 1 \\ \frac{3}{4} \end{bmatrix}$

$x_1 = V_1 e^{\lambda t}$ ➔ $x_1 = \begin{bmatrix} 1 \\ 2 \end{bmatrix} e^0 = \begin{bmatrix} 1 \\ 2 \end{bmatrix}$

$x_2 = x_1 t + V_2 e^{\lambda t}$ ➔ $x_2 = \begin{bmatrix} 1 \\ 2 \end{bmatrix} t + \begin{bmatrix} 1 \\ \frac{3}{4} \end{bmatrix} e^0 = \begin{bmatrix} t+1 \\ 2t + \frac{3}{4} \end{bmatrix}$

Therefore: The General Solution is:

$$x = C_1 x_1 + C_2 x_2$$

$$x = C_1 \begin{bmatrix} 1 \\ 2 \end{bmatrix} + C_2 \begin{bmatrix} t+1 \\ 2t + \frac{3}{4} \end{bmatrix}$$

$$x_1 = \begin{bmatrix} 1 \\ 2 \end{bmatrix}$$

$$x_2 = \begin{bmatrix} t+1 \\ 2t + \frac{3}{4} \end{bmatrix}$$

NON-HOMOGENEOUS SYSTEM WITH CONST. COEFF.

Fundamental Matrix

First Order, NON-Homogeneous Linear System

With Constant Coefficients

The Fundamental Matrix can help with solving these systems.

FUNDAMENTAL MATRIX = Ψ

Let: $A = n \times n$ matrix.

Let: $\lambda_1, \lambda_2, \lambda_3, \ldots \lambda_n$ Be Eigen Values of Matrix A

Let: $V_1, V_2, V_3, \ldots V_n$ Be corresponding Eigen Vectors of Matrix A

Let: $x_i = V_i e^{\lambda_i t}$ For $i = 1, 2, 3, \ldots n$

Then: $\Psi(t)$ = Fundamental Matrix = The Solution Matrix

$$\Psi(t) = \text{Fundamental Matrix}$$
$$\Psi(t) = [\ x_1(t),\ x_2(t), x_3(t), \ldots x_n(t)\ ]$$

Example: For $A = \begin{pmatrix} -1 & -4 \\ 1 & 1 \end{pmatrix}$ Find the Fundamental Matrix.

Solution:

Previously, we found:

$$x_1 = V_1 e^{-t} \cos(2t) = \begin{pmatrix} 2i \\ 1 \end{pmatrix} e^{-t} \cos(2t)$$

$$x_2 = V_2 e^{-t} \sin(2t) = \begin{pmatrix} -2i \\ 1 \end{pmatrix} e^{-t} \sin(2t)$$

$$x_1 = \begin{pmatrix} 2i\, e^{-t} \cos(2t) \\ e^{-t} \cos(2t) \end{pmatrix} \quad \text{and} \quad x_2 = \begin{pmatrix} -2i \sin(2t) \\ e^{-t} \sin(2t) \end{pmatrix}$$

$\Psi(t) =$ Fundamental Matrix

$$\Psi(t) = \begin{bmatrix} 2i\, e^{-t} \cos(2t) & -2i \sin(2t) \\ e^{-t} \cos(2t) & e^{-t} \sin(2t) \end{bmatrix}$$

Example: Compute the Fundamental Matrix for:

$$A = \begin{bmatrix} 1 & 2 & 0 & 1 \\ 0 & 2 & 0 & 0 \\ 0 & 0 & 3 & 2 \\ 0 & 0 & 0 & 4 \end{bmatrix}$$

Note: All zeros below diagonal so this is an "Upper Triangular" matrix. So, the determinant is simply the product of the diagonal entries.

Solution:

$$\det(A - \lambda \cdot I_4) = 0$$

$$\begin{vmatrix} 1-\lambda & 2 & 0 & 1 \\ 0 & 2-\lambda & 0 & 0 \\ 0 & 0 & 3-\lambda & 2 \\ 0 & 0 & 0 & 4-\lambda \end{vmatrix} = 0$$

$(1-\lambda)(2-\lambda)(3-\lambda)(4-\lambda) = 0 \quad \rightarrow \quad \lambda = 1, 2, 3, 4$

For: $\lambda = 1$

$$\boxed{Av = \lambda v}$$

$$\begin{bmatrix} 1 & 2 & 0 & 1 \\ 0 & 2 & 0 & 0 \\ 0 & 0 & 3 & 2 \\ 0 & 0 & 0 & 4 \end{bmatrix} \cdot \begin{pmatrix} v_1 \\ v_2 \\ v_3 \\ v_4 \end{pmatrix} = (1) \begin{pmatrix} v_1 \\ v_2 \\ v_3 \\ v_4 \end{pmatrix}$$

(1.) $v_1 + 2v_2 + v_4 = v_1$

(2.) $2v_2 = v_2$ → $v_2 = 0$

(3.) $3v_3 + 2v_4 = v_3$

(4.) $4v_4 = v_4$ → $v_4 = 0$

(1.) $v_1 + 0 + 0 = v_1$ → $v_1 = Free$

(3.) $3v_3 + 0 = v_3$ → $v_3 = 0$

Select: $v_1 = 1$

$$V_1 = \begin{bmatrix} 1 \\ 0 \\ 0 \\ 0 \end{bmatrix}$$

$$x_1 = V_1 e^{\lambda_1 t} = \begin{bmatrix} 1 \\ 0 \\ 0 \\ 0 \end{bmatrix} e^t = \begin{bmatrix} e^t \\ 0 \\ 0 \\ 0 \end{bmatrix}$$

For: $\lambda = 2$

$$\boxed{Av = \lambda v}$$

$$\begin{bmatrix} 1 & 2 & 0 & 1 \\ 0 & 2 & 0 & 0 \\ 0 & 0 & 3 & 2 \\ 0 & 0 & 0 & 4 \end{bmatrix} \cdot \begin{pmatrix} v_1 \\ v_2 \\ v_3 \\ v_4 \end{pmatrix} = (2) \begin{pmatrix} v_1 \\ v_2 \\ v_3 \\ v_4 \end{pmatrix}$$

(1.) $v_1 + 2v_2 + v_4 = 2v_1$

(2.) $2v_2 = 2v_2$ ➔ $v_2 = Free$

(3.) $3v_3 + 2v_4 = 2v_3$

(4.) $4v_4 = 2v_4$ ➔ $v_4 = 0$

(3.) $3v_3 + 0 = 2v_3$ ➔ $v_3 = 0$

(1.) $v_1 + 2v_2 + 0 = 2v_1$ ➔ $2v_2 = v_1$

Select: $v_2 = 1$ ➔ $v_1 = 2$

$$V_2 = \begin{bmatrix} 2 \\ 1 \\ 0 \\ 0 \end{bmatrix}$$

$$x_2 = V_2 \, e^{\lambda_2 t} = \begin{bmatrix} 2 \\ 1 \\ 0 \\ 0 \end{bmatrix} e^{2t} = \begin{bmatrix} 2e^{2t} \\ e^{2t} \\ 0 \\ 0 \end{bmatrix}$$

For: $\lambda = 3$

$$\boxed{Av = \lambda v}$$

$$\begin{bmatrix} 1 & 2 & 0 & 1 \\ 0 & 2 & 0 & 0 \\ 0 & 0 & 3 & 2 \\ 0 & 0 & 0 & 4 \end{bmatrix} \cdot \begin{pmatrix} v_1 \\ v_2 \\ v_3 \\ v_4 \end{pmatrix} = (3) \begin{pmatrix} v_1 \\ v_2 \\ v_3 \\ v_4 \end{pmatrix}$$

(1.) $v_1 + 2v_2 + v_4 = 3v_1$

(2.) $2v_2 = 3v_2$ → $v_2 = 0$

(3.) $3v_3 + 2v_4 = 3v_3$

(4.) $4v_4 = 3v_4$ → $v_4 = 0$

(3.) $3v_3 + 0 = 3v_3$ → $v_3 = Free$

(1.) $v_1 + 0 + 0 = 3v_1$ → $v_1 = 0$

Select: $v_3 = 1$

$$V_3 = \begin{bmatrix} 0 \\ 0 \\ 1 \\ 0 \end{bmatrix}$$

$$x_3 = V_3 \, e^{\lambda_3 t} = \begin{bmatrix} 0 \\ 0 \\ 1 \\ 0 \end{bmatrix} e^{3t} = \begin{bmatrix} 0 \\ 0 \\ e^{3t} \\ 0 \end{bmatrix}$$

For: $\lambda = 4$

$$\boxed{Av = \lambda v}$$

$$\begin{bmatrix} 1 & 2 & 0 & 1 \\ 0 & 2 & 0 & 0 \\ 0 & 0 & 3 & 2 \\ 0 & 0 & 0 & 4 \end{bmatrix} \cdot \begin{pmatrix} v_1 \\ v_2 \\ v_3 \\ v_4 \end{pmatrix} = (4) \begin{pmatrix} v_1 \\ v_2 \\ v_3 \\ v_4 \end{pmatrix}$$

(1.) $v_1 + 2v_2 + v_4 = 4v_1$

(2.) $2v_2 = 4v_2$ ➔ $v_2 = 0$

(3.) $3v_3 + 2v_4 = 4v_3$

(4.) $4v_4 = 4v_4$ ➔ $v_4 = Free$

(3.) $3v_3 + 2v_4 = 4v_3$ ➔ $v_4 = \frac{1}{2} v_3$

(1.) $v_1 + 0 + v_4 = 4v_1$ ➔ $v_4 = 3v_1$

Select: $v_4 = 1$ ➔ $v_3 = 2$ $v_1 = \frac{1}{3}$

$$V_4 = \begin{bmatrix} \left(\frac{1}{3}\right) \\ 0 \\ 2 \\ 1 \end{bmatrix}$$

$$x_4 = V_4 \, e^{\lambda_4 t} = \begin{bmatrix} \left(\frac{1}{3}\right) \\ 0 \\ 2 \\ 1 \end{bmatrix} e^{4t} = \begin{bmatrix} \left(\frac{1}{3}\right) e^{4t} \\ 0 \\ 2e^{4t} \\ e^{4t} \end{bmatrix}$$

The Fundamental Matrix is:

$$\Psi = \begin{bmatrix} e^t & 2e^{2t} & 0 & \left(\frac{1}{3}\right)e^{4t} \\ 0 & e^{2t} & 0 & 0 \\ 0 & 0 & e^{3t} & 2e^{4t} \\ 0 & 0 & 0 & e^{4t} \end{bmatrix}$$

The Fundamental Matrix will be used to

Help solve a system of NON-Homogeneous D.E.'s

How to Solve Non-Homogeneous Systems (General Idea)

4 Methods to solve a Non-Homogeneous DE	
1)	Variation of Parameters (We have used this.)
2)	Undetermined Coefficients
3)	Diagonalization of Matrix A
4)	Laplace Transform

How to Solve a NON-Homogeneous System of DE's

Using the Method of Variation of Parameter

General Example: $x' = Ax + g(t)$

Where: $x, g(t)$ are $n \times 1$ vectors.

Solution:

Step	NON-Homogeneous System of DEs Solve using the Method of Variation of Parameter
1	• Solve the Homogeneous part. Ignore the $g(t)$ part. • Solve: $x' = Ax$ • Find: λ_1, λ_2 and V_1, V_2 • Compute: $x_1, x_2,$
2	• Construct the Fundamental Matrix • $\Psi(t) = [\, x_1 \quad x_2 \,]$ • Compute $\Psi^{-1}(t)$ Using any technique.
3	• Find: $\Psi^{-1}(t) \cdot g(t)$ • Then, Integrate it. • $\int \Psi^{-1}(t) \cdot g(t)\, dt$ Integrate every component
4	• Find the Specific Solution: • $X = \Psi(t) \cdot \int \Psi^{-1}(t) \cdot g(t)\, dt$
5	• Write the General Solution: • $x = C_1 x_1 + C_2 x_2 + X$

Example: Solve: $\begin{cases} x_1' = -2x_1 + x_2 + 2e^{-t} \\ x_2' = x_1 - 2x_2 + 3t \end{cases}$

Solution:

$$x' = Ax + g(t)$$

$$\begin{pmatrix} x_1' \\ x_2' \end{pmatrix} = \begin{pmatrix} -2 & 1 \\ 1 & -2 \end{pmatrix} \begin{pmatrix} x_1 \\ x_2 \end{pmatrix} + \begin{pmatrix} 2e^{-t} \\ 3t \end{pmatrix}$$

Step 1: Solve the Homogeneous System.

$$\begin{pmatrix} x_1' \\ x_2' \end{pmatrix} = \begin{pmatrix} -2 & 1 \\ 1 & -2 \end{pmatrix} \begin{pmatrix} x_1 \\ x_2 \end{pmatrix}$$

$$A = \begin{pmatrix} -2 & 1 \\ 1 & -2 \end{pmatrix}$$

$$\det(A - \lambda \cdot I_1) = 0$$

$$\begin{vmatrix} -2-\lambda & 1 \\ 1 & -2-\lambda \end{vmatrix} = 0$$

$(-2-\lambda)(-2-\lambda) - (1)(1) = 0$

$4 - (2)(-2)(\lambda) + \lambda^2 - 1 = 0$

$3 + 4\lambda + \lambda^2 = 0$

$(\lambda + 3)(\lambda + 1) = 0 \quad \rightarrow \quad \lambda = -3, -1$

Compute Eigen Vector: For $\lambda = -3$

$$Av = \lambda v$$

$$\begin{pmatrix} -2 & 1 \\ 1 & -2 \end{pmatrix} \begin{pmatrix} v_1 \\ v_2 \end{pmatrix} = (-3) \begin{pmatrix} v_1 \\ v_2 \end{pmatrix}$$

(1.) $-2v_1 + v_2 = -3v_1$

(2.) $v_1 - 2v_2 = -3v_2$

$\rightarrow v_1 = -v_2$

| Linearly Dependent. Same equation. |

Select: $v_1 = 1$ ➔ $v_2 = -1$

$$V_1 = \begin{bmatrix} 1 \\ -1 \end{bmatrix} \;\rightarrow\; x_1 = V_1 \, e^{\lambda_1 t} = \begin{bmatrix} 1 \\ -1 \end{bmatrix} e^{-3t} = \begin{bmatrix} e^{-3t} \\ -e^{-3t} \end{bmatrix}$$

Compute Eigen Vector: For $\lambda = -1$

$$Av = \lambda v$$

$$\begin{pmatrix} -2 & 1 \\ 1 & -2 \end{pmatrix} \begin{pmatrix} v_1 \\ v_2 \end{pmatrix} = (-1) \begin{pmatrix} v_1 \\ v_2 \end{pmatrix}$$

(1.) $-2v_1 + v_2 = -1\,v_1$

(2.) $v_1 - 2v_2 = -1v_2$

$\boxed{\text{Linearly Dependent. Same equation.}}$

$v_1 = v_2$

Select: $v_1 = 1$ ➔ $v_2 = 1$

$V_2 = \begin{bmatrix} 1 \\ 1 \end{bmatrix}$ ➔ $x_2 = V_2\, e^{\lambda_2 t} = \begin{bmatrix} 1 \\ 1 \end{bmatrix} e^{-t} = \begin{bmatrix} e^{-t} \\ e^{-t} \end{bmatrix}$

Construct the Fundamental Matrix

$$\Psi(t) = \begin{bmatrix} e^{-3t} & e^{-t} \\ -e^{-3t} & e^{-t} \end{bmatrix}$$

Compute the Inverse of Ψ (for a 2 × 2 matrix)

$$\boxed{\Psi^{-1}(t) = \frac{1}{determinant} \cdot \begin{bmatrix} switch & negative \\ negative & switch \end{bmatrix}}$$

$$\Psi^{-1}(t) = \frac{1}{e^{-4t} + e^{-4t}} \cdot \begin{bmatrix} e^{-t} & -e^{-t} \\ e^{-3t} & e^{-3t} \end{bmatrix}$$

$$\Psi^{-1}(t) = \frac{1}{2e^{-4t}} \cdot \begin{bmatrix} e^{-t} & -e^{-t} \\ e^{-3t} & e^{-3t} \end{bmatrix}$$

$$\Psi^{-1}(t) = \left(\frac{1}{2}\right) e^{4t} \begin{bmatrix} e^{-t} & -e^{-t} \\ e^{-3t} & e^{-3t} \end{bmatrix} = \begin{bmatrix} \frac{1}{2}e^{3t} & -\frac{1}{2}e^{3t} \\ \frac{1}{2}e^{t} & \frac{1}{2}e^{t} \end{bmatrix}$$

Compute: $\Psi^{-1}(t) \cdot g(t)$

$$\Psi^{-1}(t) \cdot g(t) = \begin{bmatrix} \frac{1}{2}e^{3t} & -\frac{1}{2}e^{3t} \\ \frac{1}{2}e^{t} & \frac{1}{2}e^{t} \end{bmatrix} \cdot \begin{pmatrix} 2e^{-t} \\ 3t \end{pmatrix}$$

$$\Psi^{-1}(t) \cdot g(t) = \begin{bmatrix} e^{2t} - \frac{3}{2}te^{3t} \\ 1 + \frac{3}{2}t\,e^{t} \end{bmatrix}$$

Integrate:

$$\int \Psi^{-1}(t) \cdot g(t)\, dt = \int \begin{bmatrix} e^{2t} - \frac{3}{2} t e^{3t} \\ 1 + \frac{3}{2} t e^t \end{bmatrix} dt$$

$$\int \Psi^{-1}(t) \cdot g(t)\, dt = \begin{bmatrix} \int e^{2t} - \frac{3}{2} t e^{3t}\, dt \\ \int 1 + \frac{3}{2} t e^t\, dt \end{bmatrix}$$

$\int e^{2t} - \frac{3}{2} t e^{3t}\, dt = \frac{1}{2} e^{2t} - \frac{3}{2} \left[\frac{1}{3} t e^{3t} - \frac{1}{9} e^{3t}\right]$

$\int e^{2t} - \frac{3}{2} t e^{3t}\, dt = \frac{1}{2} e^{2t} - \frac{1}{2} t e^{3t} + \frac{1}{6} e^{3t}$

Du		$\int dv$
t	+	e^{3t}
1	−	$\frac{1}{3} e^{3t}$
0	+	$\frac{1}{9} e^{3t}$

$\int 1 + \frac{3}{2} t e^t\, dt = t + \frac{3}{2}[t e^t - e^t]$

$\int 1 + \frac{3}{2} t e^t\, dt = t + \frac{3t e^t}{2} - \frac{3 e^t}{2}$

Du		$\int dv$
t	+	e^t
1	−	e^t
0	+	e^t

Therefore:

$$\int \Psi^{-1}(t) \cdot g(t)\, dt = \begin{bmatrix} \frac{1}{2}e^{2t} - \frac{1}{2}t e^{3t} + \frac{1}{6}e^{3t} \\ t + \frac{3te^t}{2} - \frac{3e^t}{2} \end{bmatrix}$$

Get the Specific Solution:

$$X = \Psi(t) \cdot \int \Psi^{-1}(t) \cdot g(t)\, dt$$

$$X = \begin{bmatrix} e^{-3t} & e^{-t} \\ -e^{-3t} & e^{-t} \end{bmatrix} \cdot \begin{bmatrix} \frac{1}{2}e^{2t} - \frac{1}{2}t e^{3t} + \frac{1}{6}e^{3t} \\ t + \frac{3te^t}{2} - \frac{3e^t}{2} \end{bmatrix}$$

$$X = \begin{bmatrix} \frac{e^{-t}}{2} - \frac{t}{2} + \frac{1}{6} + te^{-t} + \frac{3t}{2} - \frac{3}{2} \\ -\frac{e^{-t}}{2} + \frac{t}{2} - \frac{1}{6} + te^{-t} + \frac{3t}{2} - \frac{3}{2} \end{bmatrix}$$

$$X = \begin{bmatrix} \frac{e^{-t}}{2} + te^{-t} + t - \frac{8}{6} \\ -\frac{e^{-t}}{2} + te^{-t} + 2t - \frac{10}{6} \end{bmatrix}$$

$$X = \begin{bmatrix} \frac{e^{-t}}{2} + te^{-t} + t - \frac{4}{3} \\ -\frac{e^{-t}}{2} + te^{-t} + 2t - \frac{5}{3} \end{bmatrix}$$

$$X = \begin{bmatrix} \frac{e^{-t}}{2} + te^{-t} + t - \frac{4}{3} \\ -\frac{e^{-t}}{2} + te^{-t} + 2t - \frac{5}{3} \end{bmatrix}$$

$$X = \begin{bmatrix} 1 \\ -1 \end{bmatrix} \frac{e^{-t}}{2} + \begin{bmatrix} 1 \\ 1 \end{bmatrix} te^{-t} + \begin{bmatrix} 1 \\ 2 \end{bmatrix} t - \frac{1}{3} \begin{bmatrix} 4 \\ 5 \end{bmatrix}$$

General Solution: $\quad x = C_1 x_1 + C_2 x_2 + X$

Example: Solve: $\begin{cases} x_1' = x_1 + 2x_2 + e^{2t} \\ x_2' = 3x_2 + 3e^{4t} \end{cases}$

Solution:

$$\begin{bmatrix} x_1' \\ x_2' \end{bmatrix} = \begin{bmatrix} 1 & 2 \\ 0 & 3 \end{bmatrix} \cdot \begin{bmatrix} x_1 \\ x_2 \end{bmatrix} + \begin{bmatrix} e^{2t} \\ 3e^{3t} \end{bmatrix} \quad \rightarrow \quad A = \begin{bmatrix} 1 & 2 \\ 0 & 3 \end{bmatrix}$$

Find Eigen Values: $\det(A - \lambda_2) = 0$

$$\begin{vmatrix} (1-\lambda) & 2 \\ 0 & (3-\lambda) \end{vmatrix} = 0$$

$(1-\lambda)(3-\lambda) - 0 = 0 \qquad \rightarrow \lambda = 1, 3$

For $\lambda = 3$

$Av = \lambda v$

$\begin{bmatrix} 1 & 2 \\ 0 & 3 \end{bmatrix} \cdot \begin{bmatrix} v_1 \\ v_2 \end{bmatrix} = 3 \begin{bmatrix} v_1 \\ v_2 \end{bmatrix}$

$v_1 + 2v_2 = 3v_1 \qquad \rightarrow v_1 = v_2$

$3v_2 = 3v_2 \qquad \rightarrow v_2 = free$

Let $v_2 = 1 \qquad \rightarrow v_1 = 1 \qquad \rightarrow V_1 = \begin{bmatrix} 1 \\ 1 \end{bmatrix}$

For $\lambda = 1$

$Av = \lambda v$

$\begin{bmatrix} 1 & 2 \\ 0 & 3 \end{bmatrix} \cdot \begin{bmatrix} v_1 \\ v_2 \end{bmatrix} = 1 \begin{bmatrix} v_1 \\ v_2 \end{bmatrix}$

$v_1 + 2v_2 = v_1 \qquad \rightarrow v_2 = 0$

$3v_2 = v_2 \qquad \rightarrow v_2 = 0$

$v_1 = free \qquad\qquad\qquad\qquad \rightarrow V_2 = \begin{bmatrix} 1 \\ 0 \end{bmatrix}$

2 Linearly Independent Solutions:

$x_1 = V_1 e^{\lambda_1 t} \qquad\qquad x_2 = V_2 e^{\lambda_2 t}$

$x_1 = \begin{bmatrix} 1 \\ 1 \end{bmatrix} e^{3t} \qquad\qquad x_2 = \begin{bmatrix} 1 \\ 0 \end{bmatrix} e^t$

$x_1 = \begin{bmatrix} e^{3t} \\ e^{3t} \end{bmatrix} \qquad\qquad x_2 = \begin{bmatrix} e^t \\ 0 \end{bmatrix}$

Construct Fundamental Matrix, etc.

$\Psi(t) = \begin{bmatrix} e^{3t} & e^t \\ e^{3t} & 0 \end{bmatrix}$

$\Psi^{-1}(t) = \frac{1}{0 - e^{4t}} \begin{bmatrix} 0 & -e^t \\ -e^{3t} & e^{3t} \end{bmatrix} = \begin{bmatrix} 0 & e^{-3t} \\ e^{-t} & -e^{-t} \end{bmatrix}$

$\Psi^{-1}(t) \cdot g(t) = \begin{bmatrix} 0 & e^{-3t} \\ e^{-t} & -e^{-t} \end{bmatrix} \cdot \begin{bmatrix} e^{2t} \\ 3e^{4t} \end{bmatrix} = \begin{bmatrix} 3e^t \\ e^t - 3e^{3t} \end{bmatrix}$

Integrate:

$$\int \Psi^{-1}(t) \cdot g(t)\, dt = \int \begin{bmatrix} 3e^t \\ e^t - 3e^{3t} \end{bmatrix} dt = \begin{bmatrix} 3e^t \\ e^t - e^{3t} \end{bmatrix}$$

Specific Solution:

$$X = \Psi(t) \cdot \int \Psi^{-1}(t) \cdot g(t)\, dt = \begin{bmatrix} e^{3t} & e^t \\ e^{3t} & 0 \end{bmatrix} \cdot \begin{bmatrix} 3e^t \\ e^t - e^{3t} \end{bmatrix}$$

$$X = \begin{bmatrix} 3e^{4t} + e^{2t} - e^{4t} \\ 3e^{4t} \end{bmatrix} = \begin{bmatrix} 2e^{4t} + e^{2t} \\ 3e^{4t} \end{bmatrix}$$

General Solution:

$$x = C_1 x_1 + C_2 x_2 + X$$

$$x = C_1 \begin{bmatrix} 1 \\ 1 \end{bmatrix} e^{3t} + C_2 \begin{bmatrix} 1 \\ 0 \end{bmatrix} e^t + \begin{bmatrix} 2e^{4t} + e^{2t} \\ 3e^{4t} \end{bmatrix}$$

Example: Solve the System: $\begin{cases} x' = -x - y + e^t \\ y' = -4x - y \end{cases}$

Using Variation of Parameters.

Solution:

Step 1: Find the Homogeneous Solution:

$\begin{bmatrix} x' \\ y' \end{bmatrix} = \begin{bmatrix} -1 & -1 \\ -1 & -1 \end{bmatrix} \begin{bmatrix} x \\ y \end{bmatrix} + \begin{bmatrix} e^t \\ 0 \end{bmatrix}$ Rewrite the system.

$\det(A - \lambda I_2) = 0$ Use this to get Eigen Values.

$\begin{vmatrix} (-1-\lambda) & -1 \\ -4 & (-1-\lambda) \end{vmatrix} = 0$

$(1+\lambda)(1+\lambda) - 4 = 0$

$\lambda^2 + 2\lambda - 3 = 0$

$(\lambda - 3)((\lambda - 1) = 0$ ➔ $\lambda_1 = 1 \quad \lambda_2 = 3$

Now, find Eigen Vectors for each Eigen Value: $\lambda_1 = 1 \quad \lambda_2 = 3$

(Next Page)

Find the Eigen Vectors for each Eigen Value: $\lambda_1 = 1$ and $\lambda_2 = 3$

For: $\lambda = 1$ Solve: $Av = \lambda v$	$\begin{bmatrix} -1 & -1 \\ -1 & -1 \end{bmatrix} \begin{bmatrix} v_1 \\ v_2 \end{bmatrix} = (1) \begin{bmatrix} v_1 \\ v_2 \end{bmatrix}$ $-v_1 - v_2 = v_1 \rightarrow v_2 = -2v_1$ $-4v_1 - v_2 = v_2$ (Same Eqn.) Let: $v_1 = 1 \rightarrow v_2 = -2 \rightarrow V_1 = \begin{bmatrix} 1 \\ -2 \end{bmatrix}$ 1st Solution: $x_1 = V_1 e^{(1)t} = \begin{bmatrix} 1 \\ -2 \end{bmatrix} e^t = \begin{bmatrix} e^t \\ -2t^t \end{bmatrix}$
For: $\lambda = -3$ Solve: $Av = \lambda v$	$\begin{bmatrix} -1 & -1 \\ -1 & -1 \end{bmatrix} \begin{bmatrix} v_1 \\ v_2 \end{bmatrix} = (-3) \begin{bmatrix} v_1 \\ v_2 \end{bmatrix}$ $-v_1 - v_2 = -3v_1 \rightarrow v_2 = 2v_1$ $-4v_1 - v_2 = -3v_2$ (Same Eqn.) Let: $v_1 = 1 \rightarrow v_2 = 2 \rightarrow V_1 = \begin{bmatrix} 1 \\ 2 \end{bmatrix}$ 2nd Solution: $x_2 = V_2 e^{(-3)t} = \begin{bmatrix} 1 \\ 2 \end{bmatrix} e^{-3t} = \begin{bmatrix} e^{-3t} \\ 2e^{-3t} \end{bmatrix}$

Step 2: Find Fundamental Matrix and its inverse.

$$\psi(t) = [x_1 \; x_2] = \begin{bmatrix} e^t & e^{-3t} \\ -2e^t & 2e^{-3t} \end{bmatrix}$$

$$\psi^{-1}(t) = \frac{1}{det(\psi)} \begin{bmatrix} 2e^{-3t} & -e^{-3t} \\ 2e^t & e^t \end{bmatrix}$$

With: $det(\psi(t)) = 2e^{2t} + 2e^{-2t} = 4e^{-2t}$

$$\psi^{-1}(t) = \frac{e^{2t}}{4} \begin{bmatrix} 2e^{-3t} & -e^{-3t} \\ 2e^t & e^t \end{bmatrix} = \begin{bmatrix} \dfrac{e^{-t}}{2} & -\dfrac{e^{-t}}{4} \\ \dfrac{e^{3t}}{2} & \dfrac{e^{3t}}{4} \end{bmatrix}$$

$$\psi^{-1}(t) \cdot g(t) = \begin{bmatrix} \dfrac{e^{-t}}{2} & -\dfrac{e^{-t}}{4} \\ \dfrac{e^{3t}}{2} & \dfrac{e^{3t}}{4} \end{bmatrix} \cdot \begin{bmatrix} e^t \\ 0 \end{bmatrix} = \begin{bmatrix} \left(\dfrac{e^0}{2}\right) \\ \left(\dfrac{e^{4t}}{2}\right) \end{bmatrix} = \begin{bmatrix} \left(\dfrac{1}{2}\right) \\ \left(\dfrac{e^{4t}}{2}\right) \end{bmatrix}$$

Step 3: Integrate and generate the General Solution.

Compute:

$$\int \psi^{-1}(t) \cdot g(t)\, dt = \begin{bmatrix} \int \left(\frac{1}{2}\right) dt \\ \int \left(\frac{e^{4t}}{2}\right) dt \end{bmatrix} = \begin{bmatrix} \left(\frac{t}{2}\right) \\ \left(\frac{e^{4t}}{8}\right) \end{bmatrix}$$

Specific Solution:

$$X = \psi(t) \cdot \int \psi^{-1}(t) \cdot g(t)\, dt$$

$$X = \begin{bmatrix} e^t & e^{-3t} \\ -2e^t & 2e^{-3t} \end{bmatrix} \cdot \begin{bmatrix} \left(\frac{t}{2}\right) \\ \left(\frac{e^{4t}}{8}\right) \end{bmatrix} \quad \text{Note: 2x2 * 2x1} \rightarrow \text{2x1 Matrix}$$

$$X = \begin{bmatrix} \left(\frac{te^t}{2} + \frac{e^t}{8}\right) \\ \left(-te^t + \frac{e^t}{4}\right) \end{bmatrix}$$

The General Solution is: $\quad x = C_1 x_1 + C_2 x_2 + X$

Note: If we had initial conditions, we could solve for C_1 and C_2 to find the Particular Solution (PS). But, we don't have initial conditions.

THE PHASE PLANE

THE PHASE PLANE

Remarks:

1) The solution of $x' = 0$ is called the <u>Equilibrium Solution</u> of the system $x' = Ax$

Example: Compute Equilibrium Solution

for: $x' = \begin{bmatrix} 1 & 2 \\ 5 & 7 \end{bmatrix} x$

Solution:

$\begin{bmatrix} 1 & 2 \\ 5 & 7 \end{bmatrix} \cdot \begin{bmatrix} x_1 \\ x_2 \end{bmatrix} = 0$

$x_1 + 2x_2 = 0 \quad \rightarrow \quad x_1 = -2x_2$

$5x_1 + 7x_2 = 0 \quad \rightarrow \quad x_2 = 0 \quad \rightarrow \quad x_1 = 0$

Note: $\begin{bmatrix} 0 \\ 0 \end{bmatrix}$ is ALWAYS the equilibrium solution for a homogeneous system of first-order linear DEs.

---

2) Matrix A must be invertible or singular $\rightarrow$ det$[A] \neq 0$

3) The curve, representing the general solution is called the path, solution trajectories, or <u>Phase Portrait</u>. The plane, containing the solution, is called the "<u>Phase Plane</u>." To analyze the solution, consider the Eigen Values and Eigen Vectors.

PHASE PLANE -- Real & Distinct Eigen Values

Phase Plane	Notes
Real & Distinct Eigen Values Same sign Both Negative $\lambda_1 < \lambda_2 < 0$	$x = C_1 V_1 e^{\lambda_1 t} + C_2 V_2 e^{\lambda_2 t}$ $x = e^{\lambda_2 t}[C_1 V_1 e^{(\lambda_1 - \lambda_2)t} + C_2 V_2]$ Note: $(\lambda_1 - \lambda_2) < 0$ $t \to \infty \Rightarrow x \to 0$ (both) All trajectories approach 0 thru V_2 Equilibrium Point: Node Stability: Asymptotically Stable Solution
Real & Distinct Roots Same sign Both Positive $\lambda_1 > \lambda_2 > 0$	$x = C_1 V_1 e^{\lambda_1 t} + C_2 V_2 e^{\lambda_2 t}$ $x = e^{\lambda_2 t}[C_1 V_1 e^{(\lambda_1 - \lambda_2)t} + C_2 V_2]$ Note: $(\lambda_1 - \lambda_2) > 0$, $\lambda_2 > 0$ $t \to \infty \Rightarrow x \to \infty$ (both) All trajectories approach ∞ thru V_2 Equilibrium Point: Node Stability: Unstable
Real & Distinct Roots Opposite Signs $\lambda_1 > 0$ and $\lambda_2 < 0$	$x = C_1 V_1 e^{\lambda_1 t} + C_2 V_2 e^{\lambda_2 t}$ Positive exponent diverges Negative exponent converges Equilibrium Point: Saddle Node Stability: Unstable

PHASE PLANE – Repeated Roots

Phase Plane	Notes
Repeated Roots $\lambda = \lambda_1 = \lambda_2 < 0$ Two L.I. Eigen Vectors (e.g. Both components free)	Two L.I. solutions $x = C_1 V_1 e^{\lambda t} + C_2 V_2 e^{\lambda t}$ All solutions converge. Equilibrium Point: "Proper Node" Stability: Asymptotically Stable
Repeated Roots $\lambda = \lambda_1 = \lambda_2 < 0$ One L.I. Eigen Vector Construct V_2 with special equation. (e.g. One component free.)	$\det(A - \lambda I_2) = 0 \rightarrow \lambda$ $Av = \lambda v \rightarrow V_1$ $(A - \lambda I_2)V_2 = V_1 \rightarrow V_2$ $x = C_1 V_1 e^{\lambda t} + C_2 (V_1 t e^{\lambda t} + V_2 e^{\lambda t})$ Equilibrium Point: "Improper Node" Stability: Asymptotically Stable. Phase Portrait: Swirl Appearance
Repeated Roots $\lambda = \lambda_1 = \lambda_2 > 0$ Two L.I. Eigen Vectors	Same as for first case BUT Unstable. Equilibrium Point: "Proper Node" Also called "Star Node" Stability: Unstable.
Repeated Roots $\lambda = \lambda_1 = \lambda_2 > 0$ One L.I. Eigen Vector Construct V_2 with special equation.	Same as second case BUT Unstable.

Phase Portrait – Diverging and Converging solutions

For second order DE, the general solution is: $x = x_1 + x_2$
Where: x_1 = the first solution and x_2 = the second solution.

When solving systems of DE in the form: $\begin{cases} x' = f(x,y) \\ y' = f(x,y) \end{cases}$

The solutions may be in a format, similar to:

$$x_1 = C_1 V_1 e^{\lambda_1 t} \quad \text{and} \quad x_2 = C_2 V_2 e^{\lambda_2 t}$$

Where:

- C_1, C_2 = Constants
- V_1, V_2 = Vectors
- λ_1, λ_2 = Eigen Values
- Note: V_n corresponds to λ_n

If the Eigen Value is negative, the solution will converge.
If the Eigen Value is positive, the solution will diverge.
As you have seen, the Eigen Values may be both negative, both positive, different signs, or zero.

Sketches of the first solutions, for various types of DE systems are provided on the following page.

General Solution ($x = x_1 + x_2$) and Graphs for a Second-Order DE

Eigen Values	Phase Portrait	x_1 VS time
Real and Distinct Both Negative $\lambda_1 < \lambda_2 < 0$ $x = C_1 V_1 e^{\lambda_1 t} + C_2 V_2 e^{\lambda_2 t}$		
Real and Distinct Both Positive $\lambda_1 > \lambda_2 > 0$ $x = C_1 V_1 e^{\lambda_1 t} + C_2 V_2 e^{\lambda_2 t}$		
Real and Distinct $\lambda_1 > 0, \lambda_2 < 0$ $x = C_1 V_1 e^{\lambda_1 t} + C_2 V_2 e^{\lambda_2 t}$		

General Solution ($x = x_1 + x_2$) and Graphs for a Second-Order DE

Eigen Values	Phase Portrait	x_1 VS time
Real and Repeating $\lambda_1 = \lambda_2 = \lambda > 0$ L.I. (2 Eigen Vectors) $x = C_1 V_1 e^{\lambda t}$ $\quad + C_2 V_2 e^{\lambda t}$		
Real and Repeating $\lambda_1 = \lambda_2 = \lambda < 0$ L.I. (2 Eigen Vectors) $x = C_1 V_1 e^{\lambda t}$ $\quad + C_2 V_2 e^{\lambda t}$		
Real and Repeating $\lambda_1 = \lambda_2 = \lambda > 0$ L.D. (1 Eigen Vector) $x = C_1 V_1 e^{\lambda t}$ $\quad + C_2 (V_1 t e^{\lambda t} + V_2 e^{\lambda t})$		
Real and Repeating $\lambda_1 = \lambda_2 = \lambda < 0$ L.D. (1 Eigen Vector) $x = C_1 V_1 e^{\lambda t}$ $\quad + C_2 (V_1 t e^{\lambda t} + V_2 e^{\lambda t})$		

General Solution ($x = x_1 + x_2$) and Graphs for a Second-Order DE

Eigen Values	Phase Portrait	x_1 VS time
Complex $\lambda = p \pm iq$ Real Part Negative $$x = C_1 e^{pt} \cos qt + C_2 e^{pt} \sin qt$$		
Complex $\lambda = p \pm iq$ Real Part Positive $$x = C_1 e^{pt} \cos qt + C_2 e^{pt} \sin qt$$		
Complex $\lambda = \pm iq$ Real Part Missing $$x = C_1 \cos qt + C_2 \sin qt$$		

General Solution ($x = x_1 + x_2$) and Graphs for a Second-Order DE		
Eigen Values	Phase Portrait	x_1 VS time
Real and Distinct One ZERO, one Neg. $\lambda_1 = 0$, $\lambda_2 < 0$ $x = C_1 V_1 + C_2 V_2 e^{\lambda_2 t}$		
Real and Distinct One ZERO, one Pos. $\lambda_1 > 0$, $\lambda_2 = 0$ $x = C_1 V_1 e^{\lambda_1 t} + C_2 V_2$		
Repeating Both ZERO $\lambda_1 = \lambda_2 = \lambda = 0$ $x = C_1 V_1 + C_2 [V_1 t + V_2]$		

Real and Distinct Eigen Values: $\lambda_1 < \lambda_2 < 0$ (Both Negative)	
General Solution (G.S.) $x_1 = V_1 e^{\lambda_1 t}$ and $x_2 = V_2 e^{\lambda_2 t}$ $x = C_1 V_1 e^{\lambda_1 t} + C_2 V_2 e^{\lambda_2 t}$	Equilibrium: Node, Stable

Example: Solve the system. $x' = \begin{pmatrix} -2 & 1 \\ 1 & -2 \end{pmatrix} x$	
Find Eigen Values: Use: $A - \lambda I_2 = 0$ $\begin{vmatrix} -2-\lambda & 1 \\ 1 & -2-\lambda \end{vmatrix} = 0$ $\lambda = -3, -1$	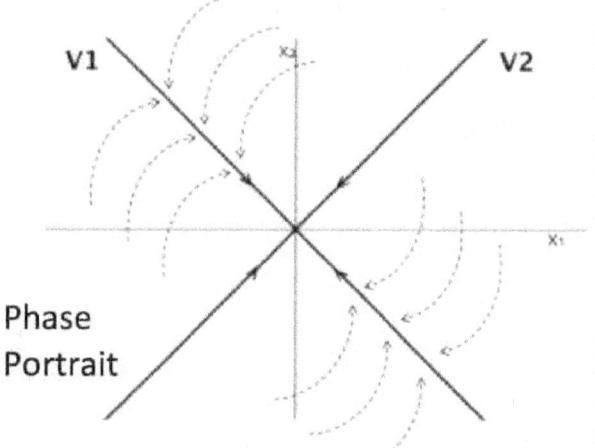 Phase Portrait
For: $\lambda = -3$ Use: $Av = \lambda v$ $\begin{pmatrix} -2 & 1 \\ 1 & -2 \end{pmatrix} \begin{pmatrix} v_1 \\ v_2 \end{pmatrix} = -3 \begin{pmatrix} v_1 \\ v_2 \end{pmatrix}$ $-2v_1 + v_2 = -3v_1 \rightarrow v_2 = -v_1$ $v_1 - 2v_2 = -3v_2 \rightarrow v_1 = -v_2$ L.D. $\rightarrow$ Let $v_2 = 1, v_1 = -1$ $V_1 = \begin{bmatrix} -1 \\ 1 \end{bmatrix}$	For: $\lambda = -1$ Use: $Av = \lambda v$ $\begin{pmatrix} -2 & 1 \\ 1 & -2 \end{pmatrix} \begin{pmatrix} v_1 \\ v_2 \end{pmatrix} = -1 \begin{pmatrix} v_1 \\ v_2 \end{pmatrix}$ $-2v_1 + v_2 = -v_1 \rightarrow v_2 = v_1$ $v_1 - 2v_2 = -v_2 \rightarrow v_1 = v_2$ L.D. $\rightarrow$ Let $v_1 = 1, v_2 = 1$ $V_2 = \begin{bmatrix} 1 \\ 1 \end{bmatrix}$
General Solution: $x = C_1 \begin{bmatrix} -1 \\ 1 \end{bmatrix} e^{-3t} + C_2 \begin{bmatrix} 1 \\ 1 \end{bmatrix} e^{-t}$	

Real and Distinct Eigen Values: $\lambda_1 > \lambda_2 > 0$ (Both Positive)	
General Solution (G.S.) $x_1 = V_1 e^{\lambda_1 t}$ and $x_2 = V_2 e^{\lambda_2 t}$ $x = C_1 V_1 e^{\lambda_1 t} + C_2 V_2 e^{\lambda_2 t}$	Equilibrium: Node, Unstable

Example: Solve the system. $x' = \begin{pmatrix} 1 & 2 \\ 0 & 3 \end{pmatrix} x$	
Find Eigen Values: Use: $A - \lambda I_2 = 0$ $\begin{vmatrix} 1-\lambda & 2 \\ 0 & 3-\lambda \end{vmatrix} = 0$ $\lambda = 3, 1$	Phase Portrait 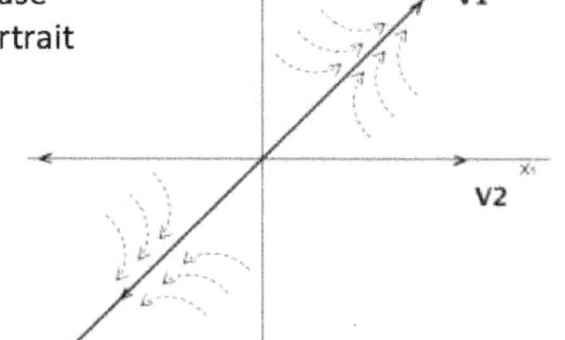
For: $\lambda = 3$ Use: $Av = \lambda v$ $\begin{pmatrix} 1 & 2 \\ 0 & 0 \end{pmatrix}\begin{pmatrix} v_1 \\ v_2 \end{pmatrix} = 3\begin{pmatrix} v_1 \\ v_2 \end{pmatrix}$ $v_1 + 2v_2 = 3v_1 \rightarrow v_2 = -v_1$ $3v_2 = 3v_2 \rightarrow v_2 = free$ Let $v_2 = 1$ $v_1 + 2(1) = 3v_1 \rightarrow v_1 = 1$ $V_1 = \begin{bmatrix} 1 \\ 1 \end{bmatrix}$	For: $\lambda = 1$ Use: $Av = \lambda v$ $\begin{pmatrix} 1 & 2 \\ 0 & 0 \end{pmatrix}\begin{pmatrix} v_1 \\ v_2 \end{pmatrix} = 1\begin{pmatrix} v_1 \\ v_2 \end{pmatrix}$ $v_1 + 2v_2 = v_1$ $3v_2 + 2v_2 = v_1$ $3v_2 = v_2 \rightarrow v_2 = 0$ $v_1 + 2(0) = v_1 \rightarrow v_1 = v_1$ (free) Let $v_1 = 1 \rightarrow V_2 = \begin{bmatrix} 1 \\ 0 \end{bmatrix}$
General Solution: $x = C_1 \begin{bmatrix} 1 \\ 1 \end{bmatrix} e^{3t} + C_2 \begin{bmatrix} 1 \\ 0 \end{bmatrix} e^t$	

Real and Distinct Eigen Values: $\lambda_1 > 0$ and $\lambda_2 < 0$ (Different Signs)	
General Solution (G.S.) $$x = C_1 V_1 e^{\lambda_1 t} + C_2 V_2 e^{\lambda_2 t}$$	Equilibrium: Saddle Node, Unstable

Example: Solve the system. $x' = \begin{pmatrix} 1 & 1 \\ 4 & 1 \end{pmatrix} x$	
Find Eigen Values: Use: $A - \lambda I_2 = 0$ $$\begin{vmatrix} 1-\lambda & 1 \\ 4 & 1-\lambda \end{vmatrix} = 0$$ $\lambda = 3, -1$	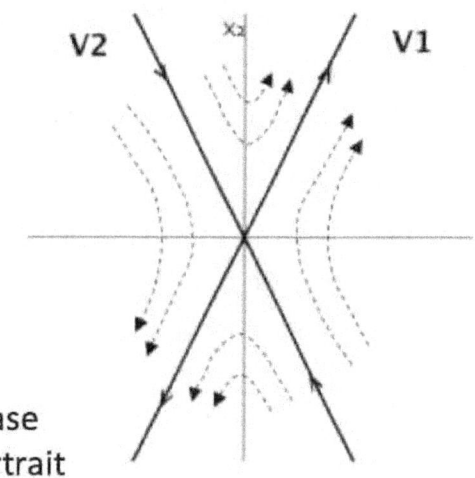 Phase Portrait
For: $\lambda = 3$ Use: $Av = \lambda v$ $\begin{pmatrix} 1 & 1 \\ 4 & 1 \end{pmatrix}\begin{pmatrix} v_1 \\ v_2 \end{pmatrix} = 3\begin{pmatrix} v_1 \\ v_2 \end{pmatrix}$ $v_1 + v_2 = 3v_1$ ➔ $v_2 = 2v_1$ $4v_1 + v_2 = 3v_2$ ➔ $4v_1 = 2v_2$ (Same) L.D. ➔ Let $v_1 = 1$ ➔ $v_2 = 2(1) = 2$ $V_1 = \begin{bmatrix} 1 \\ 2 \end{bmatrix}$	For: $\lambda = -1$ Use: $Av = \lambda v$ $\begin{pmatrix} 1 & 1 \\ 4 & 1 \end{pmatrix}\begin{pmatrix} v_1 \\ v_2 \end{pmatrix} = -1\begin{pmatrix} v_1 \\ v_2 \end{pmatrix}$ $v_1 + v_2 = -v_1$ ➔ $v_2 = -2v_1$ $4v_1 + v_2 = -v_2$ ➔ $2v_2 = -4v_1$ L.D. ➔ Let $v_1 = 1$ ➔ $v_2 = -2$ $V_2 = \begin{bmatrix} 1 \\ -2 \end{bmatrix}$
General Solution: $x = C_1 \begin{bmatrix} 1 \\ 2 \end{bmatrix} e^{3t} + C_2 \begin{bmatrix} 1 \\ -2 \end{bmatrix} e^{-t}$	

Real and REPEATED Eigen Values: $\lambda_1 = \lambda_2 = \lambda > 0$ (Both Positive)
Two L.I. Eigen Vectors

General Solution (G.S.)	Equilibrium: Star (Proper) Node, Unstable
$x = C_1 V_1 e^{\lambda t} + C_2 V_2 e^{\lambda t}$ $x = (C_1 V_1 + C_2 V_2) e^{\lambda t}$	

Example: Solve the system. $x' = \begin{pmatrix} 1 & 0 \\ 0 & 1 \end{pmatrix} x$

Find Eigen Values: Use: $A - \lambda I_2 = 0$ $\begin{vmatrix} 1-\lambda & 0 \\ 0 & 1-\lambda \end{vmatrix} = 0$ $\lambda = 1, 1$	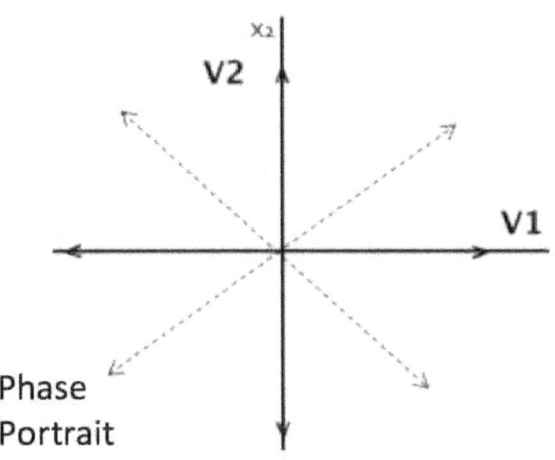 Phase Portrait
For: $\lambda = 1$ Use: $Av = \lambda v$ $\begin{pmatrix} 1 & 0 \\ 0 & 1 \end{pmatrix} \begin{pmatrix} v_1 \\ v_2 \end{pmatrix} = 1 \begin{pmatrix} v_1 \\ v_2 \end{pmatrix}$ $v_1 = v_1$ → $v_1 = free$ $v_2 = v_2$ → $v_2 = free$ Let $v_1 = 1, v_2 = 0$ (Both can't be 0) $V_1 = \begin{bmatrix} 1 \\ 0 \end{bmatrix}$	Notice v_1 and v_2 are independent and free. This gives us the opportunity to create two L.I. Eigen Vectors. Both v_1 and v_2 can't be 0. Let: $V_2 = \begin{bmatrix} 0 \\ 1 \end{bmatrix}$

General Solution: $x = C_1 \begin{bmatrix} 1 \\ 1 \end{bmatrix} e^t + C_2 \begin{bmatrix} 1 \\ 0 \end{bmatrix} e^t = \left\{ C_1 \begin{bmatrix} 1 \\ 1 \end{bmatrix} + C_2 \begin{bmatrix} 1 \\ 0 \end{bmatrix} \right\} e^t$

Real and REPEATED Eigen Values: $\lambda_1 = \lambda_2 = \lambda < 0$ (Both Negative) Two L.I. Eigen Vectors	
General Solution (G.S.) $x = C_1 V_1 e^{\lambda t} + C_2 V_2 e^{\lambda t}$ $x = (C_1 V_1 + C_2 V_2) e^{\lambda t}$	Equilibrium: Star (Proper) Node, Asymptotically Stable

Example: Solve the system. $x' = \begin{pmatrix} -1 & 0 \\ 0 & -1 \end{pmatrix} x$	
Find Eigen Values: Use: $A - \lambda I_2 = 0$ $\begin{vmatrix} -1-\lambda & 0 \\ 0 & -1-\lambda \end{vmatrix} = 0$ $\lambda = -1, -1$	Phase Portrait 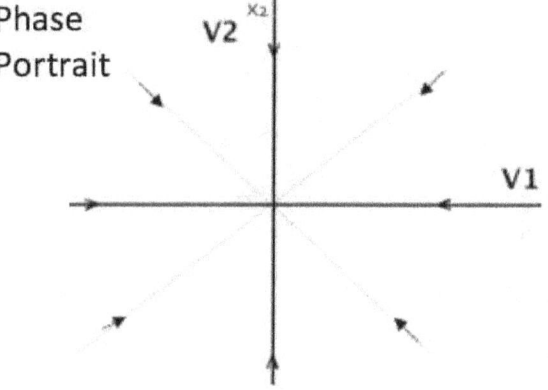
For: $\lambda = -1$ Use: $Av = \lambda v$ $\begin{pmatrix} -1 & 0 \\ 0 & -1 \end{pmatrix} \begin{pmatrix} v_1 \\ v_2 \end{pmatrix} = 1 \begin{pmatrix} v_1 \\ v_2 \end{pmatrix}$ $-v_1 = -v_1 \rightarrow v_1 = free$ $-v_2 = -v_2 \rightarrow v_2 = free$ Let $v_1 = 1, v_2 = 0$ (Both can't be 0) $V_1 = \begin{bmatrix} 1 \\ 0 \end{bmatrix}$	Notice v_1 and v_2 are independent and free. This gives us the opportunity to create two L.I. Eigen Vectors. Both v_1 and v_2 can't be 0. Let: $V_2 = \begin{bmatrix} 0 \\ 1 \end{bmatrix}$
General Solution: $x = C_1 \begin{bmatrix} 1 \\ 1 \end{bmatrix} e^{-t} + C_2 \begin{bmatrix} 1 \\ 0 \end{bmatrix} e^{-t} = \left\{ C_1 \begin{bmatrix} 1 \\ 1 \end{bmatrix} + C_2 \begin{bmatrix} 1 \\ 0 \end{bmatrix} \right\} e^{-t}$	

Real and REPEATED Eigen Values: $\lambda_1 = \lambda_2 = \lambda > 0$ (Both Positive) Two Eigen Vectors are NOT L.I.	
General Solution (G.S.) $x_1 = V_1 e^{\lambda t}$ $x_2 = V_1 t e^{\lambda t} + V_2 e^{\lambda t}$ $x = C_1 V_1 e^{\lambda t} + C_2 (V_1 t e^{\lambda t} + V_2 e^{\lambda t})$	Equilibrium: Improper Node, Unstable

Example: Solve the system. $\quad x' = \begin{pmatrix} 1 & 1 \\ 0 & 1 \end{pmatrix} x$	
Find Eigen Values: Use: $A - \lambda I_2 = 0$ $\begin{vmatrix} 1-\lambda & 1 \\ 0 & 1-\lambda \end{vmatrix} = 0$ $\lambda = 1, 1$	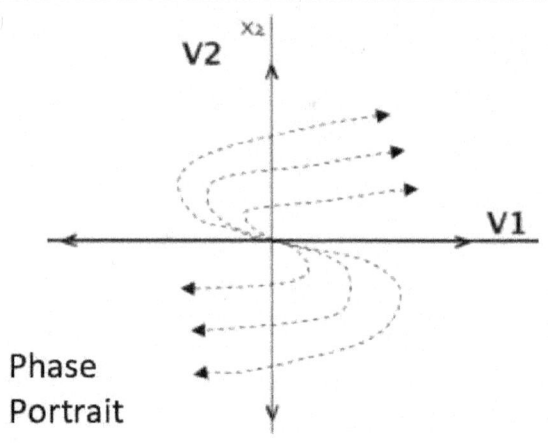 Phase Portrait
For: $\lambda = 1 \quad$ Use: $Av = \lambda v$ $\begin{pmatrix} 1 & 1 \\ 0 & 1 \end{pmatrix} \begin{pmatrix} v_1 \\ v_2 \end{pmatrix} = 1 \begin{pmatrix} v_1 \\ v_2 \end{pmatrix}$ $v_1 + v_2 = v_1 \;\rightarrow\; v_1 = free$ Let $v_1 = 1$ $(1) + v_2 = (1) \;\rightarrow\; v_2 = 0$ (not free) $V_1 = \begin{bmatrix} 1 \\ 0 \end{bmatrix}$	To find V_2 Use: $(A - \lambda I_2) V_2 = V_1$ With: $\lambda = 1$ $\begin{vmatrix} 0 & 1 \\ 0 & 0 \end{vmatrix} \begin{bmatrix} v_1 \\ v_2 \end{bmatrix} = \begin{bmatrix} 1 \\ 0 \end{bmatrix}$ $v_2 = 1$ $v_1 = free \;\rightarrow\;$ Let: $v_1 = 0$ $V_2 = \begin{bmatrix} 0 \\ 1 \end{bmatrix}$
General Solution: $\quad x = C_1 \begin{bmatrix} 1 \\ 0 \end{bmatrix} e^t + C_2 \left\{ \begin{bmatrix} 1 \\ 0 \end{bmatrix} t e^t + \begin{bmatrix} 0 \\ 1 \end{bmatrix} e^t \right\}$	

Real and REPEATED Eigen Values: $\lambda_1 = \lambda_2 = \lambda < 0$ (Both Negative) Two Eigen Vectors are NOT L.I.	
General Solution (G.S.) $x_1 = V_1 e^{\lambda t}$ $x_2 = V_1 t e^{\lambda t} + V_2 e^{\lambda t}$ $x = C_1 V_1 e^{\lambda t} + C_2(V_1 t e^{\lambda t} + V_2 e^{\lambda t})$	Equilibrium: Improper Node, Asymptotically Stable

Example: Solve the system. $x' = \begin{pmatrix} -1 & -1 \\ 0 & -1 \end{pmatrix} x$	
Find Eigen Values: Use: $A - \lambda I_2 = 0$ $\begin{vmatrix} -1-\lambda & -1 \\ 0 & -1-\lambda \end{vmatrix} = 0$ $\lambda = -1, -1$	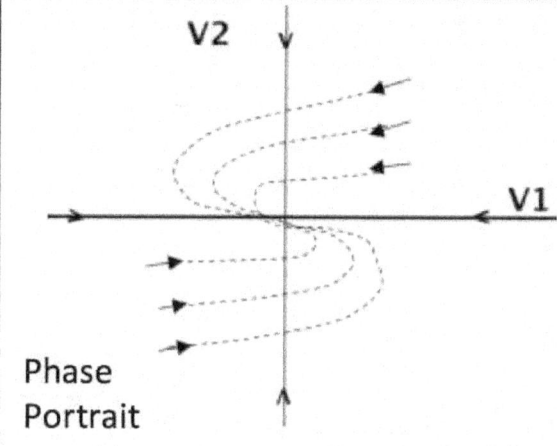 Phase Portrait
For: $\lambda = -1$ Use: $Av = \lambda v$ $\begin{pmatrix} 1 & 1 \\ 0 & 1 \end{pmatrix} \begin{pmatrix} v_1 \\ v_2 \end{pmatrix} = -1 \begin{pmatrix} v_1 \\ v_2 \end{pmatrix}$ $v_1 + v_2 = -v_1$ $v_2 = -v_2 \;\rightarrow\; v_2 = 0$ Let $v_1 = 1$ $V_1 = \begin{bmatrix} 1 \\ 0 \end{bmatrix}$	To find V_2 Use: $(A - \lambda I_2) V_2 = V_1$ With: $\lambda = -1$ $\begin{vmatrix} 0 & -1 \\ 0 & 0 \end{vmatrix} \begin{bmatrix} v_1 \\ v_2 \end{bmatrix} = \begin{bmatrix} 1 \\ 0 \end{bmatrix}$ $-v_2 = 1 \;\rightarrow\; v_2 = -1$ $v_1 = free \;\rightarrow\;$ Let: $v_1 = 0$ $V_2 = \begin{bmatrix} 0 \\ -1 \end{bmatrix}$
General Solution: $x = C_1 \begin{bmatrix} 1 \\ 0 \end{bmatrix} e^{-t} + C_2 \left\{ \begin{bmatrix} 1 \\ 0 \end{bmatrix} t e^{-t} + \begin{bmatrix} 0 \\ -1 \end{bmatrix} e^{-t} \right\}$	

Complex Eigen Values: $\lambda_1 = p + iq$ and $\lambda_2 = p - iq$ With negative real part. $(p < 0)$	
General Solution (G.S.) $x = C_1 e^{pt} \cos qt + C_2 e^{pt} \sin qt$ $x = e^{pt}[C_1 \cos qt + C_2 \sin qt]$	Equilibrium: Spiral Node, Asymptotically Stable

Example: Solve the system. $x' = \begin{pmatrix} -1 & -4 \\ 1 & -1 \end{pmatrix} x$	
Find Eigen Values: Use: $A - \lambda I_2 = 0$ $\begin{vmatrix} (-1-\lambda) & -4 \\ 1 & (-1-\lambda) \end{vmatrix} = 0$ $\lambda = -1 \pm 2i$	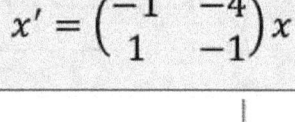 Phase Portrait
For: $\lambda_1 = -1 + 2i$ $V_1 = \begin{bmatrix} 2i \\ 1 \end{bmatrix}$	For: $\lambda_2 = -1 - 2i$ $V_2 = \begin{bmatrix} -2i \\ 1 \end{bmatrix}$
General Solution: $x = e^{-t}[C_1 \cos 2t + C_2 \sin 2t]$ Recall: $x = e^{pt}[C_1 \cos qt + C_2 \sin qt]$ With: $\lambda = p \pm iq = -1 \pm 2i$	

Complex Eigen Values: $\lambda_1 = p + iq$ and $\lambda_2 = p - iq$ With positive real part. ($p > 0$)	
General Solution (G.S.) $x = C_1 e^{pt} \cos qt + C_2 e^{pt} \sin qt$ $x = e^{pt}[C_1 \cos qt + C_2 \sin qt]$	Equilibrium: Spiral Node, Unstable

Example: Solve the system. $x' = \begin{pmatrix} 1 & 4 \\ -1 & 1 \end{pmatrix} x$	
Find Eigen Values: Use: $A - \lambda I_2 = 0$ $\begin{vmatrix} 1-\lambda & 4 \\ -1 & 1-\lambda \end{vmatrix} = 0$ $\lambda = 1 \pm 2i$	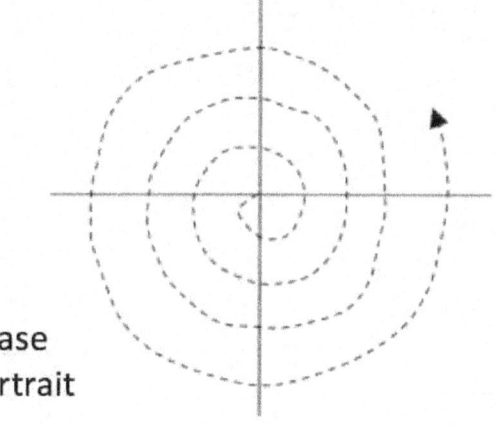 Phase Portrait
For: $\lambda_1 = 1 + 2i$ $V_1 = \begin{bmatrix} -2i \\ 1 \end{bmatrix}$	For: $\lambda_2 = 1 - 2i$ $V_2 = \begin{bmatrix} 2i \\ 1 \end{bmatrix}$
General Solution: $x = e^t[C_1 \cos 2t + C_2 \sin 2t]$ Recall: $x = e^{pt}[C_1 \cos qt + C_2 \sin qt]$ With: $\lambda = p \pm iq = 1 \pm 2i$	

Complex Eigen Values: $\lambda_1 = p + iq$ and $\lambda_2 = p - iq$ With no real part. ($p = 0$)	
General Solution (G.S.) $x = C_1 e^{0t} \cos qt + C_2 e^{0t} \sin qt$ $x = C_1 \cos qt + C_2 \sin qt$	**Equilibrium: Center, Stable**

Example: Solve the system. $x' = \begin{pmatrix} 0 & 1 \\ -1 & 0 \end{pmatrix} x$	
Find Eigen Values: Use: $A - \lambda I_2 = 0$ $\begin{vmatrix} -\lambda & 1 \\ -1 & -\lambda \end{vmatrix} = 0$ $\lambda = \pm i$	**Phase Portrait** 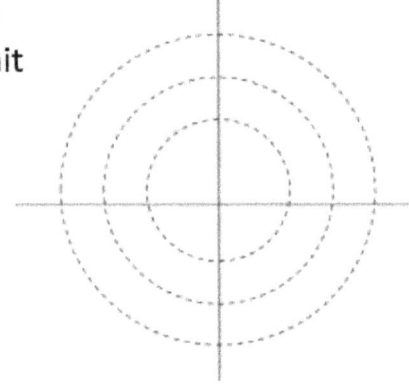
For: $\lambda_1 = i$ $V_1 = \begin{bmatrix} -i \\ 1 \end{bmatrix}$	For: $\lambda_2 = -i$ $V_2 = \begin{bmatrix} i \\ 1 \end{bmatrix}$
General Solution: $x = C_1 \cos t + C_2 \sin t$ Recall: $x = C_1 \cos qt + C_2 \sin qt$ With: $\lambda = \pm qi = \pm i$	

Real and Distinct Eigen Values: $\lambda_1 = 0 \quad \lambda_2 < 0$ (One Negative) One Eigen Value is ZERO	
General Solution (G.S.) $x = C_1 V_1 + C_2 V_2 e^{\lambda_2 t}$	Equilibrium: Stable Every point on the line containing V_1 is a critical (equilibrium) point.

Example: Solve the system. $x' = \begin{pmatrix} 1 & -2 \\ 2 & -4 \end{pmatrix} x$	
Find Eigen Values: Use: $A - \lambda I_2 = 0$ $\begin{vmatrix} 1-\lambda & -2 \\ 2 & -4-\lambda \end{vmatrix} = 0$ $\lambda = 0, -3$	Phase Portrait 
For: $\lambda = 0$ Use: $Av = \lambda v$ $\begin{pmatrix} 1 & -2 \\ 2 & -4 \end{pmatrix}\begin{pmatrix} v_1 \\ v_2 \end{pmatrix} = 0\begin{pmatrix} v_1 \\ v_2 \end{pmatrix}$ $v_1 - 2v_2 = 0 \rightarrow v_1 = 2v_2$ $2v_1 - 4v_2 = 0 \rightarrow v_1 = 2v_2$ L.D. $\rightarrow$ Let $v_2 = 1, v_1 = 2(1) = 2$ $V_1 = \begin{bmatrix} 2 \\ 1 \end{bmatrix}$	For: $\lambda = 3$ Use: $Av = \lambda v$ $\begin{pmatrix} 1 & -2 \\ 2 & -4 \end{pmatrix}\begin{pmatrix} v_1 \\ v_2 \end{pmatrix} = -3\begin{pmatrix} v_1 \\ v_2 \end{pmatrix}$ $v_1 - 2v_2 = -3v_1 \rightarrow v_2 = 2v_1$ $2v_1 - 4v_2 = -3v_2 \rightarrow v_2 = 2v_1$ Same Equation. Let $v_1 = 1$ $v_2 = 1$ L.D. $\rightarrow$ Let $v_1 = 1$, $v_2 = 2(1) = 2$ $V_2 = \begin{bmatrix} 1 \\ 2 \end{bmatrix}$
General Solution: $x = C_1 \begin{bmatrix} 2 \\ 1 \end{bmatrix} + C_2 \begin{bmatrix} 1 \\ 2 \end{bmatrix} e^{-3t}$	

Real and Distinct Eigen Values: $\lambda_1 > 0$ $\lambda_2 = 0$ (One Positive) One Eigen Value is ZERO	
General Solution (G.S.) $x = C_1 V_1 e^{\lambda_1 t} + C_2 V_2$ Equilibrium: Unstable	Equilibrium: Unstable Every point on the line containing V_2 is a critical (equilibrium) point.

Example: Solve the system. $x' = \begin{pmatrix} -1 & 2 \\ -2 & 4 \end{pmatrix} x$	
Find Eigen Values: Use: $A - \lambda I_2 = 0$ $\begin{vmatrix} -1-\lambda & 2 \\ -2 & 4-\lambda \end{vmatrix} = 0$ $\lambda = 3, 0$	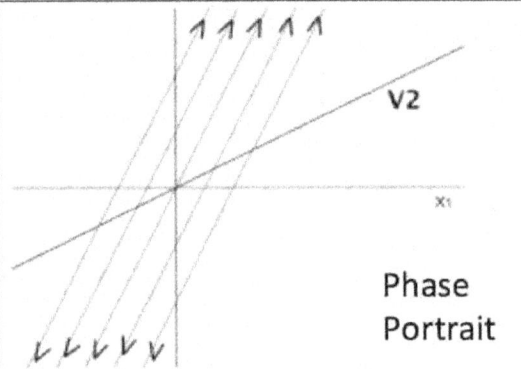 Phase Portrait
For: $\lambda = 3$ Use: $Av = \lambda v$ $\begin{pmatrix} -1 & 2 \\ -2 & 4 \end{pmatrix}\begin{pmatrix} v_1 \\ v_2 \end{pmatrix} = 3\begin{pmatrix} v_1 \\ v_2 \end{pmatrix}$ $-v_1 + 2v_2 = 3v_1 \rightarrow v_2 = 2v_1$ $-2v_1 + 4v_2 = 3v_2 \rightarrow v_2 = 2v_1$ Same Equation. Let $v_1 = 1$ $v_2 = 1$ L.D. $\rightarrow$ Let $v_1 = 1$, $v_2 = 2(1) = 2$ $V_1 = \begin{bmatrix} 1 \\ 2 \end{bmatrix}$	For: $\lambda = 0$ Use: $Av = \lambda v$ $\begin{pmatrix} -1 & 2 \\ -2 & 4 \end{pmatrix}\begin{pmatrix} v_1 \\ v_2 \end{pmatrix} = 0\begin{pmatrix} v_1 \\ v_2 \end{pmatrix}$ $-v_1 + 2v_2 = 0$ $-2v_1 + 4v_2 = 0 \rightarrow v_1 = 2v_1$ Let $v_2 = 1$ $v_2 = 1 \rightarrow v_1 = 2(1) = 2$ $V_2 = \begin{bmatrix} 2 \\ 1 \end{bmatrix}$
General Solution: $x = C_1 \begin{bmatrix} 1 \\ 2 \end{bmatrix} e^{3t} + C_2 \begin{bmatrix} 2 \\ 1 \end{bmatrix}$	

Real and Repeating Eigen Values: $\lambda_1 = \lambda_2 = \lambda = 0$
Both Eigen Values are ZERO

General Solution (G.S.)	Equilibrium: (0,0), Unstable
$x = C_1 V_1 + C_2[V_1 t + V_2]$	Every point on the line V_1 is a critical point. (Degenerate Case)

Example: Solve the system. $x' = \begin{pmatrix} 4 & -2 \\ 8 & -4 \end{pmatrix} x$

Find Eigen Values:	Phase Portrait 
Use: $A - \lambda I_2 = 0$	
$\begin{vmatrix} 4-\lambda & -2 \\ 8 & -4-\lambda \end{vmatrix} = 0$	
$\lambda = 0, 0$	

For: $\lambda = 0$ Use: $Av = \lambda v$	To find V_2 Use: $(A - \lambda I_2)V_2 = V_1$
$\begin{pmatrix} 4 & -2 \\ 8 & -4 \end{pmatrix}\begin{pmatrix} v_1 \\ v_2 \end{pmatrix} = 0 \begin{pmatrix} v_1 \\ v_2 \end{pmatrix}$	$\begin{vmatrix} 4 & -2 \\ 8 & -4 \end{vmatrix}\begin{bmatrix} v_1 \\ v_2 \end{bmatrix} = \begin{bmatrix} 1 \\ 2 \end{bmatrix}$ With $\lambda = 0$
$4v_1 - 2v_2 = 0 \rightarrow v_2 = 2v_1$	$4v_1 - 2v_2 = 1$
$8v_1 - 4v_2 = 0 \rightarrow v_2 = 2v_1$	$8v_1 - 4v_2 = 2$
Same Equation. Let $v_1 = 1$	Same Equations. Let $v_1 = 1$
$v_2 = 1$ L.D. $\rightarrow$ Let $v_1 = 1$,	$v_1 = v_2 = \frac{4v_1 - 1}{2} = \frac{3}{2}$
$v_2 = 2(1) = 2$	
$V_1 = \begin{bmatrix} 1 \\ 2 \end{bmatrix}$	$V_2 = \begin{bmatrix} 1 \\ \frac{3}{2} \end{bmatrix}$

General Solution: $x = C_1 \begin{bmatrix} 1 \\ 2 \end{bmatrix} + C_2 \left\{ \begin{bmatrix} 1 \\ 2 \end{bmatrix} t + \begin{bmatrix} 1 \\ \frac{3}{2} \end{bmatrix} \right\}$ (Straight line)

EIGEN VALUES, EIGEN VECTORS, & PHASE PORTRAIT

Example: Solve the System: $\begin{cases} x' = 3x - 2y \\ y' = 2x - 2y \end{cases}$

Compute Eigen Values, Eigen Vectors, and draw Phase Portrait.

Also, find the particular solution

For the initial conditions: $\begin{cases} x(0) = 1 \\ y(0) = 0 \end{cases}$

Solution:

This is a homogeneous system. ➜ $(0, 0)$ is an Equilibrium Solution.

Equilibrium ➜ $x' = 0$ and $y' = 0$ (Rate of change is zero.)

$$3x - 2y = 0$$
$$-(2x - 2y = 0)$$
$$\overline{}$$
$$x = 0 \quad ➜ \quad y = 0$$

➜ $(0, 0)$ is an Equilibrium Solution. (as expected)

Find Eigen Values. Rewrite equations: $\begin{bmatrix} x' \\ y' \end{bmatrix} = \begin{bmatrix} 3 & -2 \\ 2 & -2 \end{bmatrix} \begin{bmatrix} x \\ y \end{bmatrix}$

$det(A - \lambda I_2) = 0 \rightarrow \begin{vmatrix} 3-\lambda & -2 \\ 2 & -2-\lambda \end{vmatrix} = 0$

$(3 - \lambda)(-2 - \lambda) + 4 = 0$

$(-1)(3 - \lambda)(2 + \lambda) + 4 = 0$

$(-1)(6 + \lambda - \lambda^2) + 4 = -6 - \lambda + \lambda^2 + 4 = 0$

$\lambda^2 - \lambda - 2 = (\lambda - 2)(\lambda + 1) = 0 \quad \rightarrow \lambda_1 = 2, \lambda_2 = -1$

$\rightarrow$ Saddle Node, Unstable.

Compute the two Eigen VECTORS for the Equilibrium Point (0,0) with Eigen Values $\lambda_1 = 2$, $\lambda_2 = -1$.

For: $\lambda_1 = 2$ Solve: $Av = \lambda v$

$\begin{bmatrix} 3 & -2 \\ 2 & -2 \end{bmatrix} \begin{bmatrix} v_1 \\ v_2 \end{bmatrix} = 2 \begin{bmatrix} v_1 \\ v_2 \end{bmatrix} \quad \rightarrow \begin{cases} 3v_1 - 2v_2 = 2v_1 \\ 2v_1 - 2v_2 = 2v_2 \end{cases}$ Same equation!

If your calculations are correct, the two equations should be equal
    This fact may be used to check your work.
    Or, this fact may be used to just use one of the two equations.

$3v_1 - 2v_2 = 2v_1 \quad \rightarrow v_1 = 2v_2$

Let: $v_2 = 1 \quad\quad\quad \rightarrow v_1 = 2 \quad\quad\quad \rightarrow V_1 = \begin{bmatrix} 2 \\ 1 \end{bmatrix}$

For: $\lambda_2 = -1$ Solve: $Av = \lambda v$

$$\begin{bmatrix} 3 & -2 \\ 2 & -2 \end{bmatrix} \begin{bmatrix} v_1 \\ v_2 \end{bmatrix} = (-1) \begin{bmatrix} v_1 \\ v_2 \end{bmatrix} \rightarrow \begin{cases} 3v_1 - 2v_2 = -v_1 \\ 2v_1 - 2v_2 = -v_2 \end{cases} \text{Same equation!}$$

$$3v_1 - 2v_2 = -v_1 \quad \rightarrow \quad 4v_1 = 2v_2$$
$$2v_1 = v_2$$

Let: $v_1 = 1$ $\rightarrow$ $v_2 = 2$ $\rightarrow$ $V_2 = \begin{bmatrix} 1 \\ 2 \end{bmatrix}$

Now, we have two vectors: $V_1 = \begin{bmatrix} 2 \\ 1 \end{bmatrix}$ and $V_2 = \begin{bmatrix} 1 \\ 2 \end{bmatrix}$

1st Solution: $x_1 = \begin{bmatrix} 2 \\ 1 \end{bmatrix} e^{2t} = \begin{bmatrix} 2e^{2t} \\ e^{2t} \end{bmatrix}$

2nd Solution: $x_2 = \begin{bmatrix} 1 \\ 2 \end{bmatrix} e^{-t} = \begin{bmatrix} e^{-t} \\ 2e^{-t} \end{bmatrix}$

General Solution:

$$\begin{bmatrix} x \\ y \end{bmatrix} = C_1 x_1 + C_2 x_2$$

$$\begin{bmatrix} x \\ y \end{bmatrix} = C_1 \begin{bmatrix} 2e^{2t} \\ e^{2t} \end{bmatrix} + C_2 \begin{bmatrix} e^{-t} \\ 2e^{-t} \end{bmatrix} = \begin{bmatrix} 2C_1 e^{2t} + C_2 e^{-t} \\ C_1 e^{2t} + 2C_2 e^{-t} \end{bmatrix}$$

General Solution:

$$\begin{bmatrix} x \\ y \end{bmatrix} = C_1 \begin{bmatrix} 2e^{2t} \\ e^{2t} \end{bmatrix} + C_2 \begin{bmatrix} e^{-t} \\ 2e^{-t} \end{bmatrix} = \begin{bmatrix} 2C_1 e^{2t} + C_2 e^{-t} \\ C_1 e^{2t} + 2C_2 e^{-t} \end{bmatrix}$$

Particular Solution:

Solve for C_1 and C_2. Use $x(0) = 1$ and $y(0) = 0$

$$\begin{bmatrix} 1 \\ 0 \end{bmatrix} = \begin{bmatrix} 2C_1(1) + C_2(1) \\ C_1(1) + 2C_2(1) \end{bmatrix} = \begin{bmatrix} 2C_1 + C_2 \\ C_1 + 2C_2 \end{bmatrix}$$

$2C_1 + C_2 = 1$
$-(2C_1 + 4C_2 = 0\ )$
$\overline{}$
$\quad -3C_2 = 1 \qquad\qquad \rightarrow C_2 = -\frac{1}{3}$

$C_1 + 2\left(-\frac{1}{3}\right) = 0 \qquad \rightarrow C_1 = \frac{2}{3}$

Particular Solution:

$$\begin{bmatrix} x \\ y \end{bmatrix} = \begin{bmatrix} \left(\frac{4}{3}\right)e^{2t} - \left(\frac{1}{3}\right)e^{-t} \\ \left(\frac{2}{3}\right)e^{2t} - \left(\frac{4}{3}\right)e^{-t} \end{bmatrix}$$

Phase Portrait

for Equilibrium Point (0,0) with $V_1 = \begin{bmatrix} 2 \\ 1 \end{bmatrix}$ and $V_2 = \begin{bmatrix} 1 \\ 2 \end{bmatrix}$

Recall:

$$x_1 = \begin{bmatrix} 2 \\ 1 \end{bmatrix} e^{2t} = \begin{bmatrix} 2e^{2t} \\ e^{2t} \end{bmatrix} \qquad \text{For: } \lambda_1 = 2$$

$$x_2 = \begin{bmatrix} 1 \\ 2 \end{bmatrix} e^{-t} = \begin{bmatrix} e^{-t} \\ 2e^{-t} \end{bmatrix} \qquad \text{For: } \lambda_2 = -1$$

Equilibrium Point (0,0) is a Saddle Node, Unstable.

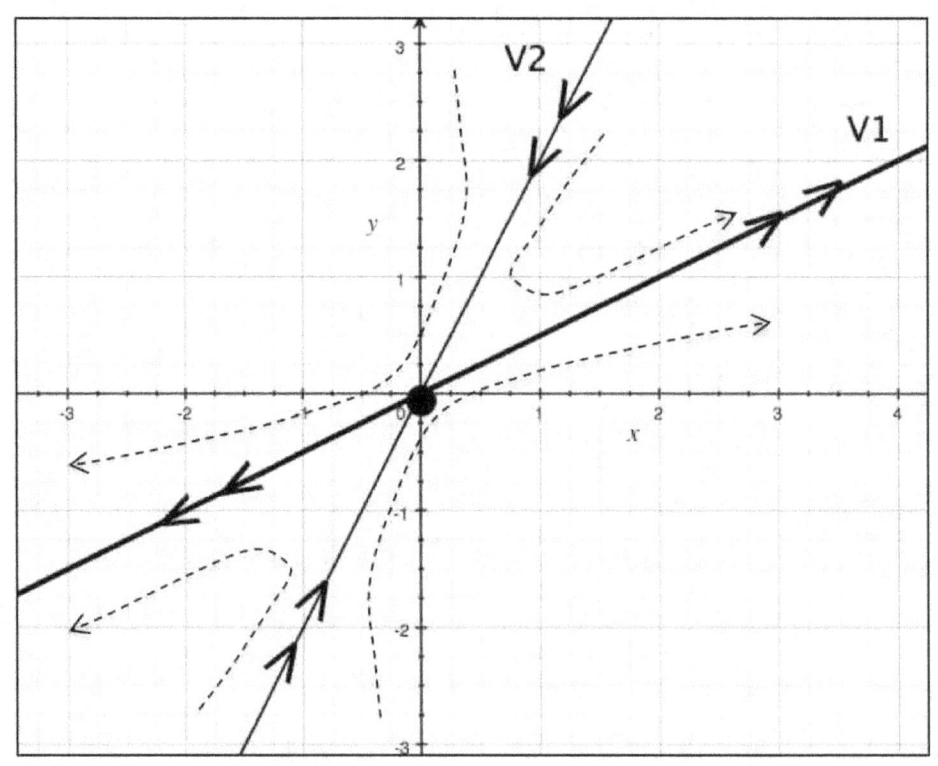

Eigen Values		Type of Equilibrium (Critical Point)	Stability
Real & distinct roots of same sign	$\lambda_1, \lambda_2 < 0$	Node	Asy. Stable
	$\lambda_1, \lambda_2 > 0$	Node	Unstable
Real & distinct roots of opposite signs.	$\lambda_1 \neq \lambda_2$	Saddle Node	Unstable
Real repeated roots 2 L.I. roots	$\lambda > 0$	Proper Node (Star)	Unstable
	$\lambda < 0$	Proper Node	Asy. Stable
Real repeated roots 1 L.I. roots	$\lambda > 0$	Improper Node	Unstable
	$\lambda < 0$	Improper Node	Asy. Stable
Complex roots with non-zero real part	$Re(\lambda) > 0$	Spiral Point	Unstable
	$Re(\lambda) < 0$	Spiral Point	Stable
Complex roots with zero real part	$\lambda = \pm qi$	Center	Stable
One zero root.	With $\lambda_2 < 0$	Infinite Equilibrium pts.	(0,0) is Stable
	With $\lambda_2 > 0$	Infinite Equilibrium pts.	(0,0) is Unstable
Two zero roots	$\lambda_1 = \lambda_2 = 0$	Infinite Equilibrium pts.	(0,0) is Unstable

Stability in NON-Homogeneous Systems:

Remark: The previous tables of Eigen Values and Phase Plane Notes can be used to determine the stability of Non-Homogeneous systems.

Non-Homogeneous System:

$$x' = Ax + k \quad \text{With: } A = \begin{pmatrix} a & b \\ c & d \end{pmatrix} \quad k = \begin{pmatrix} e \\ f \end{pmatrix}$$

	NON-Homogeneous System – How to Solve it.
Step 1	• Compute critical points (equilibrium points) • Set $x' = 0$ and $y' = 0$ Then solve for (x, y) pairs.
Step 2	• Compute Eigen Values of A. Use: $\det(A - \lambda I_2) = 0$ → λ_1, λ_2 • For Repeated Roots: Also compute Eigen Vectors.
Step 3	• Use tables to determine Phase Type and Stability for all critical points.
Step 4	• Find the General Solution, using the Eigen Values found in Step 2. Use $Av = \lambda v$ to find V_1 and V_2. Also, draw the phase portrait.

Examples: For each of the following three systems, compute the equilibrium solution and classify its type and stability. Also, find the general solution (G.S.) and draw the phase portrait.

Equilibrium Example #1: $x' = \begin{bmatrix} 0 & 1 \\ 1 & 0 \end{bmatrix} x + \begin{bmatrix} 2 \\ 3 \end{bmatrix}$

Solution:

Rewrite it: $\begin{bmatrix} x' \\ y' \end{bmatrix} = \begin{bmatrix} 0 & 1 \\ 1 & 0 \end{bmatrix} \begin{bmatrix} x \\ y \end{bmatrix} + \begin{bmatrix} 2 \\ 3 \end{bmatrix}$ → $\begin{cases} x' = y + 2 \\ y' = x + 3 \end{cases}$

Step 1: Critical Points: → $\begin{bmatrix} x' \\ y' \end{bmatrix} = \begin{bmatrix} 0 \\ 0 \end{bmatrix}$

$y + 2 = 0$ → $y = -2$
$x + 3 = 0$ → $x = -3$ → Critical Point: $(-3, -2)$

Step 2: Compute Eigen Values of A. Use: $\det(A - \lambda I_2) = 0$

$\begin{vmatrix} (-\lambda) & 1 \\ 1 & (-\lambda) \end{vmatrix} = 0$

$\lambda^2 - 1 = 0$ → $\lambda = \pm 1$

Step 3: Use Tables to determine Phase type and Stability
Opposite Signs → Saddle Node, Unstable.

Step 4: Find the G.S. and draw the Phase Portrait. Use: $Av = \lambda v$

For $\lambda = 1$	For $\lambda = -1$
$\begin{bmatrix} 0 & 1 \\ 1 & 0 \end{bmatrix} \begin{bmatrix} v_1 \\ v_2 \end{bmatrix} = (1) \begin{bmatrix} v_1 \\ v_2 \end{bmatrix}$	$\begin{bmatrix} 0 & 1 \\ 1 & 0 \end{bmatrix} \begin{bmatrix} v_1 \\ v_2 \end{bmatrix} = (-1) \begin{bmatrix} v_1 \\ v_2 \end{bmatrix}$
$v_2 = v_1$	$v_2 = -v_1$
$v_1 = v_2$ Same equation.	$v_1 = -v_2$ Same equation.
Let $v_1 = 1 \rightarrow v_2 = 1$	Let $v_1 = 1 \rightarrow v_2 = -1$
$\rightarrow V_1 = \begin{bmatrix} 1 \\ 1 \end{bmatrix}$	$\rightarrow V_2 = \begin{bmatrix} 1 \\ -1 \end{bmatrix}$

General Solution: $\quad x = C_1 \begin{bmatrix} 1 \\ 1 \end{bmatrix} e^t + C_2 \begin{bmatrix} 1 \\ -1 \end{bmatrix} e^{-t}$

Phase Portrait: Critical (Equilibrium) point is at $(-3, -2)$.

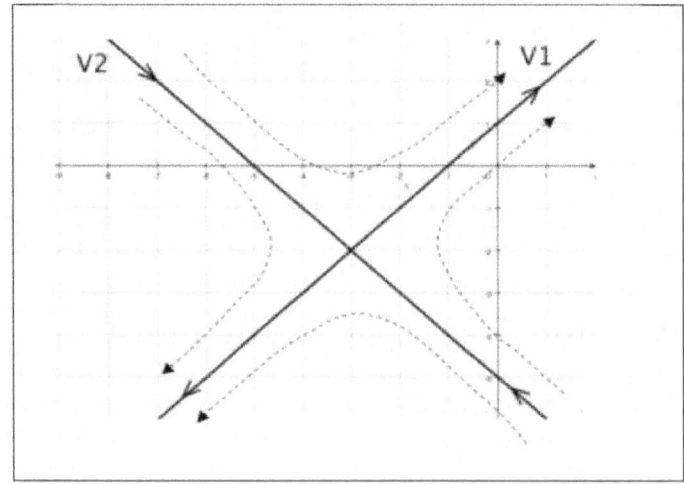

Equilibrium Example #2: $\begin{cases} x' = -2x + y - 3 \\ y' = x - 2y - 6 \end{cases}$

Solution:

$$\begin{pmatrix} x' \\ y' \end{pmatrix} = \begin{pmatrix} -2 & 1 \\ 1 & -2 \end{pmatrix} \begin{pmatrix} x \\ y \end{pmatrix} + \begin{pmatrix} -3 \\ -6 \end{pmatrix}$$ We need this form too.

Step 1: Compute Equilibrium Point(s)

$-2x + y - 3 = 0$ → $-2x + y - 3 = 0$

$x - 2y - 6 = 0$ $\qquad\quad$ $2x - 4y - 12 = 0$

$\quad$ sum $\qquad\qquad\qquad\quad$ $-3y - 15 = 0$ → $y = -5$

$\qquad\qquad\qquad\qquad\qquad\qquad\qquad\qquad$ → $x = -4$

Equilibrium Point: $(-4, -5)$

Step 2: Find Eigen Values. Use $\det(A - \lambda I_2) = 0$

$\begin{vmatrix} (-2 - \lambda) & 1 \\ 1 & (-2 - \lambda) \end{vmatrix} = 0$

$(-2 - \lambda)(-2 - \lambda) - 1 = 0$

$(2 + \lambda)^2 = 1$

$2 + \lambda = \pm 1$ → $\lambda = -2 \pm 1$ → $\lambda_1 = -3 \qquad \lambda_2 = -1$

Step 3: Use Tables to determine Phase type and Stability

Same Signs, Both Negative → Asymptotically Stable at Node $(-4, -5)$

Step 4: Find the G.S. and draw the Phase Portrait. Use: $Av = \lambda v$

For $\lambda = -3$	For $\lambda = -1$
$\begin{bmatrix} -2 & 1 \\ 1 & -2 \end{bmatrix} \begin{bmatrix} v_1 \\ v_2 \end{bmatrix} = (-3) \begin{bmatrix} v_1 \\ v_2 \end{bmatrix}$	$\begin{bmatrix} -2 & 1 \\ 1 & -2 \end{bmatrix} \begin{bmatrix} v_1 \\ v_2 \end{bmatrix} = (-1) \begin{bmatrix} v_1 \\ v_2 \end{bmatrix}$
$-2v_1 + v_2 = -3v_1$	$-2v_1 + v_2 = -v_1$
$v_1 - 2v_2 = -3v_2 \rightarrow v_2 = -v_1$	$v_1 - 2v_2 = -v_2$ Same
Let $v_1 = 1 \rightarrow v_2 = -1$	$v_1 = v_2$
$\rightarrow V_1 = \begin{bmatrix} 1 \\ -1 \end{bmatrix}$	Let $v_1 = 1 \rightarrow v_2 = 1$
	$\rightarrow V_2 = \begin{bmatrix} 1 \\ 1 \end{bmatrix}$

General Solution: $x = C_1 \begin{bmatrix} 1 \\ -1 \end{bmatrix} e^{-3t} + C_2 \begin{bmatrix} 1 \\ 1 \end{bmatrix} e^{-t}$

Phase Portrait: Critical (Equilibrium) point is at $(-4, -5)$.

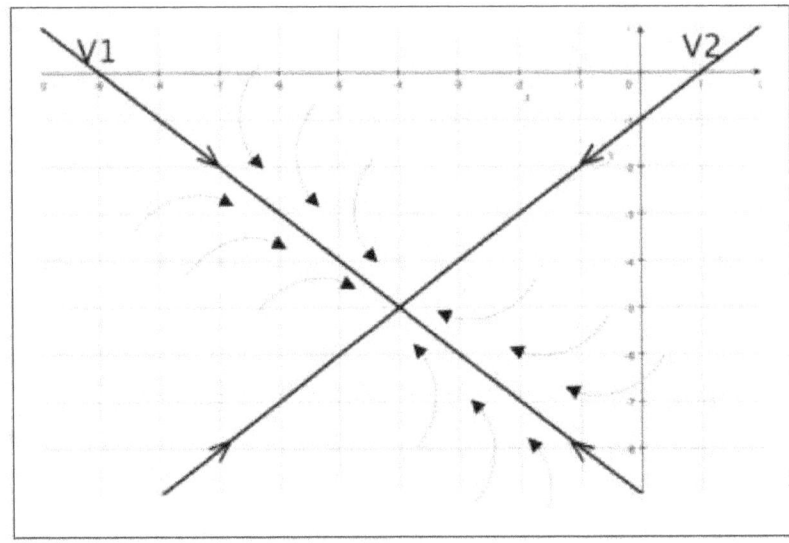

Equilibrium Example #3: $\begin{cases} x' = -x - 4y + 6 \\ y' = x - y + 4 \end{cases}$

Solution:

$$\begin{pmatrix} x' \\ y' \end{pmatrix} = \begin{pmatrix} -1 & -4 \\ 1 & -1 \end{pmatrix} \begin{pmatrix} x \\ y \end{pmatrix} + \begin{pmatrix} 6 \\ 4 \end{pmatrix}$$ We need both forms.

Step 1: Compute Equilibrium Point(s)

$-x - 4y + 6 = 0$

$x - y + 4 = 0$

sum $-5y + 10 = 0$ ➜ $y = -2$

➜ $x = -2$

Equilibrium Point: $(-2, -2)$

Step 2: Find Eigen Values.   Use $\det(A - \lambda I_2) = 0$

$\begin{vmatrix} (-1 - \lambda) & -4 \\ 1 & (-1 - \lambda) \end{vmatrix} = 0$

$(-1 - \lambda)(-1 - \lambda) + 4 = 0$

$(1 + \lambda)^2 = -4$

$1 + \lambda = \pm 2i$ ➜ $\lambda = -1 \pm 2i$ ➜ $\lambda_1 = -1 + 2i$

$\lambda_2 = -1 - 2i$

Step 3: Use Tables to determine Phase type and Stability

Real Part Negative ➜ Asymptotically Stable at Node $(-2, -2)$

Step 4: Find the G.S. and draw the Phase Portrait. Use: $Av = \lambda v$

For $\lambda = -1 - 2i$	For $\lambda = -1 + 2i$
$\begin{bmatrix} -1 & -4 \\ 1 & -1 \end{bmatrix} \begin{bmatrix} v_1 \\ v_2 \end{bmatrix} = (-1-2i) \begin{bmatrix} v_1 \\ v_2 \end{bmatrix}$	$\begin{bmatrix} -1 & -4 \\ 1 & -1 \end{bmatrix} \begin{bmatrix} v_1 \\ v_2 \end{bmatrix} = (-1-2i) \begin{bmatrix} v_1 \\ v_2 \end{bmatrix}$
Verify: $V_1 = \begin{bmatrix} -2i \\ 1 \end{bmatrix}$	Verify: $V_2 = \begin{bmatrix} 2i \\ 1 \end{bmatrix}$

General Solution: $\quad x = e^{-t} \left\{ C_1 \begin{bmatrix} -2i \\ 1 \end{bmatrix} \cos t + C_2 \begin{bmatrix} 2i \\ 1 \end{bmatrix} \sin t \right\}$

Phase Portrait: Critical (Equilibrium) point is at $(-2, -2)$.

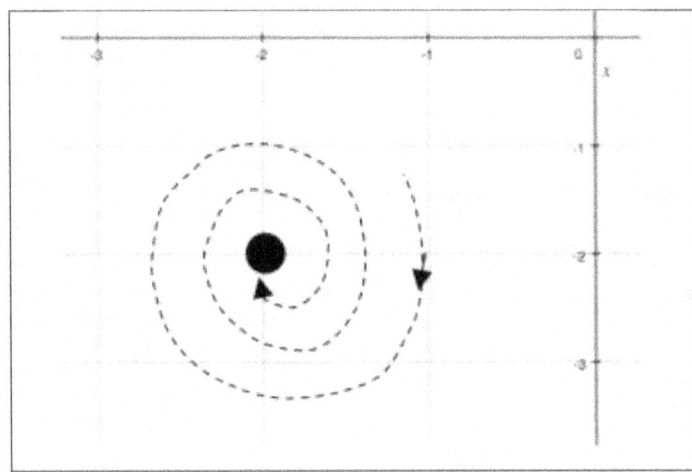

LOCALLY LINEAR SYSTEMS (GENERAL IDEA)

LOCALLY LINEAR SYSTEMS (Non-Linear Systems)

Autonomous System of DE

$$\begin{cases} x' = \frac{dx}{dt} = F(x,y) & With: \ x(t_0) = x_0 \\ y' = \frac{dy}{dt} = G(x,y) & With: \ y(t_0) = y_0 \end{cases}$$

	NON-Linear Systems – How to Solve it.
Step 1	Get Equilibrium Points: • Set: $F(x,y) = 0$ and $G(x,y) = 0$ • Solve for (x,y) • Often multiple Critical Points (Equilibrium Points).
Step 2	Get Type and Stability of <u>EACH Critical Point</u>. • This must be done locally. (small neighborhood) • Need to use Jacobian Matrix to do this analysis.. • Use: $J(x,y) = \begin{bmatrix} F_x & F_y \\ G_x & G_y \end{bmatrix}$ Partial derivatives.

Example: Compute the equilibrium points of each of the following NON-Linear systems and classify their types and stability.

$$\begin{cases} x' = x - y & = F(x, y) \\ y' = x^2 + y^2 - 8 & = G(x, y) \end{cases}$$

Solution:

Equilibrium Points: → Set RHS = 0

$x - y = 0$ → $x = y$

$x^2 + y^2 - 8 = 0$ → $x^2 + x^2 = 8$ → $x = \pm 2$

$x = 2$ → $y = 2$ → $(2, 2)$

$x = -2$ → $y = -2$ → $(-2, -2)$

Now, Compute type and stability.

Compute $J(x, y)$ The Jacobian Matrix

$$J(x, y) = \begin{bmatrix} F_x & F_y \\ G_x & G_y \end{bmatrix} = \begin{bmatrix} 1 & -1 \\ 2x & 2y \end{bmatrix}$$

(Continued ...)

Recall: $J(x, y) = \begin{bmatrix} 1 & -1 \\ 2x & 2y \end{bmatrix}$

For Point: $(2, 2)$ ➔ $J(2, 2) = \begin{bmatrix} 1 & -1 \\ 4 & 4 \end{bmatrix}$

$\begin{vmatrix} (1 - \lambda) & -1 \\ 4 & (4 - \lambda) \end{vmatrix} = 0$

$(1 - \lambda)(4 - \lambda) + 4 = 0$ ➔ $\lambda = \frac{5 \pm \sqrt{25 - 40}}{2} = \frac{5}{2} \pm \frac{\sqrt{15}}{2} i$

Imaginary Roots.

Real part positive ➔ Spiral Node, Unstable.

For Point: $(-2, -2)$ ➔ $J(-2, -2) = \begin{bmatrix} 1 & -1 \\ -4 & -4 \end{bmatrix}$

$\begin{vmatrix} (1 - \lambda) & -1 \\ -4 & (-4 - \lambda) \end{vmatrix} = 0$

$(1 - \lambda)(-4 - \lambda) - 4 = 0$ ➔ $\lambda = \frac{-3 \pm \sqrt{41}}{2} = -\frac{3}{2} \pm \frac{\sqrt{41}}{2}$

Real Roots

Opposite Signs ➔ Saddle Node, Unstable.

Type of Equilibrium Point and Stability are based on Eigen Values

Eigen Values		Type	Stability
Real & Distinct	Both > 0	Node	Unstable
	Both < 0	Node	Asymptotically Stable
	Opposite Signs	Saddle Node	Unstable
Repeated Roots 2 L.I. Eigen Vectors	Both > 0	Star /Proper Node	Unstable
	Both < 0	Star /Proper Node	Asymptotically Stable
Repeated Roots 1 L.I. Eigen Vector	Both > 0	Improper Node	Unstable
	Both < 0	Improper Node	Asymptotically Stable
Complex Roots $x = p \pm qi$	$p > 0$	Spiral Node	Unstable
	$p < 0$	Spiral Node	Stable
	$p = 0$	Center (Ovals)	Stable Node

Taylor Series representation for functions of one and two variables.

Recall: Taylor Series representation of $f(x)$ near $x = a$.

$$f(x) = f(a) + \frac{f'(a)}{1!}(x-a) + \frac{f''(a)}{2!}(x-a)^2 + \frac{f'''(a)}{3!}(x-a) + \cdots$$

Similarly:

Taylor Series representation of $F(x,y)$ near $(x,y) = (a,b)$.

$$F(x,y) = F(a,b) + F_x(a,b)(x-a) + F_y(a,b)(y-b) +$$

$$+ \frac{F_{xx}(a,b)}{2!}(x-a)^2 + \frac{2 F_{xy}(a,b)}{2!}(x-a)(y-b) + \frac{F_{yy}(a,b)}{2!}(y-b)^2 \ldots$$

Example of a Non-Linear System: $\begin{cases} x' = \frac{dx}{dt} = F(x,y) \\ y' = \frac{dy}{dt} = G(x,y) \end{cases}$

The system is called "Locally Linear" if the system is close to the Linear system. Eigen value tables can be used to determine the stability and types of equilibrium solutions. For classification of equilibrium points, approximate F and G using the first order partial derivatives. Since the system is locally linear, the higher order partial derivatives are very small or zero. This idea is discussed on the following pages.

Approximation of a Non-Linear System using Taylor Series

General Approach

For the Non-Linear System:
$$\begin{cases} x' = \frac{dx}{dt} = F(x,y) \\ y' = \frac{dy}{dt} = G(x,y) \end{cases}$$

Let: $(x, y) = (a, b)$ be any equilibrium solution.

→ $F(a, b) = 0$ and $G(a, b) = 0$

The Taylor Series Approximations
near the equilibrium Solution are:

$$F(x, y) = F(a, b) + F_x(a, b)(x - a) + F_y(a, b)(y - b) + \cdots$$
$$F(x, y) = \quad 0 \quad + F_x(a, b)(x - a) + F_y(a, b)(y - b) + \cdots$$
$$x' = F(x, y) \approx F_x(a, b)(x - a) + F_y(a, b)(y - b)$$

$$G(x, y) = G(a, b) + G_x(a, b)(x - a) + G_y(a, b)(y - b) + \cdots$$
$$G(x, y) = \quad 0 \quad + G_x(a, b)(x - a) + G_y(a, b)(y - b) + \cdots$$
$$y' = G(x, y) \approx G_x(a, b)(x - a) + G_y(a, b)(y - b)$$

$$\begin{bmatrix} x' \\ y' \end{bmatrix} = \begin{bmatrix} F_x(a,b) & F_y(a,b) \\ G_x(a,b) & G_y(a,b) \end{bmatrix} \begin{bmatrix} (x-a) \\ (y-b) \end{bmatrix}$$

$$J(x, y) = \begin{bmatrix} F_x & F_y \\ G_x & G_y \end{bmatrix} \quad \text{The Jacobian Matrix is an estimate!}$$

	General Idea
	How to compute and classify equilibrium solutions.
Step 1	Compute all Equilibrium Solutions. Solve: $\begin{cases} F(x,y) = 0 \\ G(x,y) = 0 \end{cases}$
Step 2	Compute Jacobian Matrix. $J(x,y) = \begin{bmatrix} F_x & F_y \\ G_x & G_y \end{bmatrix}$
Step 3	Evaluate $J(x,y)$. Evaluate every equilibrium solution. Then compute Eigen Values (& Eigen Vectors for Repeated Roots) To determine Type and Stability.

Example:

Compute all equilibrium solutions (Critical Points) of the system.

$$\begin{cases} x' = F(x,y) = (x-2)(y-3) \\ y' = G(x,y) = xy - 12 \end{cases}$$

Solution:

Step 1: Compute Equilibrium Solutions:

$\begin{cases} (x-2)(y-3) = 0 \\ xy - 12 = 0 \end{cases}$ ➔ Start with: $x = 2$, $y = 3$

$x = 2$ ➔ $y = 6$ ➔ $(2, 6)$ 1st Eq. Soln.

$y = 3$ ➔ $x = 4$ ➔ $(4, 3)$ 2nd Eq. Soln.

Step 2: Compute the Jacobian Matrix

$$J(x,y) = \begin{bmatrix} F_x & F_y \\ G_x & G_y \end{bmatrix} = \begin{bmatrix} y-3 & x-2 \\ y & x \end{bmatrix}$$

Step 3: Evaluate $J(x, y)$ Evaluate <u>every</u> equilibrium solution.

For: (2, 6) $\quad J(2,6) = \begin{bmatrix} 3 & 0 \\ 6 & 2 \end{bmatrix}$

$\det(A - \lambda I_1) = 0 \quad \rightarrow \quad \begin{vmatrix} 3-\lambda & 0 \\ 6 & 2-\lambda \end{vmatrix} = 0$

$(3-\lambda)(2-\lambda) = 0 \quad \rightarrow \quad \lambda = 3, 2$

Real & Distinct Eigen Values, Both positive

→ Node, Unstable

For: (4, 3) $\quad J(4,3) = \begin{bmatrix} 0 & 2 \\ 3 & 4 \end{bmatrix}$

$\det(A - \lambda I_1) = 0 \quad \rightarrow \quad \begin{vmatrix} -\lambda & 2 \\ 3 & 4-\lambda \end{vmatrix} = 0$

$(-\lambda)(4-\lambda) - 6 = 0$

$\lambda^2 - 4\lambda - 6 = 0 \quad \rightarrow \quad \lambda = 2 \pm \sqrt{10}$

Real & Distinct Eigen Values, Opposite Signs

→ Saddle Node, Unstable

SKETCH: Nodes for the two equilibrium solutions.

For: (2, 6)   Real & Distinct Eigen Values, Both positive
    → Node, Unstable

For: (4, 3)   Real & Distinct Eigen Values, Opposite Signs
    → Saddle Node, Unstable

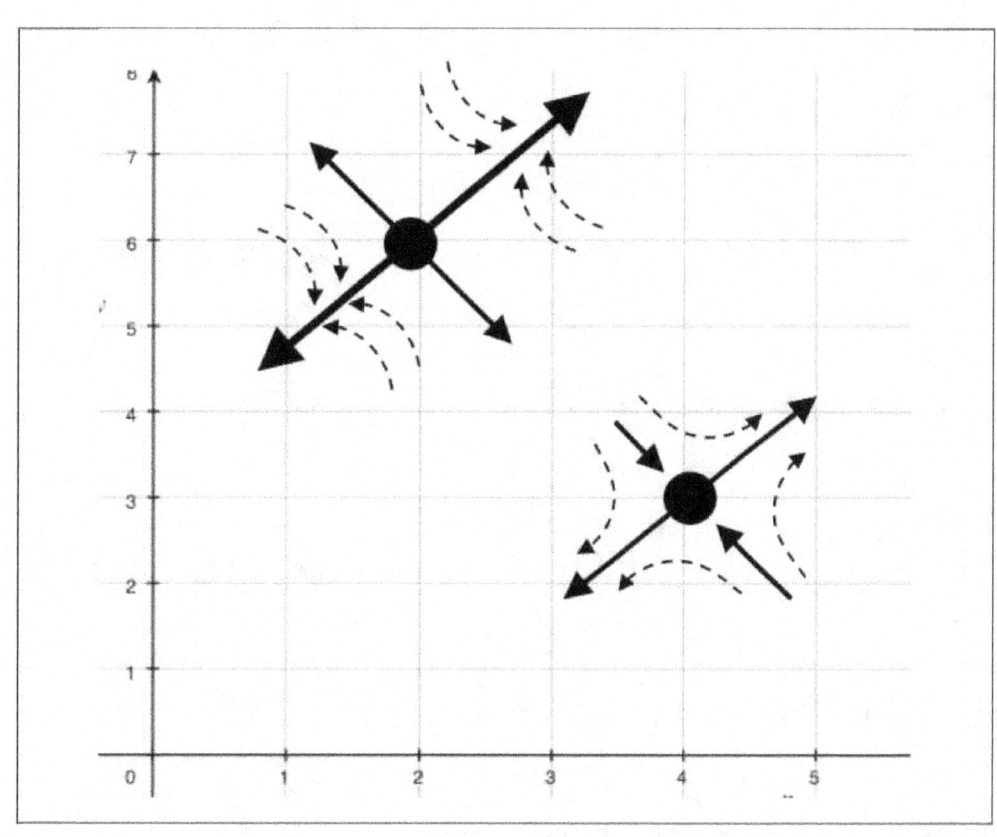

Example:

Compute all equilibrium solutions (Critical Points) of the system.

$$\begin{cases} x' = F(x,y) = 18 - x^2 - y^2 \\ y' = G(x,y) = x^2 - y^2 \end{cases}$$

Solution:

Step 1: Compute Equilibrium Solutions:

$$\begin{cases} 18 - x^2 - y^2 = 0 \\ x^2 - y^2 = 0 \end{cases}$$

Add them: $18 - 2y^2 = 0$ → $y^2 = 9$ → $y = \pm 3$

$y = 3$ → $x^2 - 3^2 = 0$

$x^2 = 9$

$x = \pm 3$ → $(3,3), (3,-3)$

$y = -3$ → $x^2 - 3^2 = 0$

$x^2 = 9$

$x = \pm 3$ → $(-3,3), (-3,-3)$

We have four equilibrium solutions.

Step 2: Compute Jacobian Matrix

$$J(x,y) = \begin{bmatrix} F_x & F_y \\ G_x & G_y \end{bmatrix} = \begin{bmatrix} -2x & -2y \\ 2x & -2y \end{bmatrix}$$

Step 3: Evaluate $J(x,y)$. Evaluate every equilibrium solution.

For: (3,3) $J(3,3) = \begin{bmatrix} -6 & -6 \\ 6 & -6 \end{bmatrix}$

$\det(A - \lambda I_2) = 0 \quad \rightarrow \quad \begin{vmatrix} (-6-\lambda) & -6 \\ 6 & (-6-\lambda) \end{vmatrix} = 0$

$(-6-\lambda)(-6-\lambda) + 36 = 0$

$(6+\lambda)(\lambda+6) + 36 = 0$

$6\lambda + 36 + \lambda^2 + 6\lambda + 36 = 0$

$\lambda^2 + 12\lambda + 72 = 0$

$\lambda^2 + 12\lambda + 36 + 36 = 0$

$(\lambda + 6)^2 = -36 \quad \rightarrow \quad \lambda = -6 \pm 6i$

Complex Roots, Negative Real Part

→ Spiral, Asymp. Stable

For: (3, −3) $J(3,-3) = \begin{bmatrix} -6 & 6 \\ 6 & 6 \end{bmatrix}$

$\det(A - \lambda I_1) = 0 \quad \rightarrow \quad \begin{vmatrix} -6-\lambda & 6 \\ 6 & 6-\lambda \end{vmatrix} = 0$

$(-6-\lambda)(6-\lambda) - 36 = 0$

$(-1)(6+\lambda)(6-\lambda) - 36 = 0$

$-36 + \lambda^2 - 36 = 0$

$\lambda^2 = 72 \quad \rightarrow \quad \lambda = \pm\sqrt{72} = \pm\sqrt{36 \cdot 2} = \pm 6\sqrt{2}$

Real Roots, Opposite Signs

→ Saddle Node, Unstable

For: $(-3, 3)$ $J(-3, 3) = \begin{bmatrix} 6 & -6 \\ -6 & -6 \end{bmatrix}$

$\det(A - \lambda I_2) = 0 \rightarrow \begin{vmatrix} (6-\lambda) & -6 \\ -6 & (-6-\lambda) \end{vmatrix} = 0$

$(6-\lambda)(-6-\lambda) - 36 = 0$

$(-1)(6-\lambda)(6+\lambda) - 36 = 0$

$-36 + \lambda^2 - 36 = 0$

$\lambda^2 - 72 = 0 \quad \rightarrow \lambda = \pm\sqrt{72} = \pm 6\sqrt{2}$

Real Roots, Opposite Signs

→ Saddle Node, Unstable

For: $(-3, 3)$ $J(-3, -3) = \begin{bmatrix} 6 & 6 \\ -6 & 6 \end{bmatrix}$

$\det(A - \lambda I_1) = 0 \rightarrow \begin{vmatrix} 6-\lambda & 6 \\ -6 & 6-\lambda \end{vmatrix} = 0$

$(6-\lambda)^2 + 36 = 0 \rightarrow \lambda = 6 \pm 6i$

Complex Roots, Positive Real part

→ Spiral, Unstable

The four equilibrium points are:

For: (3,3)   Complex Roots, Negative Real Part
  → Spiral, Asymptotically Stable

For: (3, −3)   Real Roots, Opposite Signs
  → Saddle Node, Unstable

For: (−3,3)   Real Roots, Opposite Signs
  → Saddle Node, Unstable

For: (−3,3)   Complex Roots, Positive Real part
  → Spiral, Unstable

SKETCH:

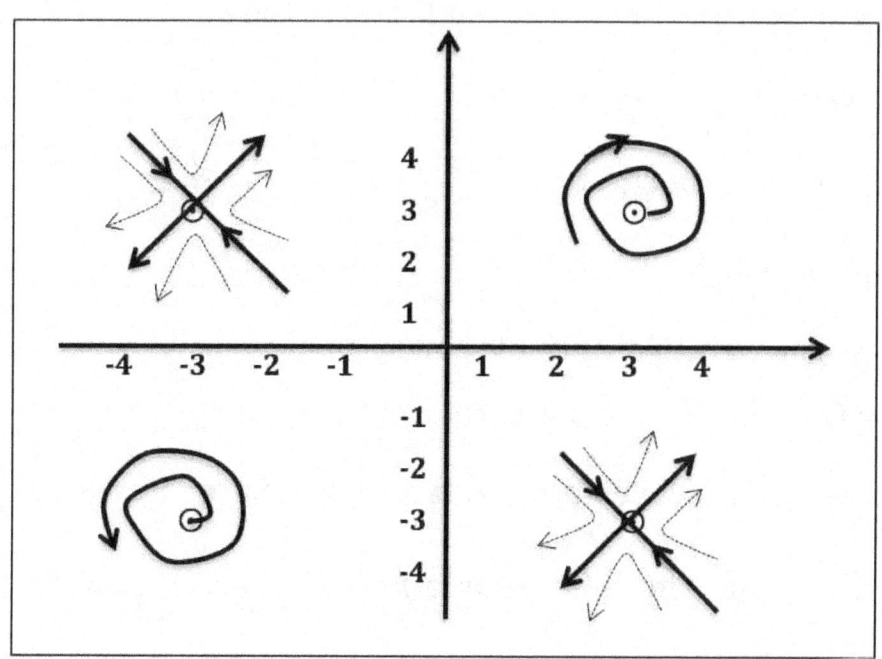

COMPETING SPECIES

COMPETING SPECIES

Two Populations of two species
that interact with other for the limited food supply.

Example: Two fish populations in a pond that don't eat each other.
but, they both eat the same food.

Recall: Population Growth (from Calculus II class, Logistic Model)

$$\frac{dP}{dt} \propto P$$

$$\frac{dP}{dt} \propto \left(1 - \frac{P}{M}\right)$$

$$\frac{dP}{dt} = kP\left(1 - \frac{P}{M}\right)$$

Where:
k = growth rate,
M = Carrying Capacity
or Saturation Point

Let x and y represent the population of the species at time t

In absence of species "y" → $\frac{dx}{dt} = kx\left(1 - \frac{x}{M}\right)$

In absence of species "y" → $\frac{dx}{dt} = x(k_1 - \sigma_1 x)$

In absence of species "x" → $\frac{dy}{dt} = y(k_2 - \sigma_2 y)$

k_1 = Growth rate for population "x"

k_2 = Growth rate for population "y"

$\frac{k_1}{\sigma_1}$ = Saturation level for population "x" (Carrying capacity)

$\frac{k_2}{\sigma_2}$ = Saturation level for population "y" (Carrying capacity)

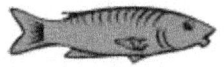 COMPETING SPECIES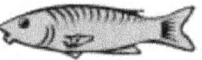

If both populations are present in the same environment, then they tend to reduce the available food supply for the other. In effect, they reduce each other's growth rates and saturation level. This can be described using the following modified system of DE.

$$\begin{cases} \frac{dx}{dt} = x(k_1 - \sigma_1 x - \alpha_1 y) \\ \frac{dy}{dt} = y(k_2 - \sigma_2 y - \alpha_2 x) \end{cases}$$

Where:

$\alpha_1 =$ measure of degree to which population "y" interferes with population "x"

$\alpha_2 =$ measure of degree to which population "x" interferes with population "y"

Note:

All constants k_1, k_2, σ_1, σ_2, α_1, α_2 are POSITIVE

And must be determined by OBSERVATION.

COMPETING SPECIES – Additional Notes

- We are interested in the realistic solution where $x > 0$ and $y > 0$.
- If the equilibrium point (x, y) with $x > 0$, $y > 0$ is an asymptotically stable node, it is a point of "Coexistence."
- If the equilibrium point (x, y) with $x > 0$, $y > 0$ is a saddle node, then it is unstable. This means that all trajectories go away from the node. In this case, only one population survives.
- The "Separatrix" divides the first quadrant into two parts for systems where only one species survives.
- The initial population, plotted on the phase portrait, assigns it to one side of the separatrtix which determines which species will survive.

Remarks:

- A solution using linearization is an approximation of the actual solution so it might be slightly different than the actual solution.

Recall:

How to Compute all equilibrium solutions (Critical Points) of the system. (General Idea)	
Step 1	Compute all Equilibrium Solutions. Solve: $\begin{cases} F(x,y) = 0 \\ G(x,y) = 0 \end{cases}$
Step 2	Compute Jacobian Matrix. $J(x,y) = \begin{bmatrix} F_x & F_y \\ G_x & G_y \end{bmatrix}$
Step 3	Evaluate $J(x,y)$. Evaluate every equilibrium solution. Then compute Eigen Values (& Eigen Vectors for Repeated Roots) To determine Type and Stability.

Example: Discuss the quantitative behavior of the system of two populations:
$$\begin{cases} x' = x(1 - x - y) \\ y' = y(0.75 - y - 0.5x) \end{cases}$$

Solution:

Step 1: Compute all Equilibrium Solutions.

$x(1 - x - y) = 0$ → $x = 0$, $(1 - x - y) = 0$

$y(0.75 - y - 0.5x) = 0$ → $y = 0$, $(0.75 - y - 0.5x) = 0$

$x = 0$ → $y(0.75 - y - 0) = 0$

→ $y = 0, 0.75$ → $(0,0), (0, .75)$

$y = 0$ → $x(1 - x - 0) = 0$

→ $x = 0, 1$ → $(0,0), (1,0)$

Now, we have three Equilibrium Solutions. Keep looking...

$(1 - x - y) = 0$ → $y = 1 - x$

$(0.75 - y - 0.5x) = 0$

Substitute: $0.75 - (1 - x) - 0.5x = 0$

$-.25 + 0.5x = 0$

$x = 0.5$

$x = 0.5$ → $y = 1 - 0.5 = 0.5$ → $(0.5, 0.5)$

We found four Equilibrium Solutions:

$(0,0), (1,0), (0,.75), (0.5, 0.5)$

Recall the System is:

$$\begin{cases} x' = F(x,y) = x(1 - x - y) = x - x^2 - xy \\ y' = G(x,y) = y(0.75 - y - 0.5x) = 0.75y - y^2 - 0.5xy \end{cases}$$

Find the Jacobian Matrix:

$$J(x,y) = \begin{bmatrix} F_x & F_y \\ G_x & G_y \end{bmatrix} = \begin{bmatrix} (1 - 2x - y) & (-x) \\ (-0.5y) & (0.75 - 2y - 0.5x) \end{bmatrix}$$

Determine the Type and Stability for all four Equilibrium Solutions.

Sol'n	Analysis
(0,0)	$J(0,0) = \begin{bmatrix} (1) & (0) \\ (0) & (0.75) \end{bmatrix}$ Eigen Values: $\begin{vmatrix} (1-\lambda) & (0) \\ (0) & (0.75-\lambda) \end{vmatrix} = 0$ $(1-\lambda)(0.75-\lambda) = 0 \quad \rightarrow \quad \lambda = 1, 0.75$ 2 Real and Distinct λ, Both Positive → Node, Unstable
(1,0)	$J(1,0) = \begin{bmatrix} (-1) & (-1) \\ (0) & (0.25) \end{bmatrix} \quad \rightarrow \quad \lambda = -1, 0.25$ 2 Real and Distinct λ, Opposite Signs → Saddle Node, Unstable
(0, .75)	$J(1, .75) = \begin{bmatrix} (.25) & (0) \\ (.125) & (-.75) \end{bmatrix} \quad \rightarrow \quad \lambda = 0.25, -0.75$ 2 Real and Distinct λ, Opposite Signs → Saddle Node, Unstable
(0.5, 0.5)	$J(.5, .5) = \begin{bmatrix} (-.5) & (-.5) \\ (-.25) & (-.25) \end{bmatrix} \quad \rightarrow \quad \lambda = -\frac{1}{2} \pm \frac{\sqrt{2}}{4}$ $\lambda \approx -0.5 \pm .27$ 2 Real and Distinct λ, Both Negative → Node, Asymptotically Stable. All trajectories converge to (0.5, 0.5) where both species coexist.

Sol'n	Summary Analysis
(0,0)	2 Real and Distinct λ, Both Positive → Node, Unstable
(1,0)	2 Real and Distinct λ, Opposite Signs → Saddle Node, Unstable
(0, .75)	2 Real and Distinct λ, Opposite Signs → Saddle Node, Unstable
(0.5, 0.5)	2 Real and Distinct λ, Both Negative → Node, Asymptotically STABLE

SKETCH:

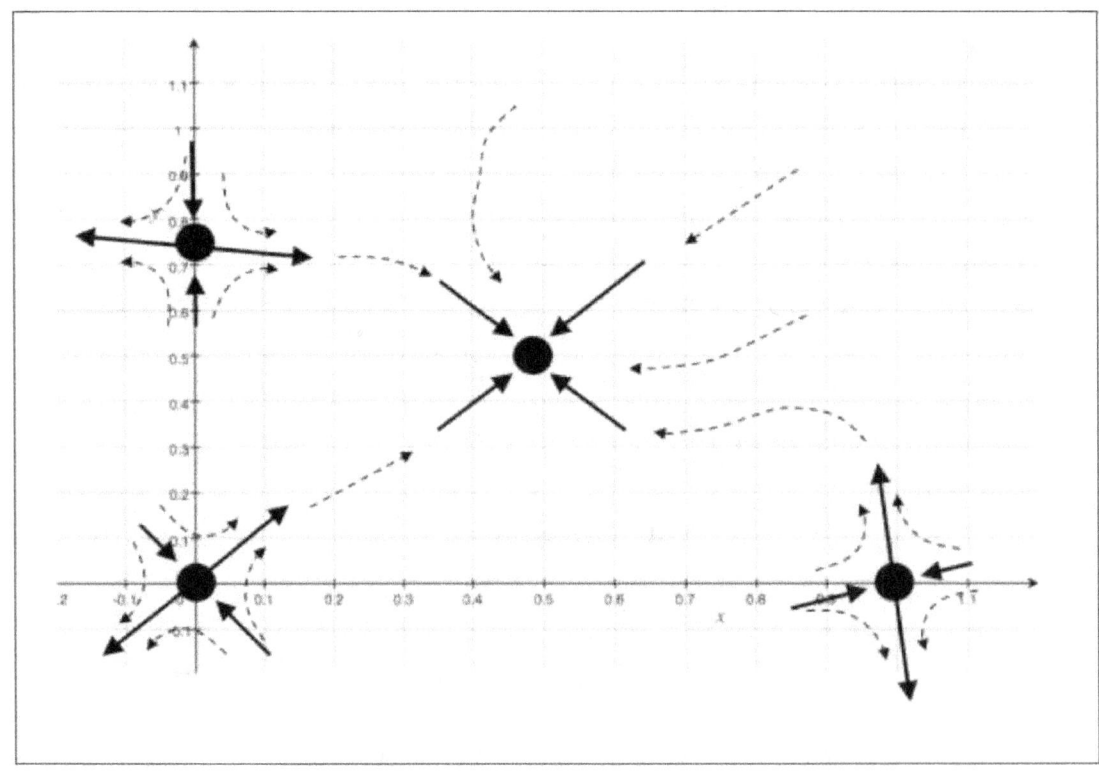

COMPETING SPECIES

Example: Consider the competing species model:

$$\begin{cases} \frac{dx}{dt} = 4x(1-x-y) \\ \frac{dy}{dt} = y(2-3x-y) \end{cases}$$

Compute all Equilibrium Solutions and find the type and stability for each solution. Also, draw the solution trajectories near each equilibrium solution (critical point).

Solution:

Equilibrium ➔ $\frac{dx}{dt} = 0$ AND $\frac{dy}{dt} = 0$

$F(x,y) = 4x(1-x-y) = 0$ ➔ $x = 0$
                                        OR ➔ $(1-x-y) = 0$

$G(x,y) = y(2-3x-y) = 0$ ➔ $y = 0$
                                        OR ➔ $(2-3x-y) = 0$

$x = 0$ ➔ $y = 0, 2$ ➔ C.P. $(0,0), (0,2)$

$y = 0$ ➔ $x = 0, 1$ ➔ C.P. $(0,0), (1,0)$

Look for another Critical Point:

$$(1 - x - y) = 0$$
$$\underline{-(2 - 3x - y) = 0}$$
$$-1 + 2x = 0 \quad\quad \rightarrow \quad x = \frac{1}{2}$$

$$x = \frac{1}{2} \quad \rightarrow \quad \left(1 - \frac{1}{2} - y\right) = 0 \quad \rightarrow \quad y = \frac{1}{2} \quad \rightarrow \quad \text{C.P.} \left(\frac{1}{2}, \frac{1}{2}\right)$$

Four Equilibrium Points:

$(0,0)$	Both populations zero.
$(0, 2)$	Only "y" population present.
$(1, 0)$	Only "x" population present.
$\left(\frac{1}{2}, \frac{1}{2}\right)$	Both populations present.

Recall:
$$\begin{cases} \dfrac{dx}{dt} = F(x,y) = 4x(1 - x - y) = 4x - 4x^2 - 4xy \\ \dfrac{dy}{dt} = G(x,y) = y(2 - 3x - y) = 2y - 6x - y^2 \end{cases}$$

Compute the Jacobian Matrix:

$$J(x,y) = \begin{bmatrix} F_x & F_y \\ G_x & G_y \end{bmatrix} = \begin{bmatrix} (4 - 8x - 4y) & (-4x) \\ (-3y) & (2 - 3x - 2y) \end{bmatrix}$$

Jacobian Matrix: $J(x,y) = \begin{bmatrix} (4-8x-4y) & (-4x) \\ (-3y) & (2-3x-2y) \end{bmatrix}$

Classify Each Equilibrium Point. $(0,0), \ (0,2), \ (1,0), \ \left(\frac{1}{2}, \frac{1}{2}\right)$

$J(x,y)$	$\det(A - \lambda I_2) = 0$
$J(0,0) = \begin{bmatrix} 4 & 0 \\ 0 & 2 \end{bmatrix}$	$\begin{vmatrix} (4-\lambda) & 0 \\ -6 & (2-\lambda) \end{vmatrix} = 0$ $(4-\lambda)(2-\lambda) = 0 \ \to \ \lambda = 4, 2$ → Node, Unstable
$J(0,2) = \begin{bmatrix} -4 & 0 \\ -6 & -2 \end{bmatrix}$	$\begin{vmatrix} (-4-\lambda) & 0 \\ 0 & (-2-\lambda) \end{vmatrix} = 0$ $(4+\lambda)(2+\lambda) = 0 \ \to \ \lambda = -4, -2$ → Node, Stable
$J(1,0) = \begin{bmatrix} -4 & -4 \\ 0 & -1 \end{bmatrix}$	$\begin{vmatrix} (-4-\lambda) & -4 \\ 0 & (-1-\lambda) \end{vmatrix} = 0$ $(4+\lambda)(1+\lambda) = 0 \ \to \ \lambda = -4, -1$ → Node, Stable
$J\left(\frac{1}{2}, \frac{1}{2}\right) = \begin{bmatrix} -2 & -2 \\ -\frac{3}{2} & -\frac{1}{2} \end{bmatrix}$	$\begin{vmatrix} (-2-\lambda) & -2 \\ -\frac{3}{4} & \left(-\frac{1}{2}-\lambda\right) \end{vmatrix} = 0$ $(2+\lambda)\left(\frac{1}{2}+\lambda\right) - 3 = 0 \ \to \ \lambda = -\frac{5}{4} \pm \frac{\sqrt{57}}{4}$ → Opposite Signs → Saddle Point, Unstable

Sketch PHASE PORTRAIT – Non-Coexisting Case

$J(0,0)$ →	$\lambda = 4, 2$ →	Node, Unstable
$(0, 2)$ →	$\lambda = -4, -2$ →	Node, Stable
$(1, 0)$ →	$\lambda = -4, -1$ →	Node, Stable
$\left(\dfrac{1}{2}, \dfrac{1}{2}\right)$ →	$\lambda = -\dfrac{5}{4} \pm \dfrac{\sqrt{57}}{4}$ →	Saddle Point, Unstable

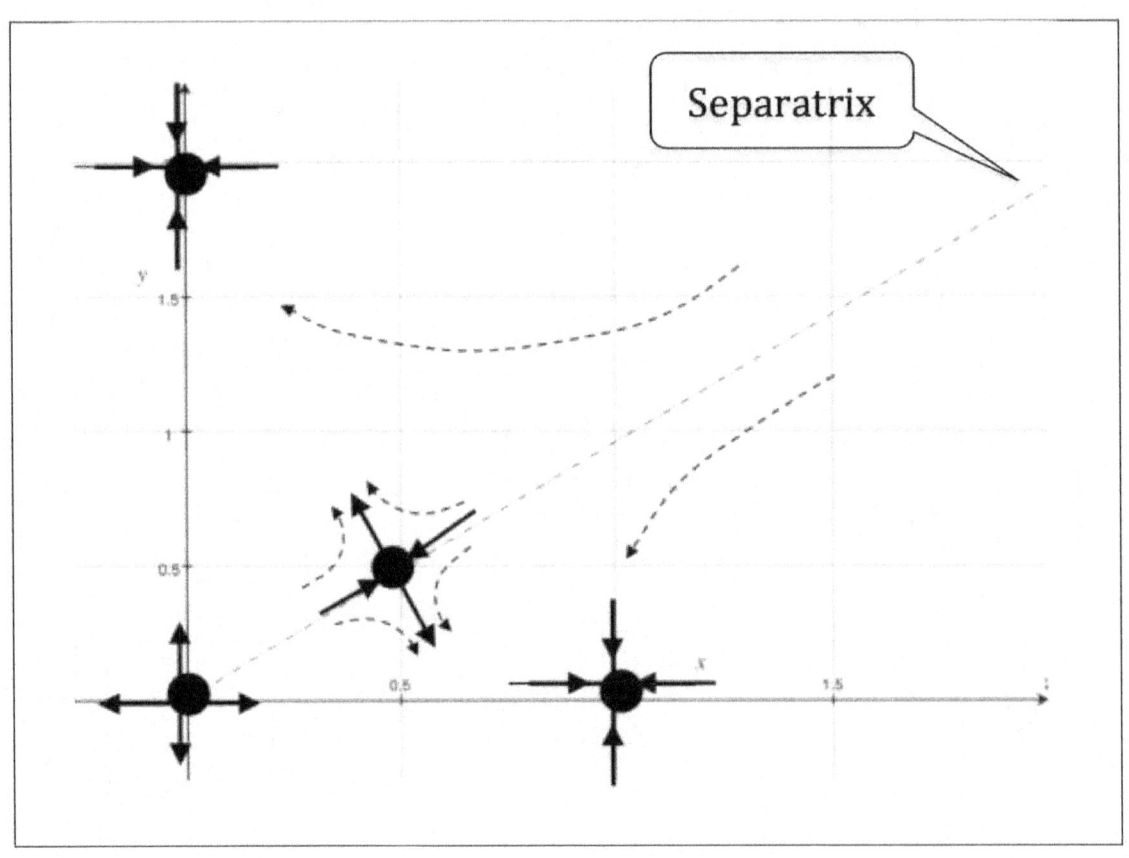

Example: Consider the Competing Species Model:

$$\frac{dx}{dt} = x(1 - x + 0.5y) = F(x,y)$$

$$\frac{dy}{dt} = y(2.5 - 1.5y + 0.25x) = G(x,y)$$

Compute all equilibrium solutions and find their type and stability. Draw the trajectories near each equilibrium solution.

Solution:

For equilibrium points, set: $\frac{dx}{dt} = 0$ and $\frac{dy}{dt} = 0$ Solve for (x, y).

$x(1 - x + 0.5y) = 0$ ➔ $x = 0$ or $(1 - x + 0.5y) = 0$

$y(2.5 - 1.5y + 0.25x) = 0$ ➔ $y = 0$ or $(2.5 - 1.5y + 0.25x) = 0$

Equilibrium (Critical) Points:

(0,0) (1)

$x = 0$ ➔ $2.5 - 1.5y + 0.25(0) = 0$ ➔ $y = \frac{2.5}{1.5} = \frac{5}{3}$ ➔ $\left(0, \frac{5}{3}\right)$ (2)

$y = 0$ ➔ $1 - x + 0.5(0) = 0$ ➔ $x = 1$ ➔ $(1,0)$ (3)

$1 - x + 0.5y = 0$ $1 - x + 0.5y = 0$
$2.5 - 1.5y + 0.25x = 0$ ➔ $\underline{10 - 6y + x = 0}$

$11 - 5.5y = 0$ ➔ $y = \frac{11}{5.5} = \frac{110}{55} = 2$

$1 - x + 0.5(2) = 0$ ➔ $x = 2$ ➔ $(2,2)$ (4)

To determine type and stability, we need to compute the Jacobian.

$F = x - x^2 + 0.5xy$ $\rightarrow$ $F_x = 1 - 2x + 0.5y$, $F_y = 0.5x$

$G = 2.5y - 1.5y^2 + 0.25xy$ $\rightarrow$ $G_x = 0.25y$, $G_y = 2.5 - 3y + 0.25x$

$$J(x,y) = \begin{bmatrix} (1 - 2x + 0.5y) & (0.5x) \\ (0.25y) & (2.5 - 3y + 0.25x) \end{bmatrix}$$

Find the Jacobian for each Critical Point.

$J(0,0) = \begin{bmatrix} (1) & (0) \\ (0) & (2.5) \end{bmatrix}$ $\rightarrow$ Eigen Values: $\lambda_1 = 2.5$, $\lambda_2 = 1$

$\rightarrow$ (0,0) is an unstable node.

$J\left(0,\frac{5}{3}\right) = \begin{bmatrix} \left(\frac{11}{6}\right) & (0) \\ \left(\frac{5}{12}\right) & \left(-\frac{5}{2}\right) \end{bmatrix}$ $\rightarrow$ Eigen Values: $\lambda_1 = \frac{11}{6}$, $\lambda_2 = -\frac{5}{2}$

$\rightarrow$ $\left(0,\frac{5}{3}\right)$ is a saddle node.

$J(1,0) = \begin{bmatrix} (-1) & (0.5) \\ (0) & (2.75) \end{bmatrix}$ $\rightarrow$ Eigen Values: $\lambda_1 = 2.75$, $\lambda_2 = -1$

$\rightarrow$ (1,0) is a saddle node.

$J(2,2) = \begin{bmatrix} (-2) & (1) \\ (0.5) & (-3) \end{bmatrix}$ $\rightarrow$ $\lambda_1, \lambda_2 = \frac{-5 \pm \sqrt{3}}{2}$ (Both negative)

$\rightarrow$ (2,2) is asymptotically stable node.

Summary of Critical Points

Critical Point	Eigen Values	Stability
$(0,0)$	$\lambda_1 = 2.5$, $\lambda_2 = 1$	Unstable Node
$\left(0, \frac{5}{3}\right)$	$\lambda_1 = \frac{11}{6}$, $\lambda_2 = -\frac{5}{2}$	Saddle node
$(1,0)$	$\lambda_1 = 2.75$, $\lambda_2 = -1$	Saddle node
$(2,2)$	$\lambda_1, \lambda_2 = \frac{-5 \pm \sqrt{3}}{2}$ (Both negative)	Asymptotically Stable Node

Phase Portrait:

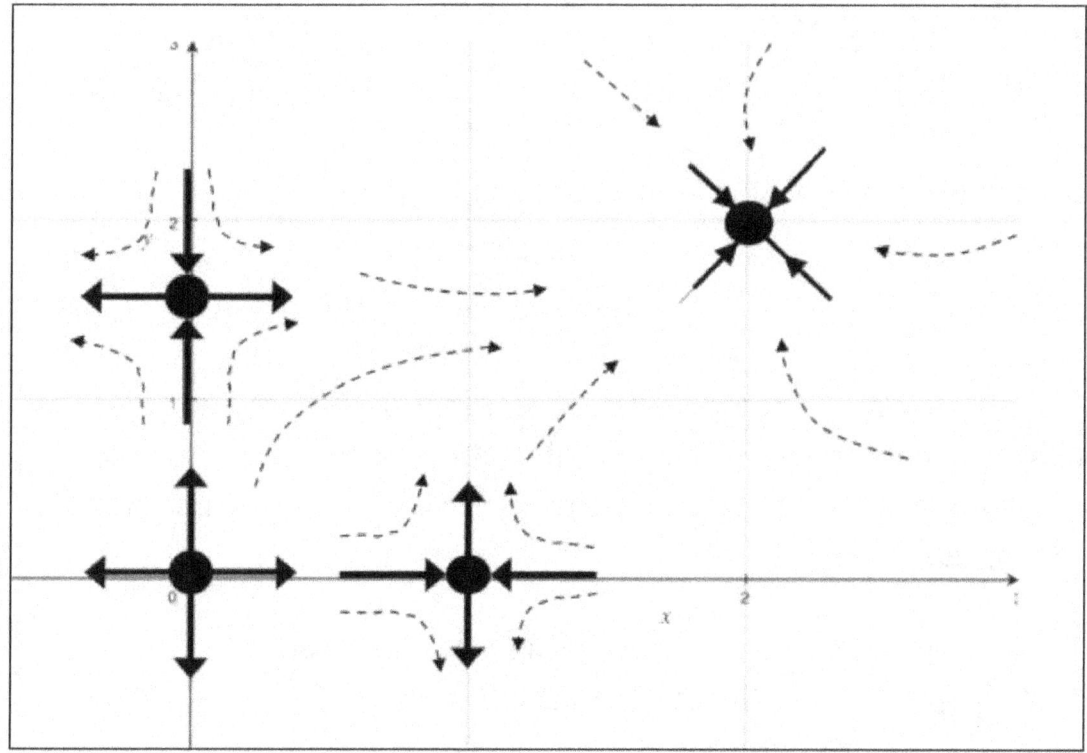

PREDATOR-PREY SYSTEMS

PREDATOR-PREY MODEL

In this section, the situation where one species (the predator) preys on the other species (the prey) is discussed. Here, prey lives on a different source of food.

Example: Consider a population of foxes and rabbits in a closed environment. The foxes prey on the rabbits and the rabbits live on vegetation.

Let x be the population of the prey.
And y be the population of the predator.

In the absence of a predator, the prey grows at the rate proportional to the current population. i.e.

$$\frac{dx}{dt} \propto x \;\Rightarrow\; \frac{dx}{dt} = ax \qquad a > 0 \;\text{ when } y = 0$$
$$a = \text{relative growth rate for the prey}$$

In the absence of prey, the predator dies out. i.e.

$$\frac{dy}{dt} = -cy \qquad c = 0 \;\text{ when } x = 0$$
$$c = \text{relative growth rate for predator}$$

The number of encounters between predator and prey is proportional to the product of their populations. Each encounter promotes the growth rate of the predator and reduces the growth rate of the prey.

Thus, the predator's growth rate increases by: μxy.
And, the prey's growth rate decreases by: kxy.
(μ, k Are positive constants.)

For the Predator-Prey Model, the modified system of DEs is:

$$\frac{dx}{dt} = ax - kxy$$

$$\frac{dy}{dt} = -cy + \mu xy$$

The same idea to do classification of equilibrium solutions can be used in this case.

Example: Consider the predator-prey system:

$$\frac{dx}{dt} = x(1 - 0.5y) = x - 0.5xy = F(x, y)$$

$$\frac{dy}{dt} = y(-0.75 + 0.25x) = -0.75y + 0.25xy = G(x, y)$$

Compute all equilibrium points and determine their type of stability. Draw the trajectories near each equilibrium.

Solution:
For equilibrium solutions, set $\frac{dx}{dt} = 0$ and $\frac{dy}{dt} = 0$. Solve for (x, y)

$x(1 - 0.5y) = 0$ ➔ $x = 0$ or $(1 - 0.5y) = 0$
➔ $x = 0$ or $y = 2$

$y(-0.75 + 0.25x) = 0$ ➔ $y = 0$ or $(-0.75 + 0.25x) = 0$
➔ $y = 0$ or $x = 3$

Critical (Equilibrium) Points: (0,0), (3,2)

Compute the Jacobian for each Critical Point.

$$J(x,y) = \begin{bmatrix} F_x & F_y \\ G_x & G_y \end{bmatrix} = \begin{bmatrix} (1-.5y) & (-.5x) \\ (.25y) & (-.75+.25x) \end{bmatrix}$$

$$J(0,0) = \begin{bmatrix} 1 & 0 \\ 0 & -.75 \end{bmatrix} \rightarrow \text{Eigen Values: } \lambda_1 = 1, \lambda_2 = -0.75$$

→ (0,0) is a saddle node (unstable)

$$J(3,2) = \begin{bmatrix} 0 & \left(-\frac{3}{2}\right) \\ \left(\frac{1}{2}\right) & 0 \end{bmatrix} \rightarrow \text{Eigen Values: } \lambda_{1,2} = \pm \frac{\sqrt{3}}{2}$$

→ (3,2) is a stable center

Phase Portrait:

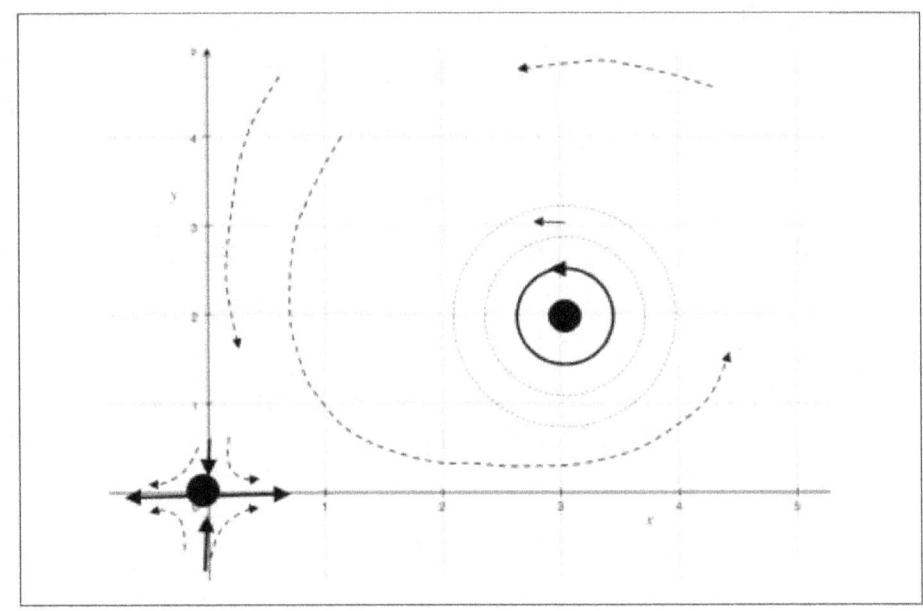

Checking for Direction in a Phase Portrait:

The direction of the center can be determined by selecting some near-by point and evaluating x' and y' at that point.

For example, in the previous diagram, x' and y' were evaluated for the point (3, 3).

$x' = x(1 - 0.5y)$
$x' = 3(1 - 0.5(3))$
$x' = 3(1 - 1.5) = 3(-0.5) = -1.5$

$y' = y(-0.75 + 0.25x)$
$y' = 3(-0.75 + 0.25(3))$
$y' = 3(-0.75 + 0.75) = 0$

Since x' is negative and y' is zero, the direction points to the left. ←

Evaluating x' and y' at point (3,0) gives:
$x' = 3$
$y' = 0$

This point also has a direction to the right. →
Which is consistent with the diagram.

Example: Consider the following model of a non-linear system, representing the relationship between two types of species (predator-prey relation).

$$x' = x(12 - x) - 4xy$$
$$y' = 2y - xy$$

(a.) Compute all equilibrium points.
(b.) Classify the type and stability of each equilibrium point.
(c.) Draw the phase portrait near each equilibrium point.

Solution – Part (a.)
$x' = x(12 - x) - 4xy = 0$
➔ $x(12 - x - 4y) = 0$ ➔ $x = 0$ or $12 - x - 4y = 0$

$y' = 2y - xy = 0$
➔ $y(2 - x) = 0$ ➔ $y = 0$ or $x = 2$

Equilibrium Points:

$x = 0, y = 0$ ➔ Point $(0,0)$ (#1)

$y = 0$ ➔ $12 - x - 4(0) = 0$ ➔ $x = 12$ ➔ Point $(12, 0)$ (#2)

$x = 2$ ➔ $12 - (2) - 4y = 0$ ➔ $y = \frac{5}{2}$ ➔ Point $\left(2, \frac{5}{2}\right)$ (#3)

(Continued ...)

Solution – Part (b.)

Compute the Jacobian for each Critical Point.
To find Eigen Values, use: $\det(J - \lambda I_2) = 0$

$$J(x,y) = \begin{bmatrix} F_x & F_y \\ G_x & G_y \end{bmatrix} = \begin{bmatrix} (12 - 2x - 4y) & (-4x) \\ (-y) & (2 - x) \end{bmatrix}$$

$J(0,0) = \begin{bmatrix} 12 & 0 \\ 0 & 2 \end{bmatrix}$ → Eigen Values: $\lambda_1 = 12$, $\lambda_2 = 2$

→ (0,0) is an unstable node.

$J(12,0) = \begin{bmatrix} -12 & -48 \\ 0 & -10 \end{bmatrix}$ → Eigen Values: $\lambda_1 = -10$, $\lambda_2 = -12$

→ (12,0) is an asymptotically stable node.

$J\left(2, \frac{5}{2}\right) = \begin{bmatrix} -2 & -8 \\ \left(\frac{5}{2}\right) & 0 \end{bmatrix}$ → Eigen Values: $\lambda_{1,2} = -1 \pm \sqrt{19}\, i$

→ $\left(2, \frac{5}{2}\right)$ asymptotically stable spiral node

Phase Portrait:

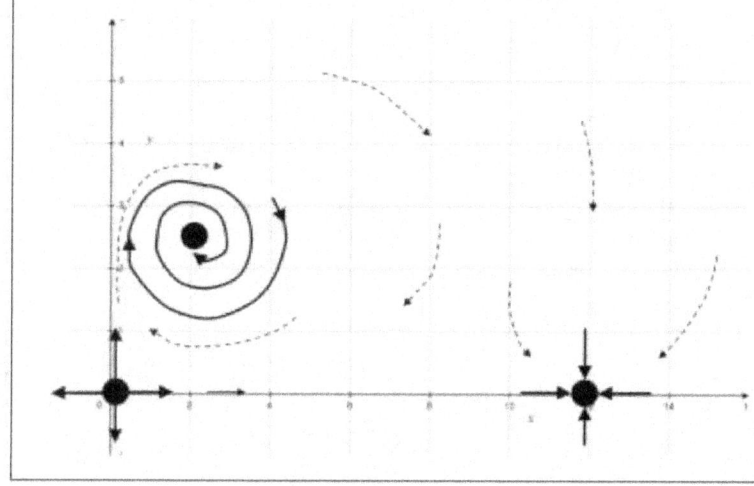

To determine the direction of the spiral, x' and y' were evaluated for point (0,2).
$x' = 0$
$y' = 4$
Upward direction.

APPENDIX

A. INTEGRATION BY PARTS (TABULAR METHOD)

Integration by Parts, is a technique that is taught, and hopefully mastered, in a second course in calculus (Integral Calculus).

Integration by Parts - Tabular Method is a very similar to regular Integration by Parts, but is faster and more efficient to use. This variant of Integration by Parts is not always taught in a second course in calculus. Since this technique is used often in this book, a description of this technique is included here.

INTEGRATION BY PARTS – TABULAR METHOD

First, we will review <u>Integration by Parts</u>, an integration technique and show some examples. Then, we will do the same examples, using <u>Integration by Parts – Tabular Method</u>.

When to use Integration by Parts: If the integral is a product of two different types of functions, and U-substitution doesn't work, try integration by parts.

How to use Integration by Parts: Break the integrand into two parts. One part will be the "u" part and the other will be the "dv" part. The general equation is:

$$\int u\, dv = uv - \int v\, du$$

Carefully select which part is which. Use the acronym "LIATE" to determine the "u" and "dv" parts. The first function, listed within the acronym, is the "u" part. The function listed second is the "dv" part.

	L	Logarithmic Functions
	I	Inverse Trig Functions
	A	Algebraic Functions
	T	Trig Functions
	E	Exponential Functions

INTEGRATION BY PARTS

Example #1: $\int (2x + 3) \cos(9x)\, dx$

Solution:

$\int (2x + 3) \cos(9x)\, dx = \int u\, dv$ "LIATE"

Let $u = (2x + 3)$ Algebraic Function

Let $dv = \cos(9x)\, dx$ Trigonometric Function

Then:

$du = $ The derivative of $u = \dfrac{d}{dx} u = \dfrac{d}{dx}(2x + 3) = (2)\, dx$

$v = $ The integral of $dv = \int dv = \int \cos 9x\, dx = \left(\dfrac{1}{9}\right) \sin 9x$

$$\boxed{\int u\, dv = uv - \int v\, du}$$

$\int (2x + 3) \cos(9x)\, dx = (2x + 3)\left(\dfrac{1}{9}\right) \sin(9x) - \int \left(\dfrac{1}{9}\right) \sin 9x\, (2)\, dx$

$ = (2x + 3)\left(\dfrac{1}{9}\right) \sin(9x) - (2)\left(\dfrac{-1}{81}\right) \cos(9x)$

$ = \left(\dfrac{1}{9}\right)(2x + 3) \sin(9x) + \left(\dfrac{2}{81}\right) \cos(9x) + C$

INTEGRATION BY PARTS (IBP) – Example #2: $\int x^2 e^x \, dx$

$$\int u\,v' = uv - \int v\,u' \qquad \text{LIATE}$$

Solution:

$\int x^2 e^x \, dx = \int (x^2)(e^x) \, dx$

Let: $u = (x^2)$ \hfill Algebraic Function

Let: $dv = v' = (e^x)dx$ \hfill Exponential Function

$u = (x^2)$	$v' = (e^x)\,dx$
$u' = 2x\,dx$	$v = e^x$

$\int u\,v' = uv - \int v\,u'$

$\int x^2 e^x \, dx = (x^2) \cdot (e^x) - \int (e^x)(2x)dx$

$\qquad = x^2 e^x - 2\int e^x x \, dx$ \hfill Must do IBP again.

$u = (x)$	$v' = (e^x)\,dx$
$u' = (1)\,dx$	$v = e^x$

$\qquad = x^2 e^x - 2[\,xe^x - \int e^x \, dx\,]$

$\qquad = x^2 e^x - 2xe^x + 2e^x + C$

Question: For $\int x^5 e^{2x} \, dx$, how many times should IBP be used?
Answer: 5 Times.

INTEGRATION BY PARTS - Example #3: $\int_0^1 \tan^{-1} x \, dx$

$$\int u\, v' = uv - \int v\, u' \qquad \text{LIATE}$$

Solution:

$\int \tan^{-1} x \, dx = \int (1) \tan^{-1} x \, dx \, dx$ NOTE: Compute $\int_0^1$ later.

Let: $u = (\tan^{-1} x)$ Inverse Trig. Function

Let: $dv = v' = (1)$ Algebraic Function

$u = (\tan^{-1} x)$	$v' = (1)\, dx$
$u' = \dfrac{1}{1+x^2} dx$	$v = x$

$\int u\, v' = uv - \int v\, u'$

$\int \tan^{-1} x \, dx = (\tan^{-1} x) \cdot (x) - \int (x)\left(\dfrac{1}{1+x^2}\right) dx$

$= x \tan^{-1} x - \int \left(\dfrac{1}{1+x^2}\right)(x\, dx)$

$= x \tan^{-1} x - \dfrac{1}{2}\int \left(\dfrac{1}{1+x^2}\right)(2x\, dx)$

$= x \tan^{-1} x - \dfrac{1}{2} \ln|1 + x^2|$

$\int_0^1 \tan^{-1} x \, dx = \left[x \tan^{-1} x - \dfrac{1}{2}\ln|1 + x^2| \right]_0^1$

$= \left[(1)\left(\dfrac{\pi}{4}\right) - \left(\dfrac{1}{2}\right)\ln(2) \right] - \left[0 - \left(\dfrac{1}{2}\right)\ln(1) \right]$

$= \dfrac{\pi}{4} - \dfrac{1}{2}\ln(2)$

INTEGRATION BY PARTS – TABULAR METHOD

Example #4: $\int (2x + 3) \cos(9x)\, dx$

Solution:

$\int (2x + 3) \cos(9x)\, dx = \int u\, dv$

Let $u = (2x + 3)$ Algebraic Function

Let $dv = \cos(9x)\, dx$ Trigonometric Function

Du		$\int dv$
$2x + 3$	$+$	$\cos(9x)\, dx$
2	$-$	$\frac{1}{9} \sin(9x)$
0	$+$	$\frac{-1}{81} \cos(9x)$
Take derivative of this side until you can't or it's $= 0$.		Take integral of this side.

Alternate Signs

Multiply Diagonally

Last Horizontal arrow ➔ Integral

$\int (2x + 3) \cos(9x)\, dx$

$= (2x + 3)\left(\frac{1}{9}\right) \sin(9x) - (2)\left(\frac{-1}{81}\right) \cos(9x) + \int \left(-\frac{1}{81}\right)(0)\, dx$

$= \left(\frac{1}{9}\right)(2x + 3) \sin(9x) + \left(\frac{2}{81}\right) \cos(9x) + C$

INTEGRATION BY PARTS – TABULAR METHOD

Example #5: $\int x^2 e^x \, dx$ (previously done)

Solution:

$\int x^2 e^x \, dx = \int (x^2)(e^x) \, dx$

Let: $u = (x^2)$ Algebraic Function   (LIATE)

Let: $dv = v' = (e^x)$ Exponential Function

Du		$\int dv$
x^2	+	e^x
$2x$	−	e^x
2	+	e^x
0	−	e^x
Take derivative of this side until you can't or it's = 0.		Take integral of this side.

Alternate Signs

Multiply Diagonally

Last Horizontal arrow → Integral

$\int x^2 e^x \, dx =$

$= (x^2) \cdot (e^x) - (2x)(e^x) + (2)(e^x) - \int (0)(e^x) \, dx$

$= x^2 e^x - 2x e^x + 2 e^x + C$

INTEGRATION BY PARTS -- TABULAR METHOD

Example #6: $\int_0^1 \tan^{-1} x \, dx$ (previously done)

Solution:

$\int \tan^{-1} x \, dx = \int (1) \tan^{-1} x \, dx \, dx$ NOTE: Compute $\int_0^1$ later.

Let: $u = (\tan^{-1} x)$ Inverse Trig. Function (LIATE)

Let: $dv = v' = (1)$ Algebraic Function

Du	$\int dv$
$\tan^{-1} x$ +	1
$\dfrac{1}{1+x^2}$ −	x
Take derivative of this side until you can't or it's = 0.	Take the integral of this side.

Alternate Signs

Multiply Diagonally

Last Horizontal arrow → Integral

$\int \tan^{-1} x \, dx = (\tan^{-1} x) \cdot (x) - \int (x) \left(\dfrac{1}{1+x^2}\right) dx$

$\qquad = x \tan^{-1} x - \int \left(\dfrac{1}{1+x^2}\right)(x \, dx)$

$\qquad = x \tan^{-1} x - \dfrac{1}{2} \int \left(\dfrac{1}{1+x^2}\right)(2x \, dx)$

$\int_0^1 \tan^{-1} x \, dx = \left[x \tan^{-1} x - \dfrac{1}{2} \ln|1+x^2| \right]_0^1$

$\qquad = \left[(1)\left(\dfrac{\pi}{4}\right) - \left(\dfrac{1}{2}\right) \ln(2) \right] - \left[0 - \left(\dfrac{1}{2}\right) \ln(1) \right]$ 0

$\qquad = \dfrac{\pi}{4} - \dfrac{1}{2} \ln(2)$

INTEGRATION BY PARTS – TABULAR METHOD

Example #7: $\int x^5 e^{2x}\, dx$

Solution:

$\int x^5 e^{2x}\, dx = \int u\, dv$ (LIATE)

Du		$\int dv$
x^5	+	$e^{2x}\, dx$
$5x^4$	−	$\frac{1}{2}e^{2x}$
$20x^3$	+	$\frac{1}{4}e^{2x}$
$60x^2$	−	$\frac{1}{8}e^{2x}$
$120x$	+	$\frac{1}{16}e^{2x}$
120	−	$\frac{1}{32}e^{2x}$
0	+	$\frac{1}{64}e^{2x}$
Take derivative of this side until you can't or it's $= 0$.		Take integral of this side.

Alternate Signs

Multiply Diagonally

$\int x^5 e^{2x}\, dx = x^5 \left(\frac{1}{2}\right) e^{2x} - 5x^4 \left(\frac{1}{4}\right) e^{2x} + 20x^3 \left(\frac{1}{8}\right) e^{2x} \dots$
$\qquad - 60x^2 \left(\frac{1}{16}\right) e^{2x} + 120x \left(\frac{1}{32}\right) e^{2x} - 120 \left(\frac{1}{64}\right) e^{2x} + C$

INTEGRATION BY PARTS – TABULAR METHOD

Example #8: $\int 36x^5 \ln x$ (Example with logs)

Solution:

$\int 36x^5 \ln x \, dx = 36 \int x^5 \ln x \, dx = 36 \int u \, dv$ (LIATE)

Du	$\int dv$
$\ln x$ +	x^5
$\dfrac{1}{x}$ −	$\left(\dfrac{1}{6}\right) x^6$
STOP at 2nd step with LOGS.	Take integral of this side.

Alternate Signs

Multiply Diagonally

Last Horizontal arrow ➔ Integral

$36 \int x^5 \ln x \, dx = 36 \left[\ln x \left(\dfrac{1}{6}\right) x^6 - \dfrac{1}{6} \int x^5 \, dx \right]$

$ = 6 \left[x^6 \ln x - \int x^5 \, dx \right]$

$ = 6 \left[x^6 \ln x - \left(\dfrac{1}{6}\right) x^6 \right]$

$ = 6 x^6 \ln x - x^6 + C$

INTEGRATION BY PARTS – TABULAR METHOD

Example #9: $\int e^{2x} \cos 3x \, dx$ (Example with infinite loop)

Solution:

$\int e^{2x} \cos 3x \, dx = \int u \, dv$ (LIATE)

Du		$\int dv$
$\cos 3x$	+	e^{2x}
$-3 \sin 3x$	−	$\left(\frac{1}{2}\right) e^{2x}$
$-9 \cos 3x$	+	$\left(\frac{1}{4}\right) e^{2x}$
STOP when back to a multiple of the original.		

Alternate Signs

Multiply Diagonally

Last Horizontal arrow → Integral

I = The original integral.

$I = \int e^{2x} \cos 3x \, dx = \cos 3x \left(\frac{1}{2}\right) e^{2x} - (-3\sin 3x)\left(\frac{1}{4}\right) e^{2x}$

$\qquad\qquad\qquad\qquad\qquad + \int \left(-\frac{9}{4}\right) \cos 3x \, e^{2x} \, dx$

$I = \left(\frac{1}{2}\right) e^{2x} \cos 3x + \left(\frac{3}{4}\right) e^{2x} \sin 3x - \left(\frac{9}{4}\right) I$

$\left(\frac{13}{4}\right) I = \left(\frac{1}{2}\right) e^{2x} \cos 3x + \left(\frac{3}{4}\right) e^{2x} \sin 3x$

$I = \left(\frac{4}{13}\right) \left[\left(\frac{1}{2}\right) e^{2x} \cos 3x + \left(\frac{3}{4}\right) e^{2x} \sin 3x \right] + C$

B. COMPLEX NUMBERS

Complex Numbers: A Brief Review

A complex number is in the form: $Z = a + ib$
Where i is an imaginary number: $i^2 = -1$ and $i = \sqrt{-1}$

$a = Re(Z)$ ➔ a is the real part of Z

$b = Im(Z)$ ➔ b is the imaginary part

Note: a & b are real #s

Example: $Z = 3 + 5i$ ➔ $Re(Z) = 3$ and $Im(Z) = 5$

Rectangular (x, y) and Polar Coordinates (r, θ)

are two types of representation.

$\quad x = a \quad$ and $\quad y = b$

$\quad r = \sqrt{x^2 + y^2} \quad$ and $\quad \theta = \tan^{-1}\left(\frac{y}{x}\right)$

Modulus of $Z = |Z| = \sqrt{a^2 + b^2}$

$\theta = \tan^{-1}\left(\frac{b}{a}\right)$

$Z \quad = \quad Cartesian \quad = \quad Polar\ Form$

$Z \quad = \quad a + ib \quad = \quad |Z|\,e^{i\theta}$

Where: $e^{i\theta} = \cos\theta + i\sin\theta$

Example:

Compute Modulus and Argument for following complex numbers.

(1) $Z = \sqrt{3} + i$ (2) $Z = 5$ (3) $Z = 1 + 2i$ (4.) $Z = -4i$

(1) Solution:

Modulus: $|Z| = \sqrt{a^2 + b^2} = \sqrt{3 + 1} = 2$

Argument: $\theta = \tan^{-1}\left(\frac{b}{a}\right) = \tan^{-1}\left(\frac{1}{\sqrt{3}}\right) = \frac{\pi}{6}, \frac{7\pi}{6}$

(2) Solution:

Modulus: $|Z| = \sqrt{a^2 + b^2} = \sqrt{25 + 0} = 5$

Argument: $\theta = \tan^{-1}\left(\frac{b}{a}\right) = \tan^{-1}\left(\frac{0}{5}\right) = \tan^{-1}(0) = 0, \pi$

(3) Solution:

Modulus: $|Z| = \sqrt{a^2 + b^2} = \sqrt{1 + 4} = \sqrt{5}$

Argument: $\theta = \tan^{-1}\left(\frac{b}{a}\right) = \tan^{-1}\left(\frac{2}{1}\right) = \tan^{-1}(2)$

(4) Solution:

Modulus: $|Z| = \sqrt{a^2 + b^2} = \sqrt{0^2 + 4^2} = 4$

Argument: $\theta = \tan^{-1}\left(\frac{4}{0}\right) = \tan^{-1}(\infty) = \frac{\pi}{2}, \frac{3\pi}{2}$

Conjugate of Complex Number:

- If $Z = a + bi$ then it's conjugate is: $\bar{Z} = a - bi$
- When you multiply: $Z\bar{Z} = |Z|^2 = a^2 + b^2$
- Note: This is always Real

Example:

Rewrite the complex number in standard form. $Z = i^{25}$

Solution: $i^{25} = i^{24} i = (i^2)^{12} i = (1)i = i$

Example:

Rewrite the complex number in standard form. $Z = \dfrac{2 + 3i}{4 - 5i}$

Solution: $\dfrac{2 + 3i}{4 - 5i} \left(\dfrac{4 + 5i}{4 + 5i} \right) = \dfrac{8 + 10i + 12i + 15i^2}{4^2 - (5i)^2}$

$= \dfrac{-7 + 22i}{16 + 25} = \dfrac{-7 + 22i}{41} = -\dfrac{7}{41} + \dfrac{22}{41} i$

Example:

Rewrite the complex number in standard form. $Z = \dfrac{7}{3 - 4i}$

Solution: $\dfrac{7}{3 - 4i} \cdot \left(\dfrac{3 + 4i}{3 + 4i} \right) = \dfrac{7(3 + 4i)}{(3)^2 - (4i)^2} = \dfrac{21 + 28i}{9 + 16} = \dfrac{21}{25} + \dfrac{28}{25} i$

Example: Compute: $Z = \sqrt{3 - 4i}$

Solution:

$\sqrt{3 - 4i} = a + ib$ We know the format.

$3 - 4i = a^2 + 2abi + b^2 i^2$

$3 - 4i = a^2 + 2abi - b^2$

$3 - 4i = (a^2 - b^2) + (2ab)i$

Compare Like Terms:

$(a^2 - b^2) = 3$ Equation (1)

and $2ab = -4.$

$ab = -2$ Equation (2)

$b = -\dfrac{2}{a}$

Modulus of both sides must be equal.

 Recall: For $Z = a + bi$

 The Modulus is: $|Z| = \sqrt{a^2 + b^2}$

        $|Z|^2 = a^2 + b^2$

Modulus of $(a + ib)^2$

$(a^2 + b^2)^2 = 3^2 + 4^2 = 25$

$(a^2 + b^2) = \pm\sqrt{25} = \pm 5 = 5$ Modulus is Positive

$(a^2 + b^2) = 5$ Equation (3)

Equations (1) and (3) ➔

$2a^2 = 8$

$a^2 = 4$

$a = \pm 2$ ➔ Since $b = -\dfrac{2}{a}$

$b = \pm 1$

Therefore, $a = \pm 2$ and $b = \pm 1$.

Four Possibilities but only 2 solutions because $a \cdot b = -2$

To get the product (-2) one term must be negative and one term must be is positive.

$\sqrt{3 - 4i} \;=\; a + ib \;=\; -2 + i \;=\; 2 - i$ Done.

Note: The remaining two cases do not give a product of -2. They give a product of 2 and we cannot evaluate it.

Example: Compute the square root of: $Z = 1 + 2i$

Solution:

$$\sqrt{1 + 2i} = a + bi \qquad \text{We know the format.}$$

$$1 + 2i = (a + bi)^2 = a^2 + 2abi + b^2 i^2$$

$$1 + 2i = (a^2 - b^2) + (2ab)i$$

→ $1 = (a^2 - b^2)$ → $a^2 = 1 + b^2$

→ $2 = 2ab$ → $1 = ab$ → $a = \dfrac{1}{b}$

Find the modulus of both sides.

$\text{Mod}[1 + 2i] = \sqrt{1^2 + 2^2} = \sqrt{5}$
$\text{Mod}[(a^2 - b^2) + (2ab)i] = \sqrt{(a^2 - b^2)^2 + (2ab)^2}$ $\qquad = \sqrt{a^4 - 2a^2 b^2 + b^4 + 4a^2 b^2}$ $\qquad = \sqrt{a^4 + b^4 + 2a^2 b^2}$ $\qquad = \sqrt{(a^2 + b^2)^2} = a^2 + b^2$

Equate the modulus of both sides:

$$\sqrt{5} = a^2 + b^2 = (1 + b^2) + b^2 = 1 + 2b^2$$

→ $b = \pm\sqrt{\dfrac{\sqrt{5} - 1}{2}}$

→ $a = \dfrac{1}{b} = \pm\sqrt{\dfrac{2}{\sqrt{5} - 1}\left(\dfrac{\sqrt{5} + 1}{\sqrt{5} + 1}\right)} = \pm\sqrt{\dfrac{2(\sqrt{5} + 1)}{5 - 1}} = \pm\sqrt{\dfrac{\sqrt{5} + 1}{2}}$

(Continued...)

So far we know the format: $a + bi$

We calculated two possible values for a and for b.

$$a = \pm\sqrt{\frac{1+\sqrt{5}}{2}} \quad \text{and} \quad b = \pm\sqrt{\frac{\sqrt{5}-1}{2}}$$

We also know: $1 = ab$ (positive)

We must select one conjugate pair from the 4 possibilities.

a and b are both positive or both negative

$$a + bi = \sqrt{\frac{\sqrt{5}+1}{2}} + i\sqrt{\frac{\sqrt{5}-1}{2}} \quad \text{OR}$$

$$a + bi = -\sqrt{\frac{\sqrt{5}+1}{2}} - i\sqrt{\frac{\sqrt{5}-1}{2}}$$

Verification:

$$ab = \sqrt{\frac{\sqrt{5}+1}{2}} \cdot \sqrt{\frac{\sqrt{5}-1}{2}} = \sqrt{\frac{5-1}{4}} = \sqrt{\frac{4}{4}} = 1$$

→ Use the positive "a" and positive "b"

$$ab = \left(-\sqrt{\frac{\sqrt{5}+1}{2}}\right) \cdot \left(-\sqrt{\frac{\sqrt{5}-1}{2}}\right) = \sqrt{\frac{5-1}{4}} = \sqrt{\frac{4}{4}} = 1$$

→ Use the negative "a" and negative "b"

C. LINEAR INDEPENDENCE

LINEARLY INDEPENDENT VECTORS – Review

The set of vectors $x_1, x_2, x_3, \ldots x_n$ are linearly independent (L.I.)

If: $\quad \alpha_1 x_1 + \alpha_2 x_2 + \alpha_3 x_3 + \ldots \alpha_n x_n = 0 \quad$ (Zero Vector)

Then: $\quad \alpha_1 = \alpha_2 = \alpha_3 = \ldots = \alpha_n = 0$

Where: $\quad \alpha_1, \alpha_2, \ldots \alpha_n$ are scalars.

Note: $\quad$ If $\alpha_1 x_1 + \alpha_2 x_2 + \alpha_3 x_3 + \ldots \alpha_n x_n = 0$

gives at least one non-zero " α_i "

then, the vectors are linearly dependent (L.D.).

Example: Check if $V_1 = \begin{bmatrix} 1 \\ 2 \end{bmatrix}$ and $V_2 = \begin{bmatrix} 2 \\ 1 \end{bmatrix}$ are L.I. or L.D.

Solution:

$\alpha_1 V_1 + \alpha_2 V_2 = 0$

$\alpha_1 \begin{bmatrix} 1 \\ 2 \end{bmatrix} + \alpha_2 \begin{bmatrix} 2 \\ 1 \end{bmatrix} = \begin{bmatrix} 0 \\ 0 \end{bmatrix}$

$\begin{bmatrix} \alpha_1 + 2\alpha_2 \\ 2\alpha_1 + \alpha_2 \end{bmatrix} = \begin{bmatrix} 0 \\ 0 \end{bmatrix} \quad \rightarrow \quad \begin{cases} \alpha_1 + 2\alpha_2 = 0 \\ 2\alpha_1 + \alpha_2 = 0 \end{cases} \quad \rightarrow \quad \begin{cases} 2\alpha_1 + 4\alpha_2 = 0 \\ 2\alpha_1 + \alpha_2 = 0 \end{cases}$

$\hspace{10cm} \text{Subtract:} \quad 3\alpha_2 = 0$

$\alpha_2 = 0 \quad \rightarrow \quad \alpha_1 + 2\alpha_2 = 0 \quad \rightarrow \quad \alpha_1 + 2(0) = 0 \quad \rightarrow \quad \alpha_1 = 0$

Thus: $\alpha_1 = \alpha_2 = 0 \quad$ So, V_1 and V_2 are Linearly Independent (L.I.)

Example: Check if $V_1 = \begin{bmatrix} 1 \\ 2 \\ 1 \end{bmatrix}$ and $V_2 = \begin{bmatrix} 2 \\ 0 \\ 3 \end{bmatrix}$ are L.I. or L.D.?

Solution:

$\alpha_1 V_1 + \alpha_2 V_2 = 0 \rightarrow \begin{bmatrix} \alpha_1 + 2\alpha_2 \\ 2\alpha_1 \\ \alpha_1 + 3\alpha_2 \end{bmatrix} = \begin{bmatrix} 0 \\ 0 \\ 0 \end{bmatrix} \rightarrow \begin{cases} \alpha_1 + 2\alpha_2 = 0 \\ 2\alpha_1 = 0 \\ \alpha_1 + 3\alpha_2 = 0 \end{cases}$

$\alpha_1 = 0$

$\alpha_1 + 2\alpha_2 = 0 \rightarrow 0 + 2\alpha_2 = 0 \rightarrow \alpha_2 = 0$

Thus: $\alpha_1 = \alpha_2 = 0$ So, V_1 and V_2 are Linearly Independent (L.I.)

Example: Check if $V_1 = \begin{bmatrix} 1 \\ 2 \end{bmatrix}$ and $V_2 = \begin{bmatrix} 3 \\ 6 \end{bmatrix}$ are L.I. or L.D.?

Solution:

$\alpha_1 V_1 + \alpha_2 V_2 = 0 \rightarrow \begin{bmatrix} \alpha_1 + 3\alpha_2 \\ 2\alpha_1 + 6\alpha_2 \end{bmatrix} = \begin{bmatrix} 0 \\ 0 \end{bmatrix} \rightarrow \begin{cases} \alpha_1 + 3\alpha_2 = 0 \\ 2\alpha_1 + 6\alpha_2 = 0 \end{cases}$

Note: Parallel Lines

$\alpha_1 = -3\alpha_2$

We can get at least one non-zero α_2 value for which $\alpha_1 \neq 0$

(e.g. $\alpha_2 = 1 \rightarrow \alpha_1 = -3$)

Thus: $\alpha_1 \neq 0$ and $\alpha_2 \neq 0$

So, V_1 and V_2 are Linearly Dependent (L.D.)

LINEARLY INDEPENDENT VECTORS – Remarks

The solution of higher-order Differential Equations (DE) requires the construction of a number of solutions equal to the degree of the DE. All constructed solutions must be different or, in other words, Linearly Independent (L.I.). This can be checked by using either the definition or by using the following results.

#	Two ways to check for Linear Independence (L.I.)
1	"n" vectors $(V_1, V_2, V_3 \ldots V_n)$ in R^n (n-dimensional space) are L.I. if their determinant $|V_1, V_2, V_3 \ldots V_n| \neq 0$ Otherwise, they are L.D. L.I. Example: $V_1 = \begin{bmatrix} 1 \\ 2 \end{bmatrix}$ and $V_2 = \begin{bmatrix} 2 \\ 1 \end{bmatrix}$ $\begin{vmatrix} 1 & 2 \\ 2 & 1 \end{vmatrix} = 1 - 4 = -3 \neq 0$ → Linearly Independent L.D. Example: $V_1 = \begin{bmatrix} 1 \\ 2 \end{bmatrix}$ and $V_2 = \begin{bmatrix} 3 \\ 6 \end{bmatrix}$ $\begin{vmatrix} 1 & 3 \\ 2 & 6 \end{vmatrix} = 6 - 6 = -3 = 0$ → Linearly Dependent
2	Checking L.I./D.I. using Wronskian Determinant. $y_1 = e^{r_1 x}$ and $y_2 = e^{r_2 x}$ are L.I. if $W(y_1, y_2) \neq 0$ In 2D: $W(y_1, y_2) = \begin{vmatrix} y_1 & y_2 \\ y'_1 & y'_2 \end{vmatrix} \neq 0$ → L.I.

Problem: Let $V_1 = \begin{bmatrix} 1 \\ 0 \\ 1 \end{bmatrix}$, $V_2 = \begin{bmatrix} -2 \\ 0 \\ 1 \end{bmatrix}$, and $V_3 = \begin{bmatrix} 0 \\ -1 \\ 0 \end{bmatrix}$

Show that $V_1, V_2, and\ V_3$ are Linearly Independent (L.I.)

Solution #1: Verify: $\det[V_1, V_2, V_3] \neq 0$

$$\det \begin{bmatrix} 1 & -2 & 0 \\ 0 & 1 & -1 \\ -1 & 0 & 1 \end{bmatrix} \neq 0$$

$1[1-0] - (-2)(0-1) = 1 + 2(-1) = -1 \neq 0$ ➔ L.I.

Solution #2: Verify: $\alpha V_1 + \beta V_2 + \mu V_3 = 0$

$$\alpha \begin{bmatrix} 1 \\ 0 \\ -1 \end{bmatrix} + \beta \begin{bmatrix} -2 \\ 1 \\ 0 \end{bmatrix} + \mu \begin{bmatrix} 0 \\ -1 \\ 1 \end{bmatrix} = \begin{bmatrix} 0 \\ 0 \\ 0 \end{bmatrix}$$

$\alpha - 2\beta = 0$

$\beta - \mu = 0$ ➔ $\mu = \beta$

$\alpha - \mu = 0$ ➔ $\mu = \alpha$

$\alpha - 2\beta = 0$

$\mu - 2\mu = 0$ ➔ $\mu = 0$ ➔ $\beta = 0$ ➔ $\alpha = 0$ ➔ L.I.

D. MATRICES REVIEW

Matrices

$A = [A_{ij}]$ $$A = \begin{bmatrix} a_{11} & a_{12} & \cdots & a_{1n} \\ a_{21} & a_{22} & \cdots & a_{2n} \\ \cdots & & & \\ a_{m1} & a_{m2} & \cdots & a_{mn} \end{bmatrix}_{m \times n}$$	m Rows n Columns Called "Matrix of size $m \times n$"
Remarks: • Square Matrix ➔ $m = n$ • Rectangular Matrix ➔ $m \neq n$	---
Transpose: A^T = Transpose of A obtained by interchanging columns by rows.	**Example:** $A = \begin{bmatrix} 1 & 7 \\ 2 & 5 \end{bmatrix} \rightarrow A^T = \begin{bmatrix} 1 & 2 \\ 7 & 5 \end{bmatrix}$
Conjugate : $\overline{A}$ = Conjugate of A	**Example:** $A = \begin{bmatrix} 2-i & 4+i \\ 3+2i & 7 \end{bmatrix}$ $\overline{A} = \begin{bmatrix} 2+i & 4-i \\ 3-2i & 7 \end{bmatrix}$
Adjoint: $A^* = (\overline{A})^T$	**Example:** $A = \begin{bmatrix} 1 & 2+i \\ 7 & 3-2i \end{bmatrix}$ $(\overline{A})^T = \begin{bmatrix} 1 & 2-i \\ 7 & 3+2i \end{bmatrix}^T$ $(\overline{A})^T = \begin{bmatrix} 1 & 7 \\ 2-i & 3+2i \end{bmatrix}$

PROPERTIES OF MATRICES	
Equality: $A = B$ IFF $a_{ij} = b_{ij}$	$\begin{bmatrix} a & b \\ c & d \end{bmatrix} = \begin{bmatrix} 2 & 3 \\ 7 & -9 \end{bmatrix}$ ➔ $a = 2,\ b = 3,\ c = 7,\ d = -9$
Zero Matrix: Matrix with all entries $= 0$ Notation: 0 Sometimes called: 0_n	$0_2 = \begin{bmatrix} 0 & 0 \\ 0 & 0 \end{bmatrix}$
Identity Matrix: Diagonal entries $= 1$ Remaining entries $= 0$ Notation: I_n	$I_3 = \begin{bmatrix} 1 & 0 & 0 \\ 0 & 1 & 0 \\ 0 & 0 & 1 \end{bmatrix}$
Addition and Subtraction: Matrices must be same size. Component-wise operation.	$A = \begin{bmatrix} 1 & 2 \\ 3 & -2 \end{bmatrix} \quad B = \begin{bmatrix} 2 & 2 \\ 1 & -5 \end{bmatrix}$ $A + B = \begin{bmatrix} 3 & 4 \\ 4 & -7 \end{bmatrix}$ $A - B = \begin{bmatrix} -1 & 0 \\ 2 & 3 \end{bmatrix}$
Multiplication (Scalar) $\alpha A = [\alpha \cdot a_{ij}]$	$A = \begin{bmatrix} 1 & 2 \\ 1 & -3 \end{bmatrix} \quad \alpha = -3$ $\alpha A = \begin{bmatrix} -3 & -6 \\ -3 & 9 \end{bmatrix}$

PROPERTIES OF MATRICES (Continued)

Multiplication of 2 Matrices: Only possible if #Columns in 1st matrix = #Rows in 2nd matrix. $$[\ ] \cdot [\ ]$$ $$m \times n \quad n \times k$$ Result is $m \times k$	$A = \begin{bmatrix} 1 & 2 \\ 0 & 5 \end{bmatrix} \quad B = \begin{bmatrix} 1 & 2 & 1 \\ 2 & 0 & 3 \end{bmatrix}$ $\quad\quad 2 \times 2 \quad\quad\quad\quad 2 \times 3$ $A \cdot B$ is possible. $\quad B \cdot A$ NOT possible. $A \cdot B = \begin{bmatrix} 1 & 2 \\ 0 & 5 \end{bmatrix} \cdot \begin{bmatrix} 1 & 2 & 1 \\ 2 & 0 & 3 \end{bmatrix}$ $A \cdot B = \begin{bmatrix} (1+4) & (2+0) & (1+6) \\ (0+10) & (0+0) & (0+15) \end{bmatrix}$ $A \cdot B = \begin{bmatrix} 5 & 2 & 7 \\ 10 & 0 & 15 \end{bmatrix}$
Inverse: $A = \begin{bmatrix} a & b \\ c & d \end{bmatrix}$ A^{-1} = Inverse of A $A^{-1} = \dfrac{\begin{bmatrix} d & -b \\ -c & a \end{bmatrix}}{\det(A)}$ $\det(A) = ad - bc$	$A = \begin{bmatrix} 2 & 1 \\ 3 & 5 \end{bmatrix}$ $\det(A) = 2 \cdot 5 - 1 \cdot 3 = 7$ $A^{-1} = \dfrac{1}{7} \begin{bmatrix} 5 & -1 \\ -3 & 2 \end{bmatrix} = \begin{bmatrix} \frac{5}{7} & -\frac{1}{7} \\ -\frac{3}{7} & \frac{2}{7} \end{bmatrix}$

End.

www.ingramcontent.com/pod-product-compliance
Lightning Source LLC
Chambersburg PA
CBHW081423220526
45466CB00008B/2250